Studies in Economic Ethics and Philosophy

Springer-Verlag Berlin Heidelberg GmbH

Agnès Labrousse · Jean-Daniel Weisz
Editors

Institutional Economics in France and Germany

German Ordoliberalism versus the French Regulation School

With 22 Figures
and 11 Tables

Springer

Agnès Labrousse
Jean-Daniel Weisz
Centre Marc Bloch
Schiffbauerdamm 19
10117 Berlin
Germany

ISSN 1431-8822
DOI 10.1007/978-3-662-04472-8

Cataloging-in-Publication Data applied for
Die Deutsche Bibliothek - CIP-Einheitsaufnahme
Institutional Economic in France and Germany: German Ordoliberalism versus the French Regulation School; with 11 tables / Agnes Labrousse; Jean-Daniel Weisz ed.. - Berlin; Heidelberg; New York; Barcelona; Hong Kong; London; Milan; Paris; Singapore; Tokyo: Springer, 2001
(Studies in Economic Ethics and Philosophy)

Originally published by Springer-Verlag Berlin Heidelberg New York in 2001.
MyCopy version of the original edition 2001

Hardcover Design: Erich Kirchner, Heidelberg

SPIN 10724818 42/2202-5 4 3 2 1 0 - Printed on acid-free paper
www.springer.com/mycopy

Preface

This volume publishes the proceedings of the conference "*Wirtschaftsordnungspolitik*: German Ordoliberalism and the French Regulation School Compared" held in Berlin on 8-9 May 1998, together with an additional text by Professor Carsten Hermann-Pillath.

The conference was organized by the Centre Marc Bloch in Berlin and the Frankfurter Institut für Transformationstudien (FIT) of the European University Viadrina (Frankfurt an der Oder) with the support of the Social Science Research Centre Berlin (WZB).

We especially would like to thank the Professor Robert Delorme in Paris and the Professor Etienne François, former director of the Centre Marc Bloch, as well as Professor Hans-Jürgen Wagener and Dr. Frank Bönker from the FIT, who supported this project from the outset. In the WZB, we wish to thank David Soskice and Bob Hancké for their assistance with the conference. The project also received friendly and active support from the members of the research group on the GDR and the new *Bundesländer* at the Centre Marc Bloch, whom we would like to thank warmly.

The publication of the book was only made possible thanks to the financial and logistic support of the Centre Marc Bloch, which bore the full costs of translation and preparation of the volume for publication. We have a great debt to Professor Colliot-Thélène, the current director of the Centre Marc Bloch, and to its general secretary, Dr. Gérard Darmon, for their personal, institutional and financial aid in the final crucial phase.

For the final correction of the book we should like to thank Dr. Juliet Vale.

Berlin, 10 July 2000

Agnès Labrousse Jean-Daniel Weisz

Contents

Part B

Evolution and Present Relevance of Ordoliberalism

Part C

The Methodological Foundations Compared: The Issue of Complexity

Note of the Editors

In order to differentiate the meaning of regulation in the French phrase "théorie de la régulation" from other meanings widespread in the English language like regulation in cybernetics and the economic regulation of sectors (opposed to deregulation) we used capital letters for "Regulation theory" and "Regulationist". However regulation is not capitalized in the adjectival form ("regulationist approach", for instance) or when it refers to central concepts of the Regulation theory where no confusion is possible (like the "mode of regulation").

Regarding Ordoliberalism, there are several possible translations for the term "zentralgeleitete Wirtschaft" used by Eucken: "centrally directed economy", "centrally managed economy", "centrally monitored economy". Following the translation of T.H. HUTCHINSON (EUCKEN 1952), we adopted the first one. The subcategory, "zentralverwaltete Wirtschaft" has been translated as "centrally administered economy".

INTRODUCTION

ROBERT DELORME

This book originates in a conference held in Berlin in 1998 the theme of which was a comparison of the German Ordoliberal school with the French Regulation school. The primary practical aim of the conference was to create an opportunity for researchers knowing almost nothing from each other to meet and exchange ideas. The distinctive characteristic of these two approaches is their initial grounding in their respective national contexts. Their origins played an important role in the trajectories they subsequently followed. This feature is abundantly illustrated throughout the book. However, the scope of these approaches to institutional economics is in no way restricted to their national contexts. And yet they remain almost ignored in books and surveys on institutional economics, although many publications and articles on both are available in English, French and German. One is therefore tempted to believe that there must be some kind of an invisible barrier preventing their due recognition to date in the literature on institutional economics, apart from some notable exceptions aside.

These two unorthodox approaches suffer from the fact that they are often perceived through clichés. Those who believe that the theory supporting Ordoliberalism can be reduced to a Hayekian type of thought and those who believe that the French Regulation school boils down to Marxist thinking will be surprised. There are also rather unexpected similarities. The aim of this volume is thus neither to produce some kind of integrated synthesis nor to set out a comprehensive presentation of both approaches. Rather, it is to compare them and to elicit information about their central tenets, what makes them original and how they address some current important economic issues.

These considerations are organized in four parts. Part I (Chapters 1-4) presents the two schools in their national contexts. Part II deals with a problem specific to Ordoliberalism. Whereas Regulation theory is relatively recent and still expanding, it might be tempting to view Ordoliberalism – and especially the theory underlying it – as belonging to the past or, at the very least, as part of the history of thought. Part II (Chapters 5-7) attempts to correct this impression and seeks to demonstrate the present relevance of

Ordoliberalism. Part III (Chapters 8 and 9) is devoted to a comparison of the methodological foundations. In the final part (Chapters 10-12), both theories are assessed and compared in the context of the issue of post-socialist transformation.

Chapter 1, by Jean-François Vidal, is entitled "Birth and Growth of the Regulation School in the French Intellectual Context (1970-1986)" and focuses on the first fifteen or so years of the school, from the beginning of the seventies to the mid-eighties. Vidal singles out three partially overlapping features. First is what he calls "pre-Regulation and the crisis of macroeconomics" (1971-1976). It covers a period during which several researchers working with the central government economic administration emphasized some shortcomings of the applied macroeconomic modelling which then prevailed and also undertook research on capital accumulation. Then came a period between 1975 and 1983 marked notably by the publication of books by Michel Aglietta, Alain Lipietz and (jointly) Robert Boyer and Jacques Mistral. Vidal calls it the "historical materialism phase", thus indicating a common background against which the theory of Regulation can be interpreted. Indeed, one of the characteristics of Regulation is revealed in this way. It is the rejection of both the quasi-official Marxist doctrine of state monopoly capitalism and of the structuralist Marxist doctrine. Vidal points out that there is regulation because the structure does not reproduce spontaneously. In this way regulationist economists were able to emphasize at a very early date that the economic instability of 1974-1975 was the start of a long-run structural crisis, not merely a business cycle or the outcome of errors in the making of economic policy. This rejection opened the way for the construction of an applied macroeconomic modelling historical and institutional macroeconomic theory that was autonomous in relation to Marxist theory. This third feature, according to Vidal, covers the period 1977-1986. He stresses four characteristics which appeared and developed during this period. First is what he calls the "agnostic turn", in other words a distancing from both Marxist and neoclassical theory. Secondly, the articulation between short-run and long-run dynamic mechanisms became influenced by the distinction between conjuncture and structure made by the historical *Annales* school. Thirdly, the method stabilized. And, finally, Regulation theory expanded through its widespread adoption and application to new areas, such as the Latin American economies and the East and Central European transformation after 1990. All in all, Regulation theory seems to stand today, in Vidal's view, as a historical and institutional macroeconomic theory oriented towards the evolution of economic systems. In this way, as Vidal suggests in his conclusion,

Regulation consists of a unique combination drawing on Karl Marx, John Maynard Keynes and Nicholas Kaldor, besides Fernand Braudel and Ernest Labrousse. Karl Polanyi and Joseph Schumpeter could also be added to this list.

In Chapter 2, Robert Boyer, one of the founders of the Regulation approach, presents it as an alternative theory of capitalism. He starts from the old and fundamental question of political economy: "what are the basic institutions which allow competition among individuals ultimately to deliver an acceptable economic order?" (p. 52) and he observes that the present victory of the capitalist system goes hand in hand with the paucity of its theory. For capitalism is not self-implementing. German reunification illustrates the difficulty of transplanting a complete set of institutions in order to build a viable market economy. Markets are neither self-equilibrating, nor self-instituting. There appears to be no theory about the process of implementing institutions required for capitalism. There are practically as many market failures as there are efficient markets in real, concrete economies. The outcomes of deregulation, privatization and liberalization are uncertain and unexpected. Boyer contends that the generalization of General Equilibrium Theory has produced failed hopes. Regulation theory provides a framework enabling to switch from the missing *secrétaire de marché* (Walrasian auctioneer) to five "institutional forms" derived in Regulation theory from a previous internal criticism of traditional Marxist theory. The very limits of Walrasian economic theory call for precisely the same basic institutions that are captured by the five institutional forms. And the notions of "regimes of accumulation" and "regulation modes" do account for the diversity of capitalist systems. Besides the gloomy record ascribed to liberal strategies, we should note Boyer's comment that "[this] approach ... is not so very far removed from Ordoliberalism, whatever the distance in terms of the implied conceptions for economic and political history, as well as economic policy strategies".

In Chapter 3, Sylvain Broyer sets out to present the central concepts, tools and methods of *Ordnungstheorie* and of Ordoliberalism understood as the political principles derived from it. Ordoliberalism develops an *Ordnungspolitik* based on the legal organization of competition through state action in order to create a framework of institutions playing a role not dissimilar to that of an economic constitution and ensuring conformity to these constitutive principles by means of a rule-oriented economic policy. Broyer outlines past criticisms of *Ordnungstheorie* and presents recent contributions which attempt to remedy some of what they identify as the shortcomings of *Ordnungstheorie*. So-called *Ordnungsökonomik* is one such contribution and offers a kind of synthesis of *Ordnungstheorie* with Hayekian and New Institutionalist notions.

Jean-Daniel Weisz's contribution in Chapter 4 stands in contrast to the preceding chapter. Whereas Broyer assimilates more or less the ordo view to neoclassical thinking, Weisz emphasizes the originality of the theoretical foundations of *Ordnungstheorie* and criticizes what he perceives to be a subsequent misinterpretation of the core methodological message forming the basis of Eucken's *Foundations of Economics*. Weisz underlines the fact that Eucken's ambition was to establish a foundation for a renewed economic analysis capable of solving the "Great Antinomy" between the German historical school and the nascent Austrian school. This ambition was based on a radical interrogation on the nature of the scientificity of economic analysis. The solution Eucken proposed was influenced by Husserl's methodology in its rejection of both Cartesianism, deductivism and empiricism. "Thinking in orders" and its systematization through morphological modelling was Eucken's answer. It was a kind of systemic thinking before systems theory became recognized in the social sciences after the 1950s. This systemic thinking underwent changes in subsequent years, both through dilution into neoclassical thinking of a reductionist type or through dichotomies of ideal-types. Weisz's contribution is an invitation to reread, or to start reading, Eucken's *Foundations of Economics*.

Part II is devoted entirely to Ordoliberalism. Here the situation is different from Regulation theory. Ordoliberalism is older, its founders are dead and ordo ideas themselves may appear obsolete or obscured by the contemporary tendency to liberalization in the wake of what one might be tempted to call "modern liberalism". Things are different with Regulation theory. Its founders are alive and apparently as active as ever, and the theory itself is expanding to some extent. There thus appears to be a definite need to address the issue of the relevance of ordo ideas to socioeconomic problems. This is done in three different ways in each of the three chapters in this part.

In Chapter 5, Frank Bönker, Agnès Labrousse and Jean-Daniel Weisz survey fifty years of the *Ordo* yearbook. Considering Ordoliberalism as a network of scholars, they suggest that it can be described as a relative close group, formed on the basis of a limited pool of editors, with dense personal and academic ties, and of a rather high concentration of contributors. However, they note a recent tendency to be receptive to external contributors. They also examine to what extent the content of the yearbook has evolved over time and identify a strong decline of articles using an ordoliberal framework over the period. However, they are careful not to conclude that there has been a decline in ordo ideas: this trend should either be viewed as an evolution towards a broader paradigm containing ordo and other forms of economic liberalism, or it can be interpreted as a symptom of a basic difficulty in the method, with a loss of methodological specificity in the ordoliberal approach.

In Chapter 6, Frank Bönker and Hans-Jürgen Wagener outline Hayek's and Eucken's thinking about the role of the state in a market economy and put it in perspective. They concentrate on differences relating to three aspects: the role of the state in constituting a market economy, the agenda of economic policy and the political preconditions for a well-functioning market economy. Thus contrasting the different positions taken by Hayek and Eucken on these topics, they contend that they were each "extremists" in their own way, with a radicalism which made for intellectual appeal but also led to some exaggeration and one-sidedness. One author's strength seems to highlight the other's weaknesses and vice versa. This view can be interpreted as evidence that Eucken's thinking is as valid today as Hayek's in the continuing debate on the topics already mentioned.

Michael Wohlgemuth discusses the current relevance of *Ordnungstheorie* for the politics and the economics of the social order in Chapter 7. He offers a broad coverage not only of the roots, historical and intellectual, of *Ordnungstheorie* but also of its content and of the extension of ordo into the German academic community. He shows that there is not just one paradigm and school of *Ordnungstheorie*. He thus distinguishes a Freiburg and a Cologne mode of ordoliberal thought, the latter being more prone to combine individual freedom in markets with social balance through government intervention as signalled by the notion of the "social market economy", which he calls a "politically effective term". As for the intellectual relevance of *Ordnungstheorie*, Wohlgemuth attributes most of the differences between Eucken, Böhm and Hayek to the ordoliberals' primary concern with the problem of private economic power and Hayek's focus on the problem of individual subjective knowledge. As far as economic theories of politics are concerned, the political relevance of *Ordnungspolitik* can be assessed in various ways. There is the success of the German economic model associated with the social market economy. There is also the European integration process which one might be tempted to interpret as a success story of the europeanization of *Ordnungspolitik*. Wohlgemuth's vision is more optimistic at the conclusion than at the outset. He maintains that the outlook for

Ordnungstheorie in its modern form, *Ordnungsökonomik*, which he has written about with Manfred Streit, is bright. Wohlgemuth emphasizes some "promising signs for the future". First, is the challenge presented by the trade-off between practical relevance and importance on the one hand and the formal rigour and precision of most theoretical work published in the discipline of economics on the other. For Wohlgemuth, the decline in demand for academic "social mathematics" is an encouraging sign that the balance is tipping towards practical relevance and political applicability. Secondly, and in conclusion, Wohlgemuth seems to consider that the wave of liberalization reforms encouraged by the International Monetary Fund and the World Bank are evidence of an impressive success on the part of "*de facto Ordnungspolitik*". This is a somewhat contentious aspect of Wohlgemuth's contribution. It is not entirely unambiguous, since it is open to question whether liberalism extended in this way is a success, whether such success can be attributed to ordo and whether the Washington consensus is, or was, an ordo type of policy.

Part III is devoted to a central problem which confronts both theories, that of how to integrate history and theory in a unified theoretical framework. The two authors of this part, Robert Delorme and Carsten Herrmann-Pillath, diagnose the same difficulty, namely complexity, either complexity taken in its own right (Delorme) or complexity considered as a property of "complex systems" (Herrmann-Pillath). The arguments concentrate on complexity *per se* in the first instance, while bearing on complex systems in the second. These paths are different. Yet they converge on the same kind of conclusion: a weakness in both theories stems from their insufficient integration of features constitutive of complexity or of complex systems. Beyond this specific case, these contributions suggest the need for more thorough reflection on the consequences of complexity for scientific investigation. Whether complex issues can be dealt with satisfactorily with tools designed for non-complex situations is a crucial question on the agenda of future research.

In Chapter 8, Robert Delorme compares *Ordnungstheorie* and the theory of Regulation from the standpoint of complexity. Both theories attempt to provide theoretical alternatives to the lack of integration of history and theory in theoretical economics. Delorme shows that although the founders of Regulation theory knew nothing of *Ordnungstheorie*, they were driven to emphasize the same kind of feature. The parallel between these theories is striking. The consequences drawn from complexity and the method elaborated to take them on board are distinctive features of these two theories. They propose a morphological analysis of the economy, each in its own way, as a solution for dealing with the particular, similar, problem which they each identify independently of the other. This problem is how to integrate the two traditionally irreducible features of history and theory in economics. Delorme defines complexity as irreducibility at a satisfactory level of reduction. The instances of irreducibility evoked in this chapter point to different degrees of complexity. Complexity is essential when it entails a change in the method of scientific enquiry, while non-essential complexity can be treated satisfactorily with the available methods. Although they never refer to this notion of essential complexity, it is clearly contained in both theoretical strategies through their assertion that there is no available satisfactory theory on their subject matter. This rejection is grounded in the absence of any satisfactory theory. But the strategy they follow seems to leave open the issue of the articulation between their systematicity and their fruitfulness. Delorme suggests in conclusion that there may be room for improvement in investigating more thoroughly the meaning of complexity and in the scope for theorizing complexity in its own right. Such an improvement becomes conditioned by, and calls for, an improvement in theorizing about complexity itself.

Carsten Herrmann-Pillath compares Ordoliberalism and Regulation theory from the standpoint of the methodological constraints associated with the study of complex systems, in Chapter 9. He relies from the start on Roy Bhaskar's criticism of current mainstream economic methodology, a methodology developed for non-complex situations and consequently inappropriate for dealing with complex systems. Thus Herrmann-Pillath, following Bhaskar's criticism, starts from properties of complex systems, derives requirements from them and examines both Ordoliberalism and Regulation theory in terms of these requirements. He emphasizes that complex systems are characterized by singularity, historicity, non-linearity, and the absence of fixed boundaries, cognitively autonomous agents and heterarchical ordering – in other words, causal relations between elements and emergent properties run in both directions. These characteristics result in problems of observation and make the identification of causal relations spectacularly difficult. Singularity and the creativity of agents appear, moreover, to be closely related. Then the challenge in theorizing about complex systems consists of devising methods for dealing with singularity and historicity, taking account of the system's boundaries and reflecting actors' creativity. The author compares both theories in terms of these features, which unsurprisingly leads him to point to several missing links, among them a theory of action, a theory of the state and the problem of the relation between system and observer. This suggests strongly that any theory which aims to deal with the subject matter of complex systems, and to offer a genuine alternative to current mainstream theory, cannot avoid reflection on the character of the object of research and the methods appropriate for such a purpose. One can discover just how problematic it can be to leave aside explicitly justified normative considerations. And this may be a weakness of Regulation theory.

The final part addresses the applicability of *Ordnungstheorie* and Regulation theory to the issue of post-socialist transformation. Bernard Chavance and Agnès Labrousse recall in Chapter 10 that regulationist research conducted in the early 1980s had already diagnosed mounting tensions in Central Europe and the Soviet Union. These tensions were analysed as expressing a qualitative turn in socialist economies analogous to a "great crisis" in regulationist parlance. The variety of national trajectories of institutional change, development modes, social trends and crisis processes, was emphasized, thus focusing on the diversity of the initial conditions for the transformation that began in 1990-1991. After setting out the basic concepts of Regulation theory as applied to post-socialist transformation, they concentrate on the transformation in East Germany. They sketch an institution and organization-centred approach based first on a multi-level conceptualization of transformation and, secondly, on the phenomena of rupture, path-dependency and hybridization. Their investigation illustrates the relevance of Regulation theory for this type of issue. But relevance does not mean monopoly and the authors point to the need to supplement Regulation theory with other institutionalist approaches, inevitably leaving on the agenda the issue of the compatibility of different theories for dealing with a given subject matter in a coherent and integrated way.

In Chapter 11 Helmut Leipold discusses some of the advantages and drawbacks of *Ordnungstheorie* in relation to the foundation of transformation policy. He asks to what degree this theory meets the requirement of providing a realistic explanation of the impact of economic institutions and policy measures. His arguments are presented as six propositions: (1) *Ordnungstheorie* provides a well-founded analysis of the impact of institutions on a market economy; (2) it provides a reliable basis for establishing and securing an efficient and acceptable market economy; (3) such a market economy is tempered by social safeguards consistent with liberal and free-market principles; (4) transformation policy in East Germany only partly corresponds to the above principles; (5) in most Central European countries only the initial reform packages were in accordance with ordoliberal principles; (6) the main deficiency of this theory, and of the principles deriving from it, is its neglect of the importance of informal rules and thus of history and path-dependence, as well as its neglect of the properties of political processes. Reform prescriptions were only oriented towards the transformation of formal, that is, legal rules. Hence the author's conclusion on the rationale for complementing what he calls "neoliberal *Ordnungstheorie*" with constitutional political economics.

Now, imagine a transcript of a panel discussion on the subject of how productive *Ordnungstheorie* and the theory of Regulation actually are, and how fruitful in relation to the transformation issue. This is what Hans-Jürgen Wagener has imaginatively devised in Chapter 12 (the discussion did not, as he tells us, actually take place). However unconventional it may appear, the seriousness of the exercise and the stimulating tone it introduces cannot be denied. Serious writing can also be lively and enjoyable. Wagener's contribution can be read as a synthesis of the previous chapters and of the colloquium that led to this volume. It provides also an open conclusion for two theories dealing with open systems.

Finally, I should like to mention the role played in this publication by its two instigators, Agnès Labrousse and Jean-Daniel Weisz. At the time of the original conference and the book's publication, they were doctoral students at the Ecole des Hautes Etudes en Sciences Sociales (EHESS) and at the University of Paris-North, respectively. Their research concerned post-socialist transformation and the German economy, and it took them to Berlin. After they had thought of bringing together scholars from the two schools who had almost no knowledge of each other, they had to organize the meeting itself and find the funding for it. The conference was such a success that then they thought they would edit a book based upon it – and once they had found the publisher, they discovered that editing a volume of contributions is really hard work. On behalf of all the conference participants and the contributors to this volume, I should like to express my thanks and best wishes to Agnès and Jean-Daniel.

Part A

The German Ordoliberalism and the French Regulation School in their National Contexts

Chapter 1

Birth and Growth of the Regulation School in the French Intellectual Context (1970-1986)

JEAN-FRANÇOIS VIDAL*

I. Introduction

The purpose of this text is to show how the principal characteristics of the French Regulation school developed. By Regulation school, we mean the bulk of work published under the direction of R. Boyer and Y. Saillard (1995), the problematic and methods of which are defined in the work of R. Boyer (1987). These studies are often termed "Paris Regulation school".

This school manifests very specific characteristics which distinguish it very clearly from other approaches in economics. It is impossible to understand its particularities with reference to the history of economic theories alone: in fact, originally, the founders of the school worked in the field of economic administration, where their task was often to employ applied mathematics in the preparation of economic decisions. Moreover, the French Regulation school rejects the notion of pure economics and, consequently, is open to social scientific methods other than those of

---

* Translated from French by Claire Laubier.

economics. This is why we must also consider the economic climate prevalent in France at the time of its conception in the mid-1970s within the climate of the social sciences as a whole. Of course, the French Regulation school evolved its problems and methods gradually over the period between 1971 and *c.*1986. We can distinguish three stages in this process, stages which overlap in time. Pre-Regulation, between 1971 and 1976 approximately, is the stage during which studies of empirical macro-economics cause the problematic of Regulation to emerge, but without its conceptualization in terms of a regime of accumulation and Regulation appearing. The second stage, roughly between 1975 and 1983, is that of a distancing of the school from Marxism and a borrowing of ideas from the historical *Annales* school: the methodological system of Regulation tends to stabilize and studies in the field begin to accumulate.

II. Pre-Regulation and the Crisis of Macroeconomics (1971-1976)

Towards the mid-1970s, when the theory of Regulation came into being, its founders were working on behalf of the central government economic administration: M. Aglietta at the *Institut National de la Statistique et des Etudes Economiques* (INSEE; National Institute of Statistics and Economic Studies), R. Boyer at the *Centre d'Etudes des Revenus et des Coûts* (CERC; Centre of Studies in Revenues and Costs), then at the *Direction de la Prévision* (DP; Central Government Economic Forecasting), where J. Mazier, J. Mistral and H. Bertrand also worked; A. Lipietz was working at CEPREMAP - *Centre d'Etudes Prospectives d'Economie Mathématique Appliquées à la Planification* (Centre of Studies in Mathematical Economic Forecasting Applied to Planning), an institute of economic research which, like the CERC, is associated with the *Commissariat Général au Plan* (Commission of Central Economic Planning). They are "economic engineers", educated at the *Ecole Polytechnique*, whose job consists mainly of applying quantitative methods of economics.

The Second World War caused a total upheaval in French economics and society. During the years 1944-1950, France's economic, social and political institutions were rebuilt and refounded, notably with the extension of voting rights to women, the implementation of a system of social security, and the creation of a public industrial and financial sector which constitutes one of the organs of state interventionism. These transformations were accompanied by a significant strengthening of economic administrative bodies. The previous *Statistique Générale de la France* (National Bureau of French Statistics) was transformed into the INSEE, with considerably increased provisions for research. The SEEF - *Service des Etudes Economiques et Financières* (Service of Economic and Financial Studies) – was created, with responsibility for the calculation of the national accounts and annual budgets, and the corner-stone of the national budget. The latter has recently become the *Direction de la Prévision*. Finally, the *Commissariat au Plan* was created, an institution with the singular role of preparing central economic planning. This institution has restricted but independent means at its disposal. It plays, however, the role of instigator or trigger. On the one hand, its role is to call on relevant research bodies to study the conditions of medium- and long-term growth and, especially, major changes in the institutions and structures of production related to growth. On the other hand, the duty of the *Commissariat au Plan* is to convince administration to respect the objectives of growth in their day-to-day management, without recourse to hierarchical privilege. In addition, a very important factor in the preparation of economic planning is that they are sanctioned by commissions which include representatives of the state, employers and trade unions. The purpose of this institutional arrangement is to raise awareness among all the relevant "social partners", in other words the heads of industrial enterprises and workers, of the economic constraints and limited scope for manoeuvre that exist, and to encourage negotiation.

It must be stressed that in the mid-1970s, when the Regulation school came into being, the INSEE, the *Direction de la Prévision* and the *Commissariat au Plan* were the only institutions which had sufficient means at their disposal to study French economic trends, using the tools of applied macroeconomics. These administrations were already constructing various applied models in order to prepare economic budgets and plans. These models largely employed the tool-box of macroeconomic functions elaborated in the English-speaking countries, and which constitute an extension of Keynesianism. However, by the end of the 1960s, the relevance of these models was called into question as a result of economic trends. This led to criticisms of these economic models by founders of the Regulation school, and the elaboration of a methodology which took their limitations into account.

At the end of the 1960s, in all industrialized countries, the crisis of macroeconomics became evident as a result of shifts in the Phillips curve which rendered empirical methods inoperable, and which were translated, of course, into a rise in inflationary tensions. In the English-speaking countries, the debate about the Phillips curve enabled economists of the monetarist school to gain ground, by calling into question the Keynesian approach and claiming that interventionist policies could only be of transitory effectiveness, that in the long-term there was a natural rate of unemployment, and that attempts to lower this by means of policies to control demand were illusory.

In France, shifts in the Phillips curve were interpreted by a section of economists in government administration as being the result of tendencies in capital accumulation and conflicts in the means of distribution. Instead of referring to M. Friedman, these authors referred to J. Robinson, M. Kalecki, R. Goodwin, even K. Marx. A first significant article was written by M. Aglietta in 1971. It criticized the use of the Phillips curve in the FIFI model (the physico-financial model), whose construction he actively took part in, and which was used to set the quantitative objectives of the sixth French plan (1971-1975).

1. Reflection on the limits of applied macroeconomics

M. Aglietta (1971) observed that econometric regression in France led to a Phillips relation characterized by a steady upward trend, and by very imprecise evaluations of the correlation between rates of growth in nominal wages and rates of unemployment. The equation was tested using data for the period 1957-1967, and its application during other periods has been problematic. This weakness in one of the vital relations of empirical macroeconomic models encouraged a more in-depth analysis of wages, regarding them as one aspect of a wider phenomenon, that of wage relations which, in capitalist economies, represent linking and opposing elements in capital–labour ratios. These relations, which swing between conflict and compromise, are fundamentally dependent on the characteristics of capital accumulation. The development of nominal and real wages depends on the way in which productivity gains are obtained, either by means of increased labour-intensiveness, or as a result of introducing new equipment, geographic relocation of premises, or by means of redistribution by sector of productivity gains. It also depends on state intervention. The introduction of the minimum wage was a recognition of the notion of a subsistence wage. The state attempted to increase mobility in order to harmonize labour markets. Policies which deliberately sought to control wage increases were put into force as a means of combating inflation. This combination of factors is so complex that it cannot be encompassed in a single macroeconomic function.

In this article, M. Aglietta began to construct a problematic of Regulation. In fact, the evolution of wage relations and the way in which these relate to capital accumulation and specific institutions played a pivotal role in this theory. But in 1971, concepts of extensive and intensive accumulation, of competitive market regulation and the regulation of trade monopolies, were completely absent.

The FIFI model, which reemerged, notably as a result of instabilities in the Phillips curve, was neither the first nor the last econometric model to be used by the administration. R. Boyer (1976) analysed successive examples of these models and came to conclusions which constitute the methodological, and even the epistemological, foundations of the French Regulation school and which, therefore, are vital to an understanding of this approach.

R. Boyer shows that French administrative bodies have used several groups of models, whose logic was related to changes in the French economy at successive periods, and to a succession of constraints encountered in the course of the process of growth. Between 1945 and 1950, the major problem was to reconstruct a coherent system of production, notably by eliminating obstacles affecting infrastructure and intermediary goods and services. At this time the administration did not have the means at its disposal to construct a truly mathematical model. But the way in which the calculations were carried out, and aims and objectives were set, indicates that the economists had conceived a model of growth of a type similar to that of von Neumann, in which maximal growth is reached as a result of the constraint that input obtained in t has been produced in t-1. This model of maximization of physical quantities chooses to ignore problems of financial distribution. On the one hand, between 1945-1948, high inflation reduced real wages; on the other, aid from the Marshall Plan helped to overcome financial constraints, notably by providing financial investment in infrastructure and the production of capital goods.

From the middle of the 1950s, reconstruction allowed a process of cumulative growth. But the French economy was faced with problems of overheating of the economy in 1957, 1962 and 1969. Global demand, fed by mechanisms of especially dynamic income formation, created inflationary pressures and a tendency to external trade deficits, necessitating the introduction of austerity measures. Such a situation lent itself well to the use of models inspired by Keynesian analysis, in order to determine global demand and inflationary or deflationary tendencies. In France, this perspective was used to construct the models ZOGOL and DECA, presented in detail by J. Benard (1972). Within this framework, the Phillips curve allowed the calculation, in principle, of the rate of inflation in relation to the pressure of global demand. It is noticeable that, in using these models, a return to a political economics based on macroeconomic equilibrium is implied, whereas reconstruction occurred as a result of political investments by the industrial sector. In addition, in the DECA model, investment by industrial enterprises depends on profits and this allows reference to be made to conflicts regarding income distribution.

In the late 1960s, the French economy began to internationalize and there was a movement away from insular economic models such as those of Keynes. The administration proceeded to construct the FIFI model, whose particular characteristic is to distinguish between the sector protected from foreign competition and the sector exposed to it. In the protected sector, the level of production and employment is determined by demand, as in Keynesianism. But in the sector exposed to foreign competition, prices have to be fixed by companies at the same rate as those of foreign competitors. For any given level of wages, profits are fixed and determine levels of investment and production. In this model the Phillips curve plays a regulatory role. If profits in the exposed sector are too low, investment is reduced and, in turn, causes unemployment to increase, wages to decrease and profits to recover. In the FIFI model, policies of budgetary intervention are ineffective, as they push wages up and reduce the competitiveness of the exposed sector. Structural policies need to be implemented which favour the transfer of resources in a skilled and qualified workforce, capital investments and technology in the exposed sector. This model is clearly better destined for a policy of structural reform within the framework of medium-term growth than a model of economic trends. It is a typical tool for medium-term planning.

R. Boyer then demonstrates how world-wide inflationary pressures from 1969-1970 onwards gave rise to the most essential hypotheses of the FIFI model: the accelerated pace of foreign inflation allows the exposed sector to raise its prices; in addition, the revaluation of the Deutsche Mark and depreciation in value of the franc resulted in a weakening or a disappearance of the fixed relation between world prices and domestic prices. As already shown, these developments mark the origin in France, as well as abroad, of the crisis of econometric models. In many countries, economists were to seek new theories and conduct new empirical trials in order to construct new models. Of course, French economists were to pursue the same course. In the early 1970s, the *Direction de la Prévision* launched new research in two different directions: the construction of a model focused on profit and capital accumulation, and one focused on the factors determining money and credit across numerous financial institutions. The Bank of France was interested in the application of the monetarist model of the Bank of St Louis to the French economy.

But, on the other hand, Regulation economists drew from their experience of econometric models a particular epistemology, which distinguishes it clearly from other approaches. According to R. Boyer, the diversity of empirical models has three causes. The first is that opposing theories exist which lead to different empirical analyses. The second is that the importance ascribed to certain variables or certain relations, and their endogenously or exogenously determined nature, depend on the political economic objectives and measures taken by the actual institutions themselves, which construct the model. Finally, and above all, in the long term, the economic structures which formed the basis of the equations entered into the models mutate, whereas by its construction a model explains a process of economic evolution based on a system of relations considered constant. In the long term, all empirical models are condemned to obsolescence. Similarly, international comparisons show that there are also huge variables in the gaps between structures and economic dynamics. Before formulating the equations that describe these types of behaviour, it is necessary first to study the characteristics of the system we hope to model, namely that of the organization of social relations and economic institutions.

2. A plethora of empirical research related to capital accumulation

Around 1973, the INSEE and *Direction de la Prévision* started to develop empirical research programmes relating to capital accumulation, income distribution, inflation and growth. These studies make reference to N. Kaldor, J. Robinson, M. Kalecki and R. Goodwin. They refer, on the one hand, to macroeconomics used for forecasting and, on the other hand to the description of an evolution of systems of production in France and other great industrial powers since the 1950s.

In 1973-1974, the STAR model (*Schéma théorique d'accumulation et de répartition*; Theoretical Scheme of Accumulation and Distribution) was constructed at the *Direction de la Prévision*. This model was to be used in the calculation of economic budgets. The team who constructed it included, most notably, R. Boyer and J. Mazier, who were joined by J. Mistral. Before detailing the econometric equations of the model, the authors presented the principal characteristics of growth in France between 1955 and 1972. The major "stylized facts" shown are the drop in added value relating to wages, the drop in consumer spending as a part of GNP, the rise in the rates of consumer debt, and the rise in the rates of company profit and rates of capital accumulation creating cyclical movements in the rates of company debt. The aim is to consider, with the aid of a dynamic model, simultaneous trends and fluctuations. As the authors indicate, the model is not exclusively focused on long-term projections of balanced growth; the path of growth, which in fact is of a cyclical nature, is followed, so that short- and long-term elements and relations are brought to light.

The STAR model largely returns to the functions of public consumption, investment, imports and employment which constitute the tool-box of applied macroeconomics associated with Keynesianism; but the elements of spending are determined as nominal values, and they depend also on the levels of debt. Above all, company spending on investment is dependent on profits, and consumer spending is especially dependent on wages. This part of the model is close to the "Cambridge" theory whereby the mechanisms for the increase in global demand are related to those of income distribution. In the STAR model, if the wage–profit ratio is taken into account, global demand at current prices depends on external factors. There is a double relation between profits and the value of GNP: for one part of the given profits, profits increase with global demand; investment, which is one contributing factor of global demand, increases with profits. These two relations determine investment, profits and the value of GNP.

The group of equations which describe the model combine an analysis of the process of production and of income distribution. In this respect it is closer to Marx than to the Cambridge neo-Keynesians. M. Fouet (1975) and Y. Barou (1976) study economic growth and fluctuations in the United States and in Great Britain, taking their inspiration from the principal variables and equations of the STAR Model. In the case of the United States, M. Fouet notes a cyclical movement in capital accumulation. During periods of recession, enterprises restructure the process of production. The restructuring allows enterprises to realize the potential productivity gains previously accumulated. The share of profit and the ratio of profits and capital accumulation increase. But the acceleration in investment is such that the growth in capital outstrips that of profits, which in turn causes a decline in cost-effectiveness and recession. In this model, recession implicitly plays the role of regulator, as it allows restructuring to take place without regard for profit. But there are considerable differences between the American and the French economic dynamics. For one thing, in the United States, tendencies in productivity gains are not linked to the pace of substitution of labour capital. Furthermore, after 1966, the capital–outlay ratio was subject to downward pressures, whereas it remained at a high level in France. The British case, analysed by Y. Barou (1976), also shows very significant differences from France: productivity gains at work are very limited there; profits as a part of added value are weak, and there is insufficient investment. The author suggests that these weaknesses can be explained in part by the nature of labour relations. But in this 1976 article, Y. Barou did not seek to give accurate definitions of the relations between economic variables and institutions. In a text written in 1978, Y. Barou reexamined the same empirical material in order to draw much more precise conclusions concerning the relationship between growth, social institutions and distribution of capital. Of course, this more in-depth study results from the fact that the basic concepts of the regimes of accumulation and regulation have been more precisely formulated.

The INSEE has also pursued statistical research into the tendencies of capital accumulation in France since the 1950s, published under the direction of C. Sautter (1974) and entitled *Fresque historique du système productif français* (*A Historic Overview of the French System of Production*). This study does not refer to the systems of behaviour of the macroeconomic models. It focuses on the measurable relationships which can be established between different productivity indicators, production factors, the distribution of productivity surplus and the capital–outlay ratio. Two main conclusions are drawn. The first is that the acceleration in the substitution of labour capital in the course of the 1960s caused an apparent drop in the capital–outlay ratio; but the negative consequences of this development for the profit ratio were offset by a decrease in the burden of company taxation, an increase in realized capital stocks, thanks to inflation, and a drop in the relative price of capital goods. These results will be interpreted by orthodox Marxist economists as demonstrating that the replacement of "living labour" by "dead labour" has begun to exert a downward pressure on the profit ratio which can only be sustained with the help of state aid to large enterprises. The second important conclusion is that the capital goods sector, defined widely to include capital goods for enterprises, as well as durable household goods, is the one which has achieved by far the best performance in terms of growth and the capital outlay ratio.

Thus we can see that the Regulation school emerged out of a prior stage of empirical macroeconomic analysis prompted by the disappointments encountered at the end of the 1960s, out of approaches adopted with regard to global demand, and out of the Phillips curve. Theoreticians often underestimate this stage. However, it is important for an understanding of Regulationist analyses. In the first instance, it helps to explain why Regulation uses a realistic methodology in which empirical observations play an important role, more important than in conventional forms of academic study. Moreover, it stresses the specificity of the French situation, namely that in French economic administrations, even at relatively high levels, a significant number of economists in the 1970s referred to J. Robinson, M. Kalecki and even to K. Marx.

During this period of pre-Regulation, empirical studies enabled the identification of the main tendencies of accumulation in France and abroad after the Second World War, and the tensions these might result in, as well as some of the causes of their disappearance at the end of the 1970s. International comparisons have shown that there are significant differences between national trends. Thus French growth during the period 1945-1975 cannot be analysed by means of one model alone. But two essential elements in the theory of Regulation are missing. First, the notion that the dynamics of capitalism can be transformed radically in the very long term has not yet been really brought to light. Secondly, the important role played by institutions in growth and fluctuations is implied, but not fully analysed.

In addition, in 1974 and 1975, the industrialized countries underwent a serious recession, accompanied by high inflation. From this period on, the founders of the Regulation school formulated the hypothesis that the economic situation of this time marked only the beginning of a long period of difficulties and that the era of strong economic growth had come to an end. This led them to study dynamics in the very long term, and the development of the industrialized world from the nineteenth century onwards. Moreover, the fact that inflation accelerated during the recession of 1974-1975, whereas price levels had slumped between 1930-1932, prompted them to examine the transformations brought about in economic institutions during the years 1930-1940.

III. The Phase of Historical and Dialectical Materialism (1975-1983)

There are very considerable differences in the approach to Marxism in Germany and in France. These differences should not be a source of misunderstanding. They can largely be understood in the light of the consequences of the Second World War. In 1945, Germany was divided into two parts. In the Eastern part, the Soviet Union imposed a Communist regime, in which Marxism played the role of a doctrine legitimizing a police-driven Regulation. When the Soviet system foundered in the 1980s, Germany could free itself.

Frances history was very different. During the Second World War, a minority of the French actively resisted National Socialism. Within this minority, Communists and Gaullists were the most numerous. After 1945, these two forces played an important role in France. They were often opposed, but in practice they reached a state of compromise. In addition, as has already been mentioned, a new economic and social constitution was established in France between 1944-1947. This reestablishment was brought about by a coalition of Gaullists, Communists, Socialists and Christian-Socialists. The participation of Communists in these decisive events greatly contributed to the importance of Marxism in the intellectual, social and political life of France between 1945 and the 1980s.

The two fundamental studies in Regulation theory are M. Aglietta's *Régulation et crises du capitalisme* [*Regulation and Crises of Capitalism*] (1976) and a collection of research papers edited by J.P. Benassy, R. Boyer, R.M. Gelpi, A. Lipietz, J. Mistral, J. Munoz, and C. Ominami entitled *Approches de l'inflation: l'exemple français* [*Approaches to Inflation, the French Example*] (1977). This collection was reedited by CEPREMAP; it was not published in its original form. The first part provided the subject of two books edited by A. Lipietz (1979-1983). The analyses of the second part were in part reused in the book by R. Boyer and J. Mistral (1978), and in the articles by J.P. Benassy, R. Boyer, R.M. Gelpi (1979) and by R. Boyer (1978; 1979).

The book by M. Aglietta, and the first part of the CEPREMAP report, make explicit use of the framework of Marxist theory. The second part of the CEPREMAP report is more empirical; it points out that its theoretical point of reference is not necessarily that of Marxism, and that it can also be justified in terms of the neo-Keynesian theories of disequilibria.

1. The tenets of Michel Aglietta

The work of M. Aglietta (1976) is the product of the post-doctoral thesis he prepared under the direction of Raymond Barre. It was discussed in a seminar at the CEPREMAP in 1975 and published in an altered form. In the introduction to the 1976 edition, M. Aglietta presents his methodology in a very explicit manner. He starts from the notion that standard analyses ignore two basic aspects of growth. Growth constitutes an upheaval in the methods of production and lifestyles. These upheavals are experienced by all those who participate in the economic process, but form no part of the theories. Economic and social changes are accompanied by social and political conflicts which are also absent from standard theory. The study of growth should be one of dynamics, in the etymological sense of the term, in other words, a study of forces and changes in conditions. The purpose of the research is to identify the forces which transform the social system and guarantee its cohesion over a long period, to explain the conditions and types of behaviour that result in qualitative changes, and to discover the role played by economic crises in these transformations. It must be stressed that these have remained the research topics of the Regulation school, even after it distanced itself from Marxism. In addition, as C. Vercelonne (in Seba?/Vercelonne 1994) emphasizes, the notion that growth is the outcome of upheavals, tensions and conflicts is also a central idea in French forms of economic planning, the purpose of which is both to stimulate change and to control its effects by means of the coordination of state intervention and negotiation between social partners. M. Aglietta demonstrates explicitly that his method is that of historical and dialectical materialism. It is necessary to identify the inner logic in the transformation of economic systems which manage to reproduce themselves while simultaneously undergoing a process of transformation. In the capitalist system, there is a dialectic of being and becoming, in which the system transforms itself while staying the same: this is Regulation. The characteristics and fundamental contradictions of capitalism need to be defined first in abstract terms; but the logic in the contradictions and transformations of these concepts can be properly understood only by studying their concrete development in the history of economic facts; the logical progression of these concepts becomes a representation of real historical movement. Here M. Aglietta appears to give Marxism a Hegelian tone.

The first chapter is an example of the method of free interchange between theoretical concepts and the history of economic facts. He starts with an analysis of the concepts of the labour theory of value, of surplus value and of capital formation. His main conclusion is that the Marxist theory of value does not logically imply a downward trend in the profit ratio. The profit ratio depends on the surplus–value ratio and the intrinsic value of the product, which, by implication, might show upward or downward trends. The identification of these trends requires an analysis of the historical dynamic of capitalism. These trends can be examined successfully, since the Marxist concepts of the labour theory of value, of surplus value and capital formation, can be transposed in the gap between the empirical variables of real wages, labour productivity and depreciation, with the help of the monetary concepts of working hours and the workforce, which in turn reflect the level of production costs and the level of consumer prices. An examination of American growth patterns shows that the dynamic of capital accumulation has changed fundamentally during the twentieth century. Until the First World War, capital accumulation was brought about mainly in the capital goods sector. It was characterized by a high level of instability, by limited productivity gains, and by inertia in patterns of working-class consumption. During the 1920s, productivity gains accelerated and a transformation in patterns of working-class consumption became apparent. After the crisis of the 1930s, a new and more stable regime of accumulation was established, the outcome of acceleration in the consumer goods sector, and accompanied by radical changes in working-class patterns of consumption. During the course of this preliminary analysis, the notion emerges of regimes of accumulation succeeding each other across great economic crises.

For M. Aglietta, these transformations have derived principally from the expansion of Taylorism, which, if present for a given period of time in the workplace, allows manpower time to increase and consequently, surplus value too. They also derive from the expansion of Fordism, which is based on the simultaneous changes in the production process in the consumer goods sector and in the lifestyle of the working class (a mechanism of relative surplus value). These transformations are not the result of exogenously determined forms of technical progress to which the American economy has adapted. They result, in the first instance, from changes in the organization of labour decided by the owners of capital. But when the changes in the production process are not accompanied by a change in working-class consumer patterns, the slowing-down in the consumer goods sector, which restricts outlets in the capital goods sector, leads to worsening profit ratios and a weakening in accumulation. The progression from Taylorism to Fordism enables this contradiction to be overcome. But the boom in the consumer goods sector, as a result of changes in working-class consumer patterns, implies a transformation of wage relations and a radical change in the wage dynamic.

At this point in his reasoning (p. 173 in the first edition of *Régulation et crises du capitalisme*), M. Aglietta returns to his article of 1971 on the development of wages in France. This article examined the question of fluctuations in the Phillips curve. The response was that behind Phillips we rediscover Marx: over very long periods, these fluctuations result from the contradictions of capitalism, and the need to transform the contradictory relationship between capitalists and workers, in order to maintain the profit ratio. It is precisely the conflicts ensuing from this contradictory relationship which will bring about its transformation and ensure its permanence, albeit in a variety of forms over long periods. Only in a stabilized regime of capital accumulation, is it possible to obtain a stable econometric equation to describe wages. Such conditions are confirmed in Fordism, where real wages are influenced only to a small degree by economic fluctuations, and where wage increases anticipate productivity gains. But by examining more deeply the conditions underlying transformations in relations between capitalists and wage-earners, M. Aglietta calls into question the orthodox version of historical and dialectical materialism adopted by the Communist parties. According to this version, changes in society result from the development of productive forces which upset social relations and institutions. The latter are described as "superstructures", in the sense that they are completely dependent on productive forces and the social relations on which these rely. According to the orthodox version of Marxism, the most important changes in twentieth-century capitalism are the development of more concentrated and growing state support state for capital accumulation. For M. Aglietta, the most important change is that of wage relations. This has resulted from a process of conflicts and compromises during the course of which wage relations have been transformed and codified into new institutions, the most important being collective bargaining agreements. The institutions are described by the term "structural forms". The term "form", which alludes to the Marxist expression "form of value", to describe labour in the abstract, indicates that institutions have the effect of rendering homogeneous concrete relations which, in themselves, are of infinite diversity, in order to give them global cohesion. For example, collective bargaining agreements harmonize wage increases allowing changes in consumer patterns within the working class as a whole. By assuring a progression from the infinite diversity of the concrete to the generality of the abstract, institutions work in the same way as concepts that ultimately end in rationalizing the diversity of the real. In addition, the term "structural" indicates that the institutions are characterized by a certain stability. In the long term, they evolve according to economic, political and ideological factors. The serious economic crisis of the 1930s created severe social and political conflicts in the United States, the outcome of which was the development of new institutions codifying new compromises and having the effect of channelling the dynamics of incomes and consumption, and thus of assuring the development of a Fordist regime of capital accumulation.

This concept of the institution presented by M. Aglietta still remains in the final version of the Regulation school. However, M. Aglietta's concept distances itself from historical and dialectical materialism, at least in its official version, for two reasons. The first is that institutions are likely to have an influence on capital accumulation; they are to a certain extent "infrastructures" which can play a pivotal role. The second is that institutions are no longer considered to be direct instruments of the dominant class. They are the result of compromise between the social classes, and they can have a profound effect on the conditions of life of the working class. Compared to traditional Marxism, this concept opens the way to reformism, reintroduces the notion of compromise, and suggests that there is no fatal determinism attached to trends in the forces of production. Furthermore, the idea that institutions have the effect of bringing together concrete behaviour patterns in a global dynamic opens the way to a regulationist concept, progressing from macroeconomics to microeconomics. Markets alone are not sufficient in assuring the global coherence of decisions at a micro-level. Institutions, and the norms of behaviour they bring about, contribute a great deal to the establishment of cohesion, and it is very important to analyse their role and their transformation in the context of long-term and short-term dynamics. These are, according to the Regulationists, the institutional and historical foundations of macroeconomics.

The second part of Aglietta's book analyses how the emergence of Fordism has transformed competition between enterprises and the financial and monetary system. The constant upheaval in the general process of production implies a high and stable level of accumulation, with a rapid pace of decommissioning of equipment. It implies competition through investment and economies of scale and, consequently, the development of large industrial groups. Stability of funding is obtained by means of the rapid development of intermediaries, to the detriment of the financial markets. In the case of intensive accumulation, the two main sources of finance for enterprises are provisions for depreciation, which generate auto-financing, and bank credits. In addition, the existence of large industrial groups stands in opposition to mobility of capital and, consequently, there can no longer be equality of profit ratios between sectors. This may result in price rates diverging from production costs. Compared to conventional Marxism, and also to neoclassical theory, this conclusion is of an agnostic nature, and it has been supported by the majority of regulationists: patterns of competition are so diverse in the real economy, that it is impossible to construct an adequate theory of relative prices.

For M. Aglietta credit should be clearly distinguished from money. Credit, for example a bill of exchange, may circulate as an instrument of payment, but only within the restricted area of industry and business. Money has the property of being a universally accepted means of payment in any national economy. Liquidity allows the purchase of any type of product; but it also acts as a constraint. Payments cannot be delayed indefinitely, and the economic agent must have sufficient cash at his disposal to pay for the merchandise required. This is the monetary constraint. Capital accumulation, where money is used in order to obtain the means of production, is financed in part by bank credits, that is to say, by the issue of private banker's debt. In order to become money, the bank credit has to be able to circulate in the economy as a whole, and thus receive social validation. The way in which this validation is effected has a considerable influence on the economic dynamics of crises. Crises initially result from the fact that areas of production in enterprises created by private labour cannot always be transformed into money through exchange based on normal prices. In this case, they are subject to severe monetary constraints. The non-realization of merchandise decreases the creditworthiness of enterprises, and that in turn of their banks. If the bank deposit money is socially validated by the possibility of converting it into gold, the growth of the monetary sum is limited, which forbids the long-term maintenance of the non-profit-making agents. This results in a form of crisis characterized by a chain reaction of bankruptcies and the depreciation of the merchandise in the form of price cuts. But if the bank deposit is validated as bank deposit money, by converting it into money issued by the Central Bank, the latter still has the possibility of increasing the value of the cash to finance the deficits of private agents, and to alleviate monetary constraint. The crisis is translated into monetary devaluation, in other words, inflation. In traditional Marxism, money is considered to be of little importance compared to the process of production and income distribution. In this approach to Regulation, which owes a great deal to S. de

Brunoff, money and financial systems play an important role in economic dynamics.

The work of M. Aglietta as a whole thus sets out the main concepts used by the Regulation school to analyse growth, the crises and changes in twentieth-century capitalism. Because of its wide scope, it enables the conceptualization of the studies in empirical macroeconomics carried out in France between 1970-1975. The trends observed between the 1950s and the 1970s correspond to a regime of intensive accumulation, established as a result of the crisis of the 1930s. The economic dynamics result largely from the fact that behaviour is directed by the institutions established in the years 1930-1940. The inconsistency in the empirical models of the late 1960s results from the fact that the regularities in the intensive regime of accumulation were in the process of weakening, a sign that a serious crisis was approaching.

The studies of M. Aglietta were discussed within the CEPREMAP in 1975, and led to two distinctive programmes of research. The first consisted in reviving Marxism, by calling on the principal results of the Regulation approach. This is the approach adopted in the first part of the CEPREMAP report on inflation. The second consists in tackling the regulationist concepts in more detail, with regard to the history of capitalist economies, without necessarily referring to Marx. This agnostic approach, in which historical and dialectical materialism becomes historical and institutional macroeconomics, is that of the second part of the CEPREMAP studies. For the moment we will present the first programme, as developed in the work of A. Lipietz (1979; 1983).

2. Regulation, Marxism and structuralism

We have shown how Regulation differs from the official Marxist theory of monopolist state capitalism. But, in the 1970s in France, there was another version of historical and dialectical materialism: the structural Marxism of L. Althusser and E. Balibar, presented notably in the work *Lire le capital* [*Reading Capital*], (1971). The works of A. Lipietz are largely a rereading of *Capital* that diverges from the structuralist reading. This leads us briefly to outline the main tenets of structuralism which exerted a profound influence on French intellectuals between 1950 and 1970.

As F. Dosse (1992) points out, the father of structuralism is the anthropologist C. Lévi-Strauss, who established its basic principles in the work, *Les structures élémentaires de la parenté* [*The Elementary Structures of Kinship*], published in 1949. This method was transformed and transposed to numerous other social sciences to such an extent that it is difficult to give its precise definition.

However, it is possible to distinguish three major methodological principles of structuralism. The first is that the relevance of an analysis does not depend on empirical facts, but on the hidden relations found between the elements of a system. It is necessary to examine the hidden logic which exists "below the surface" of the superficial and ephemeral world of events and experiences. It must be stressed here that in the formation of his ideas Lévi-Strauss recognized the influence of geology and Marxism: he conceives the idea that the geographic configuration of a given terrain can be explained by its underlying geographical layers and strata, and that the empirical variables of prices and profits reflect more profoundly theories of labour value and capital gain. Lévi-Strauss proposes, so to speak, a sociology of depth.

The second main characteristic of structuralism is to give preference to synchrony over diachrony. Here Lévi-Strauss refers to the linguist F. de Saussure and his *Cours de linguistique générale* [*Course of General Linguistics*] (1872) discussed before 1914. In the nineteenth century, the principal task of linguistics was to study the evolution of languages; by comparing different languages spoken in Europe and Asia, it could be shown that they originated from a common root, Indo-European. This type of analysis favours diachrony, in other words, evolution over a period of time, and relies largely on comparative methods. In contrast, F. de Saussure considers that a language is composed at a given time of a coherent system of interdependent symbols, and that the purpose of analysis is to bring to light the synchrony, in other words, the general system of interdependences which link the elements of a language at a given time. The notion of synchrony is, in fact, familiar to economists. In his reflections on language, F. de Saussure was probably influenced by Walras's theory of General Equilibrium, in which all markets are interdependent. Lévi-Strauss was very interested in econometric models and anticipated the use of mathematics in the social sciences. In short, the notion of synchrony, or structural causality, is close to the notion of interdependence of variables in an economic model with several equations. Structuralism is strongly in favour of an analysis of synchrony, rather than of diachrony. This leads to the analyses where social systems are presumed to reproduce themselves in an identical form with stable structures, and to underestimating the importance of evolutionary development and events.

The third characteristic of structuralism is that it is an extreme form of holism. This results from the fact that Lévi-Strauss associates the notion of the unconscious, borrowed from Freud, with the notion of the institution having a stabilizing effect on the social system. The analysis of kinship structures is revealing. For Lévi-Strauss, the prohibition of incest, which is observed in all societies in varying forms, indicates that different social groups are forced to exchange women through the union of marriage. This norm, which is an essential element of social cohesion, is not consciously known by those who practise it; it results from unconscious mental structures and is imposed on individuals through incest taboos. This concept is in total opposition to the methodology of individualism; the individual is not aware of his actions and does not act out of choice, since his behaviour is dictated by a social logic, the existence of which he does not even suspect.

Starting from these principles, L. Althusser and E. Balibar (1971) proceed to a rereading of K. Marx's *Capital*. They reach a series of conclusions which are opposed to traditional interpretations of Marxism. The first is that Marxist theory itself is a structure. This means first of all that one cannot be satisfied with a literal reading of *Capital*. It is necessary to seek the deeper truth of Marxism, for Marx himself did not completely theorize his own theory and, consequently, his writings may contain ambiguities. An in-depth reading of *Capital* enables us to see that Marx constructed a system of concepts which was radically new. The idea that Marxism resulted from a transformation of the philosophy of Hegel, in which spirit is replaced by economics, is a mistaken one. There was no progressive (diachronic) development which led to Marxism. Marx made an "epistemological break" with his predecessors. This enabled Marx to show that the truth behind capitalist modes of production lies in the hidden (theoretical) concepts of the relationship of production and surplus value.

Each mode of production is a structure which reproduces itself. Marx's "schemes of reproduction" show how the mode of capitalist production is reproduced in the process of production. Its fundamental structure is the relation of ownership of the means of production. Capitalists possess and own the means of production; they have the ownership of the finished product; this allows them to obtain its capital gain; with mechanization they control the methods of work. In the process of production, money capital enables the purchase of the means of production, that is to say, capital goods and the labour force. The labour force is paid the minimum wage for survival; it remains poor and has to sell its labour all over again. Product sales enable capital gains to be realized and capital to be accumulated. The process of production is a form of reproduction in the full sense of the word, in other words, a means of perpetuating the proletariat and capital and the relations of ownership.

In this analysis, the inherent contradiction within capitalist modes of production is minimized. Capital accumulation is accompanied by the development of mechanization and, consequently, of the rise in the organic composition of capital, which causes a downward trend in profit ratios. But the contradiction does not lie in the synchronic logic of capitalism; it is a result of its projection in time, which has a dynamic quality (in the mathematical sense of the term) and not a diachronic one. Above all, the increase in the organic composition of capital can always be offset by the rise in the ratio of capital gains. E. Balibar suggests that the contradiction always gives rise to a certain balance, even when this balance is achieved by means of a crisis; periodic crises are but "violent and momentary solutions to existing contradictions, violent eruptions which reestablish upsets in the balance for a while", as Marx says. Crisis brings about a devaluation of capital and the reestablishment of profit ratios. Crisis reproduces the structure of the mode of capitalist production.

A third important consequence of structural Marxism is to eliminate the subject: the determining element of the structure is the relationship of production, which produces social classes; social classes (and individuals) are not subjects but elements that support the relationships of production. This leads to an important implicit consequence. In traditional readings of Marxism, the social classes are also subjects, political players who revolutionize political institutions when these have become incompatible with the development of productive forces; revolution is a moment in the diachronic process which leads from one mode of production to another. In structural Marxism, where the subject is eliminated, transitions become very difficult to render into theory, and events are ignored.

The decline of structuralism in France probably began with the events of May 1968, in spite of the fact that J. Lacan claimed that "structures had taken to the streets". What is more, regulationist economists started from the idea that the years 1974-1975 marked the beginning of a big crisis which would end with a restructuring of capitalism: during the course of the crisis, Fordism would disappear and be replaced by new regimes of accumulation. This intuition is fundamentally opposed to the structuralist approach. A. Lipietz's two books contain a critique of structural Marxism, starting from the notion of Regulation. Regulation exists because structure does not reproduce itself spontaneously.

In A. Lipietz's study of 1979, he criticizes especially the idea that structures reproduce themselves without any major contradiction. He stresses that it is precisely Marx's schemes of reproduction which show that the double and circular nature of the process of production (production reproduces both the labour force and capital) are not reproduced simultaneously; it is by no means certain that all goods will be disposed of, and that all the labour force will find jobs. In reality, the structural Marxist interpretation underestimates the role of money which introduces a disparity between the supply and demand of goods. Imbalances and crises are all the more likely to happen when changes in the norms of production and consumption occur at varying rates. These contradictions can be regulated in two ways. A. Lipietz uses the results of the second part of the CEPREMAP report: in competitive Regulation distortions are regulated after they occur by falls in prices and bankruptcies; in monopolist Regulation, characteristic of Fordism, they are regulated before they occur, by mechanisms which anticipate productivity gains and which bring about an inflationary drift; monopolist Regulation also enables the coordination of the development of the capital goods sector and the consumer goods sector.

In his book published in 1983, A. Lipietz criticizes structuralist metaphysics, according to which the empirical world is deceptive and the real, visible world can only be known by theoretical means and imposes itself on human beings without them being aware of it. A. Lipietz recognizes the fact that Marx distinguished between a world of empirical exoteric appearances and a real esoteric world; but he tries to show that the esoteric world of appearances has, especially as regards monopolist Regulation, a consistency, which is more and more real, and which enables it to guide the behaviour of its economic subjects. The esoteric world is that of the exchange of goods, of money, prices, profits and wages. The exoteric world is that of the relations of production, of abstract labour, of labour value and of capital gain. The exoteric world has the "fetishism" of goods as its origin, in other words, the economic agents who exchange goods are not aware of the fact that they are exchanging abstract work: the price, which is only a form of the labour–value ratio, disguises this fact in the eyes of those exchanging the goods. The problem of the progression from value to prices has often been analysed as being the algebraic problem of the formation of "the costs of production", but this is not the major factor. The most important consideration is the progression from labour value to prices and nominal incomes. Most notably, it is necessary to understand the reasons why, in Fordism, there are significant productivity gains, which force down the labour value of the goods and continually force up the general level of prices.

The exoteric world guides the decision-making of economic agents and is of an autonomous nature compared to the esoteric world; in other words it is governed by its own laws. Added value has to be distributed between wages and profit (and in some cases annuities) which gives rise to conflict. Added value must be spent, in order to be realized. Wages are paid to workers, and then these are spent in their totality on consumer goods. In contrast, profits are not necessarily spent in their totality; the realization of profit in this case implies the creation of money. In the case of monopolist Regulation, where money is no longer linked to gold, this creation of money anticipates the growth of production and gives rise to a continuous fluctuation in the general level of prices. Labour value, transformed into nominal prices, causes an inflationary surge. In addition, in Fordism, there are significant disparities in productivity gains between sectors, which implies the existence of considerable and opposing variations in labour value in the exoteric world; but in the esoteric world, there may be restrictions which limit the variations in relative prices and cause distortions between nominal incomes and production, finally leading to a depreciation in the value of money, measured in hours of abstract labour. Here, A. Lipietz returns to the concept of equivalent labour and money used by M. Aglietta. Finally, the rigidity of relative prices in the exoteric world is reinforced by the continuous creation of money in the form of bank deposit money: the Central Bank validates nominal prices and their increase by transforming money issued by commercial banks into a universal means of payment. Thus the autonomy and reality of the exoteric world is reinforced by inflationary surges.

In the work of M. Aglietta and A. Lipietz, the Regulation approach is a new interpretation of Marxism. Compared to the traditional version of "monopolist state capitalism", it presents several major differences. It insists on the idea that the development of productive forces is directed by the social relationships of production; that is to say, the ownership of the means of production, which makes it possible to control work methods and determines the nature of productivity gains. It follows that the changes in twentieth-century capitalism were associated above all with the changes in wage relationships, which oppose and unite capitalists and wage-earners, and which have enabled the maintenance of profit ratios for a long period. Compared to structural Marxism, the Regulation school rehabilitates the diachronic nature of the economy, that is to say, evolution in its full sense. It reevaluates the contradiction which requires Regulation and rediscovers the subject and its subjectivity in price categories and nominal incomes. The Regulation school has revived Marxism in the light of the history of the economic facts of the twentieth century and advances in macroeconomic analysis.

But from the end of the 1970s onwards, Marxism experienced a brutal decline in France. No doubt there are a number of reasons for this development. To recall just one of these: the crisis in Soviet-type economies was deepening and becoming confrontational in Poland. Polish workers were opposing their Communist regime more and more openly, as it transformed itself into an official military dictatorship. The decline of the Soviet system contributed indirectly to a weakening in the research programme, the purpose of which was to revive Marxism with the help of the concept of Regulation. But it did not affect the programme of research presented in the second part of the CEPREMAP report, which involved the construction of a historical and institutional macroeconomics that no longer used Marxist theory as its point of reference.

IV. The Construction of Historical and Institutional Macroeconomics (1977-1986)

I shall demonstrate how the Regulation school distanced itself from Marxism, borrowed from the French *Annales* school and consolidated its methods to constitute the basis of the bulk of its work.

1. The agnostic turning-point

With historical and institutional macroeconomics, the Regulation approach adopted a realistic type of approach, founded on agnosticism. Agnosticism starts from the premise that it is not necessary to resort to an invisible world which is impossible to observe, in order to explain the real world. This hypothesis distinguishes the Regulation approach from both Marxist theory and the neoclassical school. In contrast to Marxist theory, the Regulation school no longer refers to the esoteric world of labour value and constructs its arguments around exoteric variables of incomes and prices, which are those of theoretical macroeconomics and national accounts.

Furthermore, the Regulation school abandons the Hegelian–Marxist dialectic, according to which contradictions can always be bypassed by means of synthesis: there is no providence to guarantee that a new, coherent and better, regime of accumulation will emerge in a serious crisis. The theory of Regulation preserves two concepts of Marx. The first is that capital needs to be defined in the wider sense of the term: it is at the same time a combination of capital goods, of monetary and financial assets, and of economic powers which enable the owners of capital to decide the rate of innovation and change in the work organization. The second is that capital accumulation is an essential factor of growth, notably because it causes changes in modes of production and lifestyles; but capital accumulation is unstable and often leads to social tensions. Marx derived a third concept from this: namely that capitalism would be overthrown by a revolution. But the Regulationists are content with saying that the contradictions of capitalism should be regulated, in other words, controlled, by means of a body of institutions which, themselves, rely on compromises between the social classes or social categories.

The agnosticism of the Regulation school is clearly evident compared to the neoclassical theory of General Equilibrium. The world of General Equilibrium is an invisible world, which is presumed better than the real world. In regulationist theory, there is no reference to the world beyond. In addition, the Regulation approach reorganizes, in principle, the diversity of the real world. Human nature cannot be defined once and for all. Human beings create tools, institutions, ideas and myths, and they are transformed by using them. Man evolves with the help of his own creations. Economic developments are, therefore, subject to path-dependency. Moreover, for very many reasons, not all markets can function in the same way, and different markets are linked to different types of institutions. These combinations vary from one country to another. For the Regulation economist, the purpose of research is not to construct a logical system of ideas which claim to explain everything. It is to gather the infinite diversity of the real world into some general types of regime of accumulation and modes of regulation, in order to explain short-term and long-term dynamics.

Marxism and the neoclassical school start from the premise that the real world is deceptive and, consequently, that one should seek a different world from the real one, and one which is more truthful. The Regulation school is not a heterodoxy which opposes these two orthodoxies, for it rests on agnostic principles.

The real world is difficult to explain, not because it is a deceptive illusion, but because it is diverse and complex. In the course of their activities, economic agents transmit messages, of which a part may be captured by statistical systems. These messages are considerable in number, incomplete and often ambiguous or distorted. The problem of economic analysis is to correct them, to filter them, and to interpret them in a logical way. There are diverse types of filters in macroeconomics, such as indexes, aggregate planning and stylized data of the Kaldor type, the relevance of which requires careful study. Logical relations between variables can be established according to different macroeconomic functions, in order to be combined into a mathematical model, after subjection to econometric tests.

2. Economic trends and structure: borrowing from the French *Annales* school

In M. Aglietta's book it was long-term growth and the changes which it brings about, which were at the centre of the analysis. The dynamics of economic trends were discussed at the end of the study, with regard to the role of money. In the second part of the CEPREMAP report, which is the foundation of historical and institutional macroeconomics, inflation and not growth is the central theme. The reason for this is that in 1975-1976, the Regulationists came to the conclusion that the instability of the years 1973-1975 marked the beginning of a serious crisis which would result in the end of the Fordist regime of accumulation. But it is also necessary to explain why the recession of 1974-1975 caused an acceleration in inflation, whereas in the nineteenth century and the first third of the twentieth century recessions were accompanied by a fall in prices. The problem, therefore, is to explain the variability over very long periods of short- to medium-term fluctuations. This shift in the problematic, from growth to inflationary trends, is accompanied by a shift in the definition of Regulation: for M. Aglietta, it is defined as the general effect of long-term transformation within an economic system that enables them to be reproduced, whereas for R. Boyer, it is "the conjunction of mechanisms contributing to the reproduction of the system as a whole, considering the conditions of economic structures and social forms. This Regulation is at the root of the dynamic of short and medium-term periods, since it defines the nature and development of economic trends." The analysis is concerned with the mechanisms of economic adjustment, studied with the aid of empirical and theoretical neo-Keynesian macroeconomics. But the Regulation school uses the macroeconomic equations differently from the way in which they are habitually used. In standard economics, relations are presumed to be basically stable. In the Regulation approach, the equations describe economic structures, and a model translates the conditions of a given moment of an economic system in a synchronic fashion. But in the real world, nothing is eternal: all structure is bound to disappear or to mutate. In the Regulation approach, the study of synchronic logic is always accompanied by a diachronic analysis, where research is carried out into how the forces which transform economic structures are translated into shifts in their macroeconomic functions, in other words, by significant changes in their parameters.

The study of wages, prices and production by R. Boyer (1978) is a clear example of this method. The empirical data point to three successive profiles related to economic trends in France. In the eighteenth century and at the beginning of the nineteenth century, it can be observed that economic crises are characterized by a fall in agricultural and industrial production, by a decrease in the nominal wage and a dramatic rise in consumer prices. After 1850, a new economic climate becomes evident: the decline in industrial production is accompanied by a fall in the nominal wage and in prices. After the 1850s, industrial capitalism becomes dominant within the liberal institutional framework established by the revolution of 1789. The imbalances between different productive sectors cause repeated crises of overproduction, prices fall and lay-offs force down wages in industry. These new forms of adjustment follow the patterns of "competitive Regulation". After 1950, the development of collective bargaining agreements and of the welfare state assure a relatively stable progression of direct and indirect wages. The slowing-down in labour adjustments limits the number of lay-offs during recessions, and the concentration of enterprises limits falls in prices. Incomes and consumption of wage-earners are stabilized, and wages and prices remain at a fixed low level. This type of economic climate follows the patterns of "monopolist Regulation". These changes occur notably as a result of shifts in wage relations, unemployment and consumer prices, which are studied in detail.

This method, which consists in showing how the transformation of economic structures modifies economic trends, is partly inspired by the French historical *Annales* school. This school of historians was founded in 1929 by M. Bloch and L. Febvre, as G. Bourde and H. Martin mention (1983). Its work is a response by historians to the development of the social sciences and structuralism. In fact, traditional history was above all a political history. Before the First World War, the sociologist and economist F. Simiand accused historians of concentrating unduly on political events, suggesting that they should study instead the rules governing causal laws, as is the case in the social sciences. Historians were sensitive to these criticisms, and as a result the journal *Annales*, which first appeared in 1929, mainly contains articles on economic and social history, with frequent reference to statistical data the construct of which homogenized the real world by eliminating experienced events. After the Second World War, C. Lévi-Strauss violently criticized historians, accusing them of putting too much emphasis on the contingency of events, without identifying the hidden structures which determine the functioning of societies. F. Braudel responded by giving a different definition of structure, as being the bulk of permanent data relating to the life of societies, which can be brought to light in long-term studies and which eventually gives rise to trends.

There is, therefore, a convergence between the Regulation school, which seeks to historicize structures, and the *Annales* historians, who seek to show the structures underlying events. For example, it can be shown that there are similarities between the Regulationalists and F. Braudel's methods of studying time. For Regulationists, there are three temporalities: the short term of economic trends, from one to three years; periods which cover several decades, characterized by the domination of a form of Regulation; the very long-term which is secular, during which there are successive regimes of accumulation and forms of Regulation during the course of great crises. F. Braudel, in his study of the Mediterranean area in the time of Phillip II of Spain (1947), throws light on the static geographical structures which determine man's conditions of existence and the strategic position of nations. The development of markets and the evolution of social groups are achieved in the long term; the short term is that of the events of political history. However, F. Braudel studied above all the periods preceding the industrial revolution, whereas the Regulationists, whose studies pertain especially to the periods succeeding this watershed, are confronted with societies with rapid developments, which have the effect of shortening all temporalities.

In fact, Regulationists have borrowed their ideas mainly from E. Labrousse. E. Labrousse was an assistant to A. Aftalion at the Faculty of Law in Paris, and he produced the main bulk of his work during the 1930s and 1940s. His aim was to study the relations between upheavals in societies and political regimes, as well as growth and crises. Over a long period, he collected and studied available statistical sources for the eighteenth and nineteenth centuries. In his work on *La crise de l'économie française à la fin de l'Ancien Régime et au début de la révolution* [*The Crisis of the French*

Economy at the End of the ancien régime and the Beginning of the Revolution] (1944), he claims that the French Revolution was caused by the conjunction of three economic cycles. From 1733 to 1778, there was a long period of agricultural expansion and the sale of land rents stimulated the growth of towns and industry, thus strengthening the influence of the bourgeoisie. From the 1770s onwards, the trend was reversed, and a long depression affected profits and wages; this difficult phase was linked to the sharp decline in public finances, which the *ancien régime* was incapable of reforming. Finally, in 1788, a subsistence crisis erupted, with the result that in 1789 all the grievances accumulated, leading to a major crisis that was to transform institutions. In the work which E. Labrousse coedited with F. Braudel (1976) he also studied the evolution of cyclical crises in the nineteenth century. He showed that in the course of this century, fluctuations in agricultural production diminished in intensity. This was the case because cereal production, which is very susceptible to climatic risks, was gradually replaced by the cultivation of potatoes and fodder and the breeding of livestock, which are much less unstable. The variation in agricultural prices diminished considerably, thanks to the development of national and international transport.

The crises of the *ancien régime* had significant repercussions until the 1860s, since the dramatic rises in cereal prices sharply reduced the incomes of the urban population and often resulted in riots. However, these *ancien-régime*-style crises gradually gave way to "industrial crises", where fluctuations in the mining and metallurgical sectors were more pronounced than in the consumer goods sector. During the period of expansion, capital accumulations increased at an accelerated pace but ultimately led to a decline in markets, resulting in industrial crisis. E. Labrousse points out that there are numerous theories of crisis, but that, in reality, one must explain why different types of crises occur.

R. Boyer and J. Mistral (1978) applied this problematic in their comparison of the economic dynamics of stagflation in 1974-1975, and of deflation in 1930-1932. They show that these differences arise from the fact that competitive Regulation was still dominant during the 1930s, whereas monopolist Regulation came into effect after the beginning of the 1960s. In 1930-1932, it is true that the real average wage rose, since prices were more flexible than salaries; but the speed of adjustment of employment was such that cutbacks in personnel reduced real wages overall. In addition, the markets were structured in such a way that companies were forced to lower their prices, to respond to the fall in demand. In 1974-1975, the speed of adjustment in employment was slow, and thus reductions in the labour force were restricted. The continued rise in the average wage, combined with the growth in indirect wages, maintained the general level of real wages so that consumption continued to rise. Companies had the option not to reduce their prices. The change in Regulation between the years 1930 and 1960 was translated into shifts in the relations between wages and unemployment, between employment and production, and between prices and demand.

The Regulationists brought to light the succession of three types of Regulation, inspired by the observations of E. Labrousse regarding crises of the *ancien régime* and crises of liberal capitalism, and elaborated and developed the Labroussian axiom according to which each economy experiences structural crises. The structure of a modern economy, which comprises the main characteristics of its productive system and the major economic institutions, determines the mass of trends related to growth, described as the "regime of accumulation". One must distinguish first between two types of crises. "Small crises" are short-term readjustments, enabling a return to the characteristic trends of the regime of accumulation and a reproduction of structures; the small crises are regulatory. There are also "great crises", where capital accumulation does not result in a spontaneous redistribution. A great crisis may result from the exhaustion of a regime of accumulation, from growing contradictions between procedures of Regulation and the regime of accumulation, in the weakening of regulations, either because of exterior shocks, or as a consequence of internal dynamics. Great crisis leads to a change in the regime of accumulation, by transforming its institutions. We can observe that great crisis is to the economy what revolution is to politics. There is, however, an important difference: a political revolution may succeed in three days, whereas a great crisis requires at least a decade to be resolved.

3. The stabilization of the regulationist method

The studies contained in the second part of the CEPREMAP report were completed with the aid of empirical analysis of the productive sections by H. Bertrand (1978), and of sectors by J. Mazier, M. Baslé, J.F. Vidal (1979, 1984). They enabled the elaboration of the characteristics of the extensive regime of accumulation, associated with competitive Regulation at the end of the nineteenth century, and those of the intensive regime of accumulation and of monopolist Regulation (Fordism), which prevailed between 1950 and the 1970s. Long-term empirical time series show that these regimes can be distinguished by differentiated values in capital accumulation rates (investment–capital ratio), rates of productivity growth and capital per capita, and by specific variations in growth between sectors. Time series in the productive sector show that the acceleration in the pace of growth in France after 1945 was linked to the fact that the consumer goods sector, which performed poorly during the first half of the twentieth century, underwent rapid modernization after 1945, with a considerable increase in investment and in the substitution of labour for capital. The progress from an extensive to an intensive regime of accumulation is also translated into a considerable augmentation in returns to scale, evident in the shifts in the law of Kaldor and Verdoorn. The same method also shows that after 1975, the abrupt slowing-down in productivity gains can be explained to a large degree by contractions in returns to scale, which constitutes an important factor in the crisis of Fordism.

The progression from competitive regulation to monopolist regulation, can be traced with the help of shifts in wage equations, employment equations (such as the functions of Brechling and O'Brien, which enable estimations of the pace of adjustments in employment) and price equations (relations between prices, average costs of production and pressure of demand). The shift in the principal functions of macroeconomic models can be used to indicate transformations in institutions, in particular those that regulate employment relations and the issue of currency.

Studies by C. André and R. Delorme (1983) have provided a regulationist approach to the state. Here it must be stressed that, paradoxically, neither M. Aglietta's book (1976) nor the CEPREMAP report (1977) contained chapters specifically devoted to the issue of the state. Given the fact that the Regulation theory starts from the premise that markets cannot regulate themselves independently, an analysis of the state would naturally follow from this. But the state is not mentioned, since it is the process of regulation which enables markets to function, and regulations cannot be mistaken for the state. Moreover, apart from economic regulations, there are regulations which have the purpose of channelling tensions or social and political contradictions to which the state also contributes. The different modes of regulation are varied; they may stem from habitus resulting from education and individual experience; they may be associated with the diverse social contracts and conventions which take place among individuals and organizations; they may stem from diverse state interventions. Types of state interventions depend first of all on the form of Regulation: R. Delorme and C. André show that competitive regulation corresponds to limited state mechanisms, and that monopolist regulation to the welfare state. This development was an outcome of the social contract of the nineteenth century, which was based on liberal principles, and which, with the increasing bargaining power of wage-earners, and following the crisis of 1929 and two world wars, became a Fordist compromise. Furthermore, R. Delorme and C. André also show that the precise terms of the social contract might vary from country to country, according to its historical and cultural traditions, as is shown by the example of the varying types of organization of educational systems in the major industrialized countries.

At the beginning of the 1980s, Regulationists succeeded in putting forward a theory of growth and of transformations in the major capitalist economies during the twentieth century. The programme of regulationist research appeared to be complete. In reality, research continued, notably because attention was directed towards countries which have not yet reached the stage of Fordism: the developing economies and the countries of Eastern Europe.

4. The expansion of approaches to Regulation

The expansion of the programme of regulationist research is clearly demonstrated in the work directed by R. Boyer and Y. Saillard (1995), which contains around fifty contributions, some of which deal with the basic concerns of macroeconomics in the major industrialized countries, with additional studies of the non-industrial sectors, such as agriculture or the service sector or international relations; the other contributions are studies of the Latin American countries, South-East Asia and the Maghreb, and cover numerous developments in the final crisis of planned economies. These studies show that the Regulation school employs intermediary concepts in its methods. Since Regulation never assumes eternal and absolute facts, the concepts which it puts forward have a marked element of relativity. To recoin a phrase of J.C. Passeron (1991), the realist approach leads to the conclusion that each concept contains spatio-temporal coordinates, which make it impossible to transfer any one of them to distant physical locations or periods of time. For regulationist economists who are, in their profession, immersed in a standard theory which takes an eternal view of things, such an approach might have presented certain difficulties. It was for this reason, as R. Boyer (1987) mentions, that research economists sought to apply the notion of Fordism to countries in the process of development, where industry and wages are limited.

At the beginning of the 1980s, regulationist analyses of the Latin American countries started to develop. These studies showed that the regulationist approach was fruitful, and that it could return to previous analyses of the structuralist type (in the Latin American sense of the term), dualism or dependency, at the same time incorporating these into its own methodology. The studies by C. Ominami (1986) on Chile are significant. They demonstrate that during the 1930s, the Chilean economy underwent a severe crisis. This crisis led to two series of changes. On the one hand, capital accumulation was achieved before 1929 in the mining sector; after the 1930s, it was partly reoriented towards industries producing goods to replace imports. On the other hand, the political regime was overthrown; the traditional state, controlled by land-owning oligarchies was replaced by an interventionist state which tried to bring about compromise between the owners of capital and middle-class wage-earners. Monopolist forms of regulation developed in the public sector and the modern industrial sector. But the importance of an intermediary sector prevented the development of a genuine wage relationship, on the macroeconomic scale. This is why the regime of accumulation established in Chile after the 1930s cannot be described as Fordism. It was a specific regime of accumulation which, after the 1960s, resulted in specific contradictions, such as the difficulty in resolving the question of the substitution of imports with capital goods and the increase in the food deficit.

Regulation approaches are equally fruitful in the case of the countries of Eastern Europe. B. Chavance (1990) shows that at the beginning of the 1980s, there was considerable debate as to the reasons why the Soviet system was unable to establish an intensive regime of accumulation, combined with a rapid growth in patterns of working-class consumption. Social relations between wage-earners and management presented significant differences from those in western countries. Soviet workers were subject to strong political pressures but, on the other hand, the lack of manpower and the high mobility of the workforce restricted the management's capacity to control the organization and the efficiency of work. The Soviet regime of accumulation could only be of an extensive nature given its low productivity gains. Management directed capital accumulation along the lines of section 1, and this prevented the development of patterns of consumption of the Fordist type. The system of central economic planning led to numerous imbalances within the framework of a "regime of shortage". From the 1960s onwards, the authorities launched a succession of reforms without, however, succeeding in rescuing the regime of accumulation from collapse. The failure of Regulation, in M. Aglietta's sense, ended in a very great crisis.

Towards the middle of the 1980s, it became evident that the methodology of the Regulation approach had become clearly defined, and that an endogenous process of cumulative research had begun. The methodological system could then be presented in a precise and detailed way by R. Boyer (1987).

V. Conclusion

The purpose of this chapter has been to examine the real and ideal context in which the Regulation school was created, and not to discuss its recent developments or its limitations. Our first conclusion would seem to be that the Regulation school evolved out of the process of transformation from historical and dialectical materialism to historical and institutional macroeconomics. The theory of Regulation transformed Marxism in two major areas. Given that institutions are not simply superstructures which reflect infrastructures, and that they can direct capital accumulation, based on social compromise, Regulation has "socio-democratized" Marxism since 1976-1977. Setting aside the issue of labour theory of value, Regulation theory examines the process of capital accumulation within the framework of theoretical and empirical macroeconomic concepts, which enable it to integrate advances in this discipline.

During the past fifteen years, the scope of neoclassical methodology has widened, to encompass facts relating to institutions or returns to scale in which Regulationists have also been interested. But Regulation preserves its specific approach, notably in the use it makes of history. An example of this relates to returns to scale. At the end of the 1970s, Regulationists employed previous studies by Kaldor and Verdoorn to attempt to explain why macroeconomic productivity gains increased at a rate of 5% per annum during the 1950s and 1960s, and at a rate of 2% after 1975. Of course, given the significant role played by returns to scale, it has to be admitted that institutions of the Fordist type, which stimulated demand, like the welfare state, were a cause and not only a result of growth. The neoclassical approach focused on returns to scale from the 1980s onwards, when it became clear that they might be compatible with the theoretical system of General

Equilibrium. But no research was undertaken into the breaks in the productivity trend observed in 1945-1950 and 1973-1975 within the framework of the theory of endogenous growth. This illustrates the difference between the Regulation theory and other neoclassical approaches. Regulation studies the development of economic systems, and seeks to locate changes in trends precisely in time and explain them.

Another conclusion of this text is that the birth of the Regulation school is closely linked both to institutional data, which are specifically French, and to the intellectual climate in Paris in the period from 1960 to 1970. But the Regulation approach was also substantially influenced by German nineteenth-century philosophy, by anglophone macroeconomics and by the problems and methods of French historians. It can thus be described as being truly European.

References

AGLIETTA, M.: "L'évolution des salaires en France au cours des vingt dernières années", *Revue économique*, January 1971.

AGLIETTA, M.: *Régulation et crises du capitalisme, l'expérience des Etats-Unis*, Paris (Calmann-Lévy) 1976.

AGLIETTA, M., SAUTTER, C. (Eds.): *Quatre économies dominantes sur longue période*, Paris (INSEE) 1978.

ALTHUSSER, L., BALIBAR, E.: *Lire le capital*, Vol. 1-2, Paris (Maspéro) 1971.

ANDRÉ, C., DELORME, R.: *L'Etat et l'économie*, Paris (Le Seuil) 1983.

BAROU, Y.: "L'économie britannique en crise, 1949-1974", *Statistiques et études financières*, série orange, 24 (1976).

BAROU, Y: "Salariat britannique et développement", M. AGLIETTA, C. SAUTTER (Eds.): *Quatre économies dominantes sur longue période*, Paris (INSEE) 1978.

BENARD, J.: *Comptabilité Nationale et modèles de politique économique*, Paris (Presses Universitaires de France) 1972.

BENASSY, J.P., BOYER, R., GELPI, R.M., LIPIETZ, A., MISTRAL, J., MUNOZ, J., OMINAMI, C.: *Approches de l'inflation: l'exemple français, Rapport de la convention de recherche*, no. 22/176 (CEPREMAP/CORDES), Vol. III, Dec. 1977, in: *Recherches économiques et sociales*, 12 (1978), Paris (La Documentation française).

BENASSY, J.P., BOYER, R: *Régulation des économies capitalistes et inflation*, Paris (CEPREMAP) 1978.

BERTRAND, H.: "Une nouvelle approche de la croissance française: l'analyse en sections productives", *Statistiques et Études financières*, série orange, 35 (1978).

BOULLÉ, J., BOYER, R., MAZIER, J., OLIVE, G.: "Le modèle STAR", *Statistiques et études financières*, série orange, 15 (1974).

BOURDE, G., MARTIN, H.: *Les écoles historiques*, Paris (Le Seuil) 1983.

BOYER, R., MAZIER, J., OLIVE G.: "Un nouveau modèle de prévision macroéconomique: STAR", *Economie et Statistique*, November (1974), pp. 29-53.

BOYER, R.: "La croissance française de l'après-guerre et les modèles macroéconomiques", *Revue économique*, (1976), pp. 882-940.

BOYER, R.: "Les salaires en longue période", *Economie et Statistique*, (1978) pp. 27-59.

BOYER, R., MISTRAL, J.: *Accumulation, inflation, crise*, Paris (Presses Universitaires de France) 1978.

BOYER, R.: *La théorie de la régulation: une analyse critique*, Paris (La Découverte) 1987.

BOYER, R., SAILLARD, Y.: *Théorie de la Régulation. L'état des savoirs*, Paris (La Découverte) 1995.

BRAUDEL, F.: *La Méditerranée et le monde méditerranéen à l'époque de Philippe II*, 2 vols., Paris (Armand Colin) 1947.

BRAUDEL, F., LABROUSSE, E.: *Histoire économique et sociale de la France*, Vol. 2, Paris (Presses Universitaires de France) 1976.

BRUNOFF, S. de: *La monnaie chez Marx*, Paris (Editions sociales) 1973. First edition: 1971.

CHAVANCE, B.: "L'analyse des systèmes économiques socialistes et la problématique de la régulation", *Revue d'Études Comparatives Est-Ouest*, (1990), pp.135-151.

DOSSE, F.: *Histoire du structuralisme*, 2 vols., Paris (La Découverte) 1992.

FOUET, M.: "1948-1974 croissance et répartition de la valeur ajoutée aux USA", *Statistiques et études financières*, 20 (1975).

FOUET, M.: "Pourquoi la récession de 1974-1975 a été la plus grave de l'après-guerre", *Statistiques et études financières*, série orange, 23 (1976).

LABROUSSE, E.: *La crise de l'économie française à la fin de l'Ancien Régime et au début de la révolution*, Paris (Presses Universitaires de France) 1990. First edition 1944.

LÉVI-STRAUSS, C.: *Les structures élémentaires de la parenté*, Paris (PUF) 1949.

LIPIETZ, A.: *Crise et inflation, pourquoi?*, Paris (Maspéro) 1979.

LIPIETZ, A.: *Le monde enchanté: de la valeur à l'envol inflationniste*, Paris (La Découverte) 1983.

MAZIER, J., BASLÉ, M., VIDAL, J.F.: "Croissance sectorielle et accumulation en longue période", *Statistiques et Etudes financières*, série orange, 40 (1979).

MAZIER, J., BASLÉ, M., VIDAL, J.F: *Quand les crises durent...*, Paris (Economica) 1993. First edition 1984.

OMINAMI, C.: "Chili: échec du monétarisme périphérique", in R. BOYER (Ed.): *Capitalisme fin de siècle*, Paris (Presses Universitaires de France) 1986.

PASSERON, J.C.: *Le raisonnement sociologique. L'espace non-poppérien du raisonnement naturel*, Paris (Nathan) 1991.

SAUSSURE, F. de: *Cours de linguistique générale*, Paris (Payot) 1986. First edition: 1971.

SAUTTER, C. (Ed.): *Fresque historique du système productif*, Paris (INSEE) 1974.

SEBAÏ, F., VERCELLONE, C.: *Ecole de la régulation et critique de la raison économique*, Futur antérieur, Paris (L'Harmattan) 1994.

Chapter 2

The Regulation Approach as a Theory of Capitalism: A New Derivation

ROBERT BOYER

I. Old Issue, Renewed Interest

The possibility, nature and evolution of capitalism have been a major concern for philosophers and political scientists since the very earliest periods of the emergence of markets. In historical retrospect an oscillation between two visions can be observed. For classical and, more recently, neoclassical economists, the market mechanism is basically self-equilibrating, the only source of crises being clumsy interventions on the part of political authorities or the legacy of inadequate and obsolete social relations. But first Marxists and then Keynesians have repeatedly argued the very opposite – that the pure market mechanism may lead to major crises, financial disequilibria, rising inequalities or stagnation and long run unemployment. Therefore state intervention in relation to the nationalization or socialization of investment decisions may help to maintain the viability of a mature capitalist system. This long swing in ideas about the role of the market has been clearly pointed out by Karl Polanyi (1944; 1983).

The second vision is of the roots of the unprecedented growth regime since the Second World War: multifaceted interventions by the state in relation to credit and monetary supply, labour legislation, competition, the exchange rate and external trade have promoted a new form of capitalism, combining both private and public firms, market mechanisms and state regulations (Shonfield 1967). During the 1960s, many social scientists concluded that a mixed economy combining private initiative and public intervention delivered better results than a typical laissez faire economy and, of course, better than a centrally planed economy. At that time, the theory of convergence of economic systems argued that both Western capitalist economies and Eastern socialist societies were gradually converging towards a somewhat similar economic system: in the Western world, more state intervention was correcting the drawbacks of market (unemployment and rising inequality), whereas in Eastern Europe, and especially in the Soviet Union, economic reforms were introducing more market mechanisms and decentralization in order to cure slow technical progress and promote a more sophisticated consumption pattern.

This very brief historical retrospect clearly shows how different contemporary analyses of the same issue can be. Many structural changes took place and the so-called Keynesian neoclassical synthesis broke down, to be replaced by quite different visions of the functioning of markets and the appropriate role of the state. The belief in markets has spread out to a large proportion of politicians and experts. The more market-oriented capitalist economies, such as the U.S., are frequently considered as a model to be imitated, especially by European countries affected by substantial long-term unemployment. Many national models which were exploring an intermediate configuration between typical state-led and purely market-led forms of capitalism have collapsed, one prominent example being Sweden to say nothing of France. Last but not least, the Soviet-type economies have dramatically collapsed and left a new space into which capitalist institutions along with democratic reforms in the political system can be introduced. A decade after the collapse of the Berlin Wall, the surprise and disappointment associated with these transformations call for a renewed interest in the theory of capitalism.

This chapter starts from a paradoxical observation: the huge gulf between the achievements of modern economic theory and the rather complex and unexpected evolution observed in the last two decades. To introduce this theme, I quote Robert L. Heilbroner (1988) who wrote the entry on "Capitalism" in the New Palgrave Dictionary of Economics:

"(The) conception of capitalism as a historical formation with distinctive political and cultural as well as economic properties derives from the work of those relatively few economists interested in capitalism as a " stage " of social evolution. In addition to the seminal work of Marx and the literature that his work has inspired, the conception draws on the writings of Smith, Mill, Veblen, Schumpeter and a number of sociologists and historians, notable among them, Weber and Braudel. The majority of present day economists do not use so broad a canvas, concentrating on capitalism as a market system, with the consequence of emphasizing its functional rather its institutional or constitutive aspects." (p. 350b).

In the conclusion of his enlightening analysis, Heilbroner wrote:

"Contemporary mainstream economists are largely uninterested in questions of historic projections, regarding capitalism as a system whose formal properties can be modelled, whether along general equilibrium or more dynamic lines, without any need to attribute to these models the properties that would enable them to be perceived as historic regimes and without pronouncements as to the likely structural or political destinations towards which they incline. At a time when the need for institutional adaptation seems pressing, such an historical indifference to the fate of capitalism, on the part of those who are professionally charged with its self-clarification, does not augur well for the future." (p. 353b).

The French theory of Regulation has been designed precisely in order to overcome this serious limitation to modern economic research. Building critically upon the Marxian legacy and the breakthroughs of Keynes and Kalecki, this research programme aims to provide exact analysis of the current stage of modern capitalism, the transformations of its basic institutions, the changing pattern of structural crises. But Regulation theory is only a component of a wider array of heterodox research, ranging from English and American radical thinking (Glyn 1988; Bowles/Gordon/Weisskopf 1983) to New Institutionalists (Powell/Di Maggio 1991), to say nothing of economic sociologists (Granovetter/Swedberg 1992), historians and political scientists (Hollingsworth/Schmitter/Streeck 1995). What contributions do these researches in socioeconomics make to the understanding of contemporary capitalist transformations? Perhaps it is to deliver a more convincing and relevant picture than is provided by the bulk of neoclassical - sophisticated but balkanized - micro-modelling of a system, the logic and dynamic of which is hardly captured. Let us take some of the major events which took place during the last decade and compare the forecast of leading economic theoreticians or practitioners with what actually happened. Many paradoxes emerge from this comparison, and they organize the way this chapter unfolds.

The most striking structural change is probably the collapse of Soviet-regime-type economies, which was interpreted as a definite victory for capitalism and democracy. But has capitalism actually emerged from the political and economic turmoil so evident in the former Soviet Union? What do economists really know about the implementation of capitalism? Probably still less than they do about its functioning (II). A second and earlier event was the demise of Keynesian orthodoxy concerning counter-cyclical policies and, more generally, the benefits which can be derived from state intervention. In the early 1980s, it was quite commonly believed that if Keynes was wrong, Milton Friedman and/or Friedrich von Hayek were necessarily right. A decade later, there is a sharp contrast between the optimistic beliefs of policy makers about the efficiency of markets and what modern microeconomic theory concludes about the strict limits of this coordinating mechanism confronted with the more typical issues of contemporary economies (III).

Since the majority of economists are contemporary, and observers of the current situation, one might anticipate an unprecedented level of expertise and knowledge about the dynamics of capitalism. Quite surprisingly, however, the numerous and sophisticated contributions of modern research deliver either elegant mathematical models or a series of scattered and frequently contradictory empirical results ... but no clear contribution to the understanding of modern society as a whole. This might be an unintentional outcome of the extreme division of labour among social scientists. Alas, there is apparently no "invisible hand" for economic theory.

It might therefore be useful to start again from the core question addressed by the founding fathers of political economy: what are the basic institutions which allow competition among individuals ultimately to deliver an acceptable economic order? In a sense, the theory of Regulation is faithful to this general objective. Basically, this chapter argues that the very limits of Walrasian economic theory call for precisely the same basic institutions that are captured by the five institutional forms, previously derived from an internal criticism of traditional Marxist theory. This is why the subtitle of this chapter puts forward a possible new derivation of Regulation theory. In so doing, it sets out an approach that is not so very far removed from

Ordoliberalism, whatever the distance in terms of the implied conceptions for economic and political history, as well as economic policy strategies (IV).

Taking into account all the conclusions gathered while discussing these issues, a short conclusion proposes an agenda for institutionalist research for the twenty-first century. It is hoped that the theory of Regulation will be part of the ambitious process to restore relevance and analytical rigour to socioeconomics, a welcome and much-needed complement and challenge to contemporary neoclassical research (V).

II. The Victory of the Capitalist System and the Paucity of Its Theory

One of the major events of the last decade has been the unexpected and dramatic collapse of Soviet-regime-type societies. Most observers have analysed this structural crisis as the direct consequence of two of the systemic features of this regime: the concentration of political power into the hands of the Communist party and related nomenklatura, and the centralization of economic decisions by the Gosplan.[1] Thus, conventional economists have been active in advising the new governments and have put forward a very simple equation:

Capitalism = Private property rights + Markets

which was supposed to replace the wrong mix of political and economic institutions:

Socialism = Collective property + Gosplan.

Both the Soviet and East German cases clearly demonstrate the failure of such a simplistic vision, however attractive, of a smooth and easy transition from one economic system to another. Unfortunately, economists now have to recognize that they have little understanding of the implementation of capitalist institutions, particularly if the transformation takes place quickly as

[1] The Gosplan was the state planning committee of the former USSR (before 1948: state planning commission). Its mission was to centralize and direct the industrial and agricultural production in the whole country. The term nomenklatura refers to the ruling elite (note of the editors).

a result of the structural crisis of a previous regime that continues to exert its influence.

1. From Soviet Union to Russia: Capitalism is not self-implementing

Privatization was supposed to institute clear property rights, by contrast with the old collective system. Thus, the previous inefficiency would be removed by the relentless profit-seeking activity of true entrepreneurs in the Schumpeterian sense. They would save scarce resources, get rich by responding to consumer demand and would be induced to innovate, merely to sustain high profits. The destruction of the planning and centralization that had previously existed under the aegis of Gosplan would develop "horizontal" transactions between independent economic units and progressively lead to the emergence of markets, conceived as fair and efficient, by contrast with the old system with its administratively established prices, chronic shortages, major sectoral unbalances and bias against consumer goods.

Consequently, the interplay of a new generation of managers with the strong competition induced by market mechanisms would be bound to increase the volume and quality of supply, and solve one of the major limitations of the Soviet regime. Simultaneously, it was assumed that this new market system would create strong incentives for workers to work harder for higher wages and move from inefficient public firms to modern private firms: rising productivity and living standards would then propel the Russian economy towards Western standards of living. This successful economic transition would benefit the democratic ideal: richer consumers and workers would become citizens eager for the continuation of the economic reforms who would therefore should vote for the new government which had manufactured such a happy and favourable transition. Market and democracy would reinforce one another and create the conditions for a fairly rapid, or at least steady, convergence towards American or European economic and political systems (Diagram 2.1).

The former Soviet Union is a clear counter-example of such a naïve vision. In fact, no "big bang" really took place, since the economic and political reforms only shifted the previous information and power system, which was by no means totally destroyed. First of all, the new businesses have been launched by individuals closely connected to the old nomenklatura, which provided political protection, information and links with foreign trade and firms. Therefore, in many East European countries formal privatizations have not always allowed the creation of a totally new class of entrepreneurs but have rather, by means of recombinant process, redistributed property among the elite (Stark 1990; 1992). On the other hand, however, much earlier state intervention in public spending, taxation and subsidies has been eroded or cancelled, and still more has become increasingly uncertain, given the unpredictability of government action.

This frequent change in the rules of the game is typical of the contemporary Russian situation and is quite harmful for the emergence and prosperity of real Schumpeterian entrepreneurs: it then becomes more rational to play the role of arbitragist or speculator between markets than to take the risk of long-run and irreversible investment. Therefore, the simultaneous collapse of the Soviet system and the transfer of property rights within various sections of the elite have benefited speculators more than they have enhanced business innovations, initiatives and productive capital accumulation (Sapir 1996). At the same time, rapid inflation and the state's inability to pay wages regularly and adjust pensions has produced a drastically unequal distribution of income and wealth. Most elderly workers have suffered from a severe decline in living standards, which has in turn resulted in a contraction in the production of most domestic – and especially public – firms. Last but not least, since the majority of the population is worse off, it comes as no surprise that the legitimacy of the market is now challenged and that this is manifested in political instability, potentially blocking the deepening of economic reforms. Since 1996, Russian elections have experienced the potential conflict between a fast implementation of market mechanisms and political stability within a democratic system (Diagram 2.2).

In more theoretical terms, the Russian transformation seems to contradict the neo-Austrian vision concerning the selection by pure market mechanisms from among alternative social and economic organizations. If the macroeconomic context is highly unstable (fluctuating inflation rate, inability to collect taxes and therefore to meet public spending including the civil servants' pay, large swings in relations with foreign markets and firms and so on), purely private adjustments cannot compensate for the lack of provision

From Soviet Union to Russia: The market does not necessarily generate the institutions required for its functioning

Diagram 2.1: Transition to a market economy: The conventional analysis

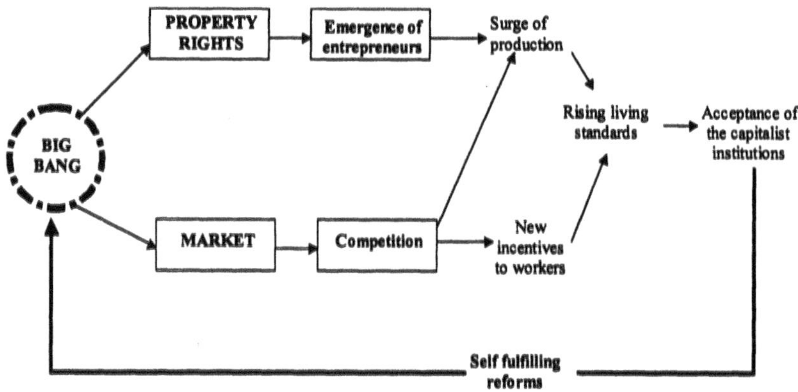

Diagram 2.2: The transformation of a Soviet-type economy: An institutional analysis

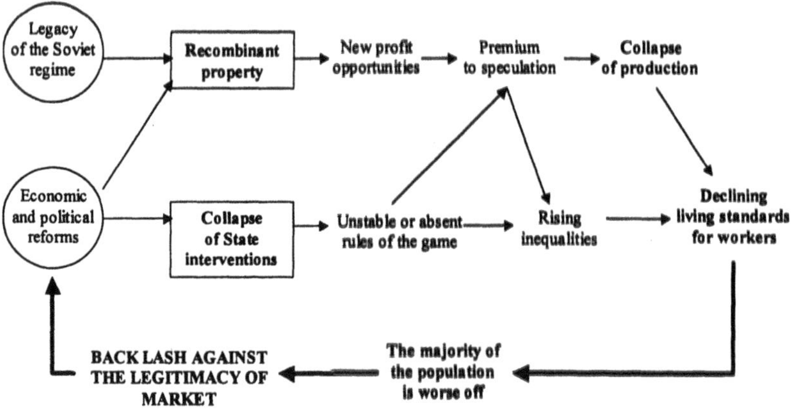

of public goods such as a securely based judicial system, accurate public and private accounting, a clear national monetary unit and an efficient payment system. The complete loss of legitimacy of the former Soviet state seems to prevent the very implementation of another modern and capitalist state, an indispensable ingredient in the implementation of a market system. Ultimately, economists come to recognize that they know very little about the necessary and sufficient institutions of capitalism, since they were considered as given or "natural" in slowly evolving Western societies. This lack of expertise is particularly serious when an old system is collapsing and requires specific formulations in order to build a new one. There is no invisible hand in the implementation and selection of basic capitalist institutions. After all, this is one of the core messages of Ordoliberalism: a market economy requires precise rules and a form of public authority is needed in order to proclaim and enforce them. But a given constitutional and institutional order that has proved coherent over one specific national space may well be inadequate when transferred to a neighbouring country. Let me give an example.

2. German reunification: Coherent but inadequate institutions

In short, the state is now so weak in Russia, that it has turned out to be virtually impossible to implement and enforce new rules of the game: **it is naïve to think that capitalism is self-implementing**, since, quite on the contrary, it requires strong and coherent intervention from political authorities, as well as a stable legal system. This strategic and structural uncertainty is in fact unique among former socialist countries. At the opposite end of the spectrum, German reunification delivers a second major lesson about the "transition to capitalism". Although all the West German institutions have been transplanted into the Eastern *Länder*, the strait-jacket imposed on firms and workers has induced an unprecedented fall in production, only compensated for by huge transfers from the West (Diagram 2.3). Despite the clear coherence of these institutions, the transformation has not been easy and is not yet complete.

First of all, the monetary integration of former Eastern *Länder* into the Western German system was implemented by the selection of a relatively high conversion rate between the Ost Mark and the Deutsche Mark. On the one hand, this decision secured a good conversion level for East German wealth and income and helped construct the homogeneity of the newly

Diagram 2.3: Transplanting a complete set of institutions is not sufficient to build a viable capitalist system

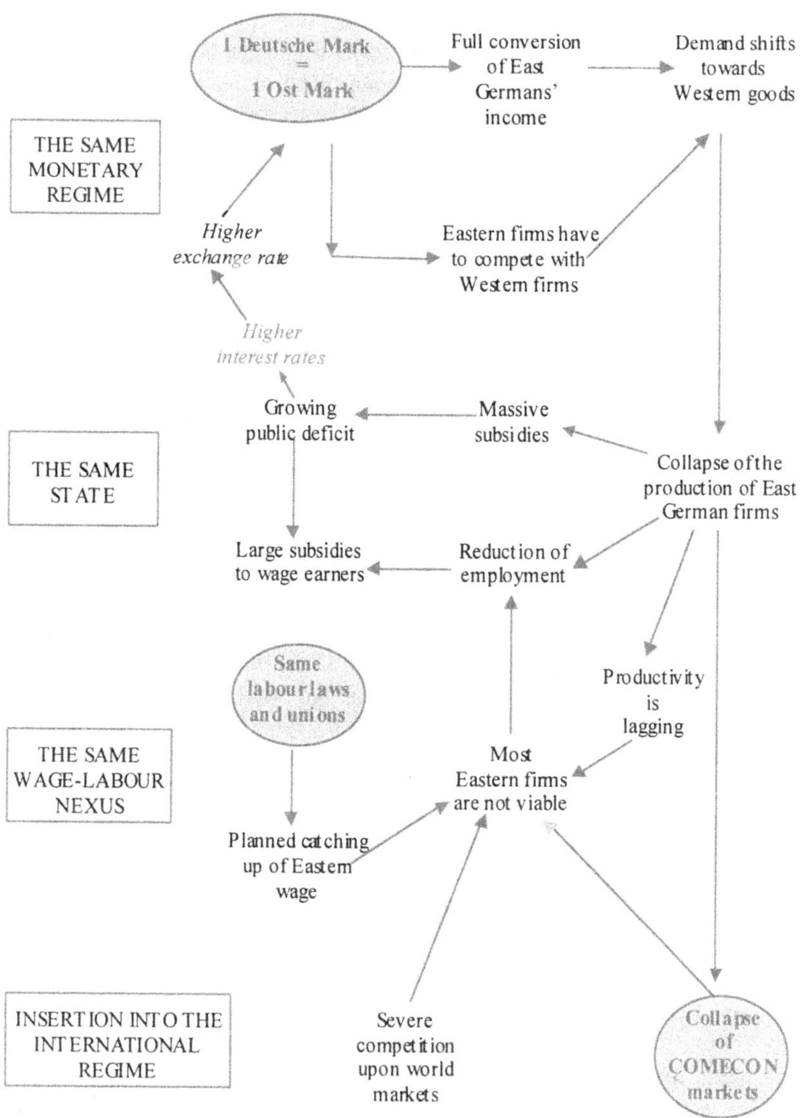

reunified Germany. On the other hand, however, few Eastern goods are competitive at this exchange rate within the new domestic territory. Furthermore, East Germans prefer Western goods and consequently the production of the former GDR has been declining very steeply as a direct consequence of the extension of the institutional architecture already in place in Western Germany to the Eastern *Länder*. The positive side of this reunification is that the poorer *Länder* have benefited from huge transfers from the richer, which are located in former West Germany. This is a significant difference by comparison with the complete decay of state and public transfers observed in Russia.

None the less, although labour laws were immediately extended from West to East, the government chose progressively to increase Eastern wages to Western levels, both in order to promote the convergence of standards of living and to prevent massive immigration from Eastern to Western *Länder*. But since Eastern production has dropped, despite privatization and restructuring, and productivity has been stagnating since the mid-1990s, this fast rise in wages has meant the levelling-off of net profits which have become negative, therefore calling for a massive and constant flow of subsidies from the West. The change in the international regime has been detrimental, too: whereas most Eastern firms used to sell to COMECON countries, the collapse of the Soviet regime destroyed the previous international division of labour among "socialist" countries and constrained the former GDR firms to compete on much more sophisticated markets, such as the single European market. They were not prepared for such a context and this shift in international trade has worsened the internal economic adjustment of the Eastern *Länder*.

Clearly, importing the institutions of the well-established and coherent Western architecture of the so-called "social market economy", or "German model", has not remotely delivered the expected results, and this disappointment calls for an explanation. Whereas in West Germany, institutions, economic specialization and the organization of firms had jointly and slowly coevolved, in East Germany, the divorce between the new institutions and the former specialization has led to a major crisis, still unsolved despite massive efforts on the part of the German authorities and an intensive learning-process. In fact, Eastern firms, managers, workers, unions, banks and local authorities could not immediately benefit from a coherent institutional design, because they had to learn how to act and how to produce efficiently within this new context. This takes time, and the process is longer than the political and legal transfer. Clearly, **institutions have to evolve**

slowly and in any case several decades are needed in order to learn how to organize production, training, innovation, credit and so on. **Capitalism cannot be implemented by treaty nor by law**. Its efficiency derives from a slow process of trial and error, which cannot be compacted into the very brief period of a few years. Similarly, some behaviours which were supposedly "normal" within old industrialized capitalist societies are not so, especially for individuals who have been socialized within a totally different political and economic order. The institution of capitalism is not a matter of the "big bang" but the final outcome of a long, painful and contradictory process.

3. Markets are not necessarily self-equilibrating, nor self-instituting

It is therefore more satisfactory to label as a "great transformation" the processes observed within former socialist countries than to consider that there is a simple transition from one regime to another. After all, who knows what precise socioeconomic system will ultimately emerge in Russia and even East Germany? But this, too, is a challenge to conventional economics and the message has been perceived by an increasingly large proportion of social scientists. Faced with these poor results, neoclassical economists are now considering a neglected issue: what are the basic institutions of capitalism? They recognize that they know very little about the so-called "optimal sequencing of structural reforms"(Dewatripont/Roland 1996).

On the one hand, theoreticians have to admit that the two welfare theorems used to prove the superiority of capitalism over the centralized system are very misleading indeed. It has been observed that the framework of General Equilibrium Theory does not correspond to a monetary decentralized economy but precisely to a Soviet-type economy, in which the auctioneer is in fact a benevolent Gosplan, centralizing all excess demands and supplies and setting prices in order to balance them. This is certainly some irony (Benassy 1982). From a theoretical standpoint, therefore, the so-called socialist economy was potentially coherent, and the general equilibrium theorems demonstrate precisely this paradox (Diagram 2.4).

Conversely, however, this means that economists have no complete and satisfactory formal demonstration of the superiority of *de facto* capitalist economies, for which of course no central planning agency operates. As soon as transactions are in fact decentralized by monetary exchange, new compatibility problems emerge due to the formation of expectations,

Diagram 2.4: For a decentralized economy: The core institution is a credit and monetary regime

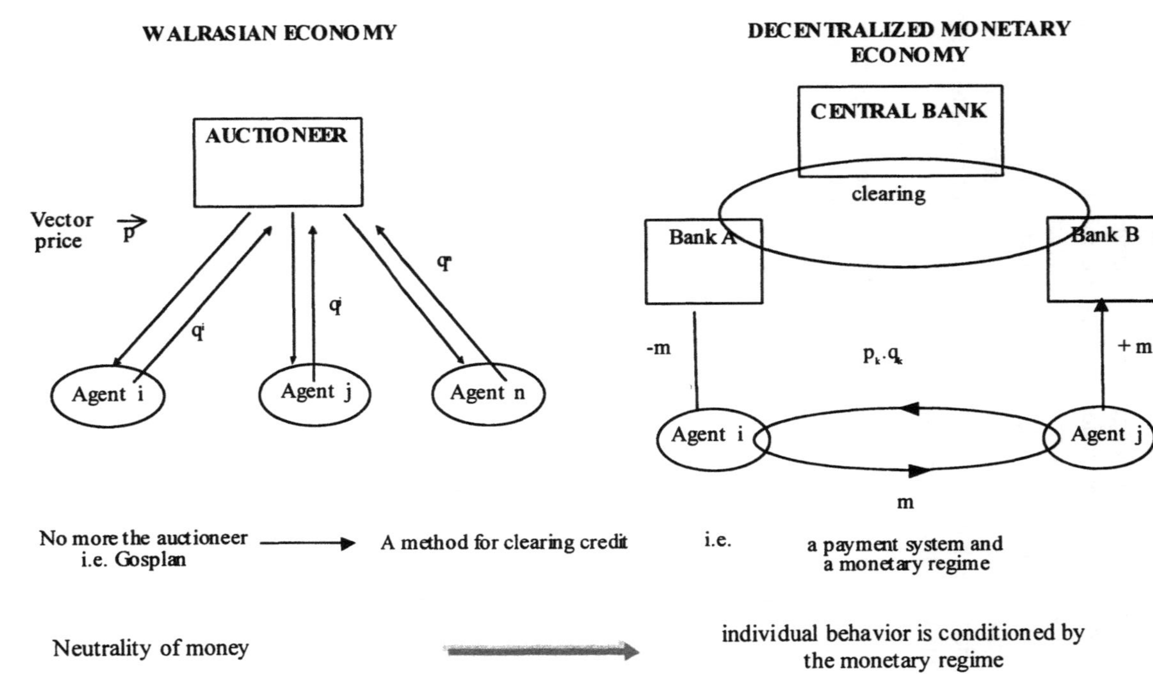

uncertainty concerning the use of money balances and financial assets, and of course the role of time in investment decisions. The equivalent of the auctioneer is then a well-organized system of payment. Ultimately, a central bank is needed in order to clear and balance the net credit and debt between banks, producers and wage earners (Diagram 2.4, left-hand side). The most crucial capitalist institution is therefore the **monetary regime**, which is actually absent in contemporary Russia, a feature which prevents the building of related and more sophisticated capitalist institutions. Economists realize that the sophisticated models they build for modern economies do not deal explicitly with one of the founding principles of capitalism: they take it for granted and have no theory about it.

If such a system is established, then economic agents will normally conduct transactions by means of bilateral exchanges. But there is a huge gulf between a series of disconnected bilateral exchanges and the constitution of a market which centralizes and makes compatible a bundle of individual demands and supplies across a huge number of individuals and goods which meet and supply them. Quite a few markets are organized according to the quotation of U.S. T-bonds on Wall Street. It is precisely highly sophisticated legislation and public intervention that are required in order to achieve the subtle properties of an integrated and efficient market. This means that **market exchange is not a natural phenomenon but, on the contrary, a social construction**, in order to assess the quality of the goods and the credit of the buyers; secure the effective delivery of the goods by the supplier, the reimbursement of credit by the debtor; ensure there is no collusion, distorted or biased information, and prevent transactions on the side (Diagram 2.5). Furthermore, the institutional arrangements designed to cope with these problems may vary dramatically from one market to another (from raw materials to financial derivatives, to say nothing of manufactured goods) and from one society to another (an African village market is not a miniature Wall Street). Historians, such as Fernand Braudel (1979), have stressed that the public markets, in contrast with private markets (where the best-informed traders gain much to the detriment of others), have resulted from struggles between traders and local public authorities. In the past, many societies arranged their exchanges without the organization of markets in the modern sense and some still persist today. Even within modern capitalist societies, it is not always necessary to build a new market, since it is the outcome of social bargaining among the relevant groups and not in the least derived automatically from any economic law (Garcia 1986).

Diagram 2.5: Market is a social construct

FROM A SERIES OF BILATERAL EXCHANGES

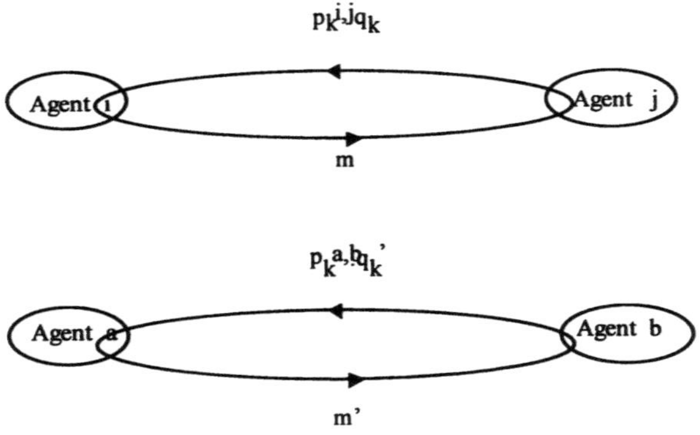

- Heterogeneous price for the same good
- It might well be a different good: asymmetric assessment of quality

TO A MARKET-PLACE

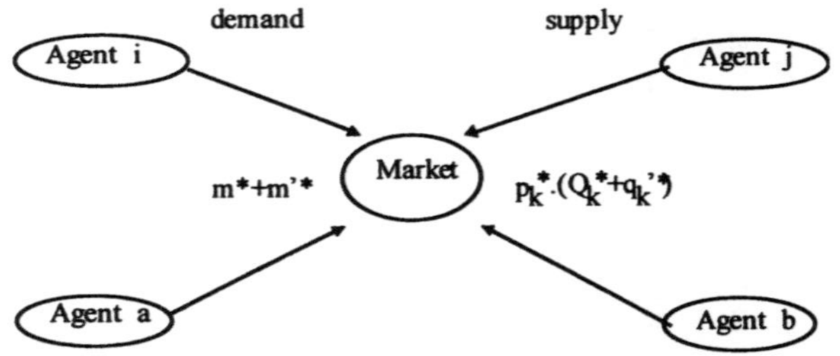

- Clear definition of the quality
- A single price whatever the buyer and the seller
- Transactions and prices now differ from the bilateral exchange case

There is a striking paradox between the achievements of modern economic theory and the contemporary issues and transformations that have to be explained: capitalism seems to have won against alternative systems but, in assessing the transition of former socialist countries, economists had to acknowledge that they knew little about the **necessary and sufficient conditions for a viable capitalist system**. Furthermore, even if this system is superior, there is no theory **about the complex process of implementation of the capitalist institutions**. This is precisely the dilemma, which was highlighted by Robert L. Heilbroner (1988) a decade ago. Will the economic profession ever change?

III. Trading Places: Politicians Trust Markets not Theoreticians

After the dramatic episode of the Great Depression between the two World Wars, the vast majority of economists agreed that pure market mechanisms were insufficient to deliver full employment, fast growth and financial stability. The so-called "Keynesian revolution" brought legitimacy to a series of state interventions in the domain of competition, labour market, finance and, of course, budgetary and monetary policies. The neoclassical-Keynesian synthesis stressed that a pure market economy was not self-equilibrating. Thus, during the "Golden Age", state intervention was accepted by a wide range of politicians, from social democrats and centrists to typical conservative parties.

After all, President Nixon declared in the early 1970s: "We are all Keynesians now". Who would dare say so in the 1990s? On the contrary, the slogan would be: "Markets we trust you!", since the general opinion seems to be that all state interventions are bound to fail or deliver counter-productive results. Such a U-turn merits explanation.

1. Why has the market ideology succeeded?

Regulation theory has provided another interpretation of the Golden Age: the triumph of Keynesian counter-cyclical policy was the most visible part of a structural transformation of absolutely all institutional forms, with the exception of competitive capitalism (Boyer 1990; Boyer/Saillard 1995). Basically, a new capital–labour compromise has promoted an unprecedented synchronization between productivity increases and real wage hikes, or more generally between production and consumption norms (Aglietta 1982). Simultaneously, competition has been mitigated by oligopolistic patterns, whereas the stability provided by a new international regime has opened a greater degree of freedom for national economic policies. Furthermore, the state has played a major role in providing the basic collective investments needed for the maturation of mass consumption and production (transport, housing, education, health). Therefore, it would be more accurate to label the **Golden Age** as **Fordist** (new productive methods), **Beveridgian** (collective provision of the goods and services necessary for the intergenerational reproduction of wage-earners) and, finally, **Keynesian** (new links between the private interests and public intervention).

But no regulation mode can last forever, since it may come into conflict with external or/and exogenous transformations affecting the world economy. But the most serious challenge is associated with the slow erosion of institutional forms, which may lose their ability to monitor the recurring imbalances associated with any capitalist regime of accumulation. Regulation theory stresses that the same factors which explain the success of a regime are at the origin of its maturation, decay and demise (Lordon 1993). This teaching applies especially to the period 1945-1996. The very success of the Golden Age institutions triggered new emerging imbalances, such as stagflation, accelerating inflation during the early 1970s, unemployment, erosion of productivity increases particularly in the U.S., destabilization of the Bretton Woods fixed exchange rate system, the growing role of exports rather than domestic demand. All these structural changes are cumulative and define unprecedented macroeconomic and sectoral evolutions which cannot be interpreted within the Fordist-Beveridgian-Keynesian mode of regulation: these novelties call for an alternative explanation, outside strictly Keynesian orthodoxy and its legacy.

For surely conventional counter-cyclical Keynesian policies have been less and less efficient during the 1970s and 1980s. Was the failure of the French 1981-1982 reflation not clear evidence of the end of an era of state intervention? Why did the highly original social democratic configuration observed in Sweden recently been unable to continue to promote full employment? If, in the past, the choice of an exchange rate was a matter of public policy, are the financial institutions not nowadays increasingly governing exchange rates and, incidentally, national monetary and budgetary policies? Since growth has slowed down, public budgets are no longer self-equilibrating and governments encounter many difficulties in reconciling the trend in spending associated with past institutionalized compromises (Delorme/André 1983) to the reduced growth of the tax base. Is it not true that in virtually every country politicians are facing hard times and extremely difficult choices?

What is more, they are left without any clear picture of the key mechanisms governing macroeconomic variables. A strong temptation can therefore be observed to focus economic policy on more sectoral and microeconomic measures, in the hope that they will ultimately solve macroeconomic imbalances. Lacking any alternative image, politicians are tempted to declare their support for the market, which is charged with solving all the problems they are unable to deal with in both the political arena and the public sector (Diagram 2.6). That is to say, all measures concerning decentralization, the liberalization of trade and finance, the flexibilization of the labour market, privatization of public enterprises, new voluntary and private schemes for welfare and so on.

Willingly or unwillingly, most governments have come to declare that they prefer market mechanisms to public intervention, that the state should be reformed and rationalized and that they experience little freedom in controlling evolution at micro- and macro-levels. Reference to the "**invisible hand**" was the only alternative remaining to provide a minimal understanding of highly complex and interdependent developments. But is this Panglossian optimism about the overwhelming efficiency of the market grounded in sound theoretical constructs? Here we meet a second surprising discrepancy between economic practices and the teachings of contemporary modern economic theory.

2. Modern economic theory versus Panglossian optimism

The second paradox is then the following: politicians and experts recommend a "market therapy" at the very point in time when the more distinguished and dedicated mathematical economists prove that the market mechanism is not efficient for most of the coordination problems arising in contemporary capitalism and which it is critical to solve (for a more complete demonstration, see Boyer 1997). One of the major achievements of modern economics is the demonstration of a set of precise conditions under which a general equilibrium exists and corresponds to a Pareto optimum (Debreu 1959). At least seven hypotheses are required. Money is only a *numéraire*

Diagram 2.6: How to explain the return to belief in self-regulating markets?

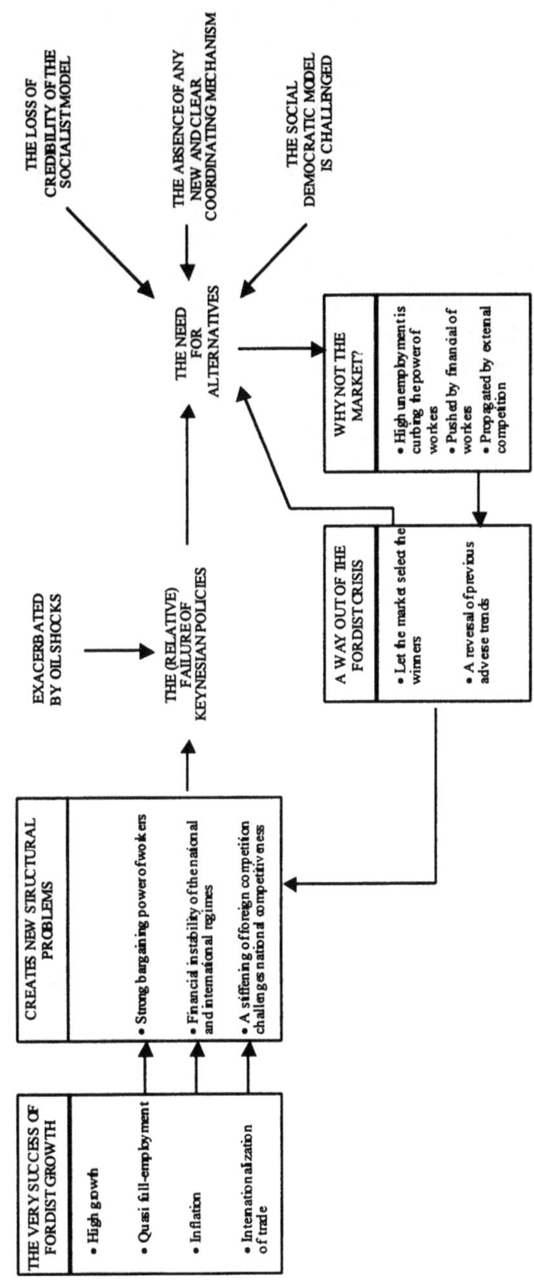

(abstract unit of account) and all transactions are centralized, and prices set, by an auctioneer. Each economic unit adopts a parametric behaviour and takes as given the strategy of other individuals. There is a complete and finite description of all the goods with no ambiguity as to quality and availability. There are no collective goods and they are unnecessary for private production. All the technologies are common knowledge, set once and for all, and they display constant returns to scale. There is a complete set of contingent future markets in which goods can be traded and prices set for each event which may happen in the future, the list of which is perfectly known at the beginning of the period. Last but not least, distribution issues are totally distinct from allocational problems or, alternatively, equity and principles of social justice can totally be disconnected from allocative efficiency (Table 2.1, column 1).

Back in the 1960s, the expectations of general equilibrium theorists were highly optimistic about the possibility of removing some of these quite restrictive hypotheses, while obtaining analytically tractable and general conclusions (Ingrao/Israel 1990). Nowadays, it is clearly established that the removal of any one of the seven hypotheses drastically affects the existence of equilibria and their Pareto optimality. First, a *decentralized monetary economy* exhibits a new variety of equilibria with some extreme cases in which no monetary equilibrium exists. The introduction of expectations, which can be fulfilled or disappointed, creates many temporary equilibria, far removed from the ideal of a single equilibrium set by reference to the fundamental variables governing the value of goods. Secondly, in any industrialized economy, *strategic behaviours* among a small number of large competitors are the rule and this feature destroys the very concept of a general equilibrium independent of some game rules concerning competition, which generally speaking is imperfect. Thirdly, *the absence of product innovation* is totally at odds with the very logic of a capitalist economy, but when they are introduced market allocations are no more efficient. Still worse, when no equilibria exist or the quality of the goods is uncertain as far as the consumer is concerned the price has to fulfil two distinct functions: revealing quality on one hand, allocating resources on the other. The same problems arise when the *production and financing of public goods* is introduced into general equilibrium models: by definition, markets alone cannot make decisions about the optimum volume to be provided by a state in charge of promoting social welfare. A fifth obstacle to the generalization of General Equilibrium theory concerns process innovations and, more generally, spill-over effects, whether positive or negative, associated with radical innovations. The allocation of investment in research by pure market

Table 2.1: As many market failures as efficient markets in real economies

THE HYPOTHESES OF THE TWO WELFARE THEOREMS	STYLIZED FEATURES OF CONTEMPORARY ECONOMIES	CONSEQUENCES UPON MARKET FUNCTIONING
1. *De facto*, complete centralization of transactions, no need for money	1. Largely decentralized exchanges allowed by money and credit	1. Multiplicity ... or absence of any equilibrium. Efficiency is no longer assured
2. Atomistic competition among very many agents	2. Imperfect competition via product differentiation is the rule	2. Market equilibria are no more efficient
3. The list of goods is finite, their quality known	3. Producers are better informed than consumers, product innovation is crucial	3. Markets do not clear: unemployment and overcapacity
4. Purely private goods, with no external effect	4. Existence of many public goods and external effects (security, education, R&D etc.)	4. Competitive markets imply an underinvestment in collective goods
5. Constant returns to scale and fixed technologies	5. Learning by doing, by using and increasing dynamic returns to scale are significant	5. Imperfect competition is the rule, inefficient techniques can persist, multiplicity of path-dependent equilibria
6. All contingent future markets exist	6. Only a few financial markets allow intertemporal transactions	6. Existing markets cannot deliver adequate coordination: inefficient equilibria are the rule
7. Equity principles have no influence upon efficiency	7. Workers' loyalty and commitment are linked to fair treatment	7. Markets do not clear; unemployment can persist

mechanisms is then not optimal. A still more devastating event takes place when the theoreticians take into account the fact that only a few contingent markets exist: then a cluster of complex dynamics may occur and produce a strong discrepancy between a market equilibrium and the optimum path for a rationally planned economy. Still more fundamentally, much empirical evidence and experimental economics convincingly show that value judgements about income distribution do have a part to play in production efficiency, since workers make their efforts and commitments proportionate to the reward they consider as fair (Table 2.1, column 3).

The dilemma of modern economic theory is precisely that all these configurations are not the exception but the rule for contemporary developed countries. Thus, since all these features are quite frequent in really existing economy, the superiority of the market only results by default of alternative coordinating mechanisms. Economic theory does not prove that markets are always, and in any circumstances, the most efficient way of allocating resources. In some cases, public intervention improves the outcome for all economic agents. For instance, contemporary growth theory (Romer 1986; Lucas 1988; well summarized in Barro/Sala-I-Martin 1995) shows that a private economy equilibrium is not Pareto optimal, as soon as innovation exerts significant external effects (positive or negative). Similarly, the strong positive externalities associated with education call for public intervention, the asymmetry of information between producers and consumers can be corrected by independent rating agencies or some public norms. But conversely, public intervention even if it corrects market failures may in turn introduce distortions. There is rarely a first-best solution, and the problem is to choose from among second- or third-best coordinating mechanisms (Wolf 1990). Nevertheless, during the Golden Age, the more egalitarian societies seem to have been more efficient than those that were more unequal, whereas some indicative planning has helped in synchronizing a series of individual investment decisions, merely to compensate for the absence of contingent markets (Table 1, column 2).

In other words, this second paradox has two sides. First, **modern economic theory contradicts the Panglossian optimism of conservative politicians**. Secondly, the very maturing and sophistication of capitalist economies make belief in the self-equilibrating role of markets and their efficiency **less and less realistic**. There is further evidence of this discrepancy between academic representations and government strategies.

3. Deregulation, privatization, liberalization: Uncertain and unexpected results

Since the end of the 1970s, the majority of economic policies have been devoted to a tentative reduction in the role of the state and increasing reliance upon market mechanisms. Financial deregulation has replaced detailed national regulations, many public firms have been sold through financial markets; some parts of welfare systems have been privatized; labour market regulations relaxed; and external trade has, of course, benefited from a new wave of tariff reductions; and exchange rates are no longer set by public authorities but are derived from the day-to-day quotations on currency markets. According to the free marketers, macroeconomic performances should therefore be far better: lower interest rates, more efficient health and education systems, greater productivity and increased global welfare as a result of the deepening of the international division of labour and, finally, no more inflation, a return to full employment, budgetary equilibrium and external trade balance (Table 2.2).

In the light of the general argument presented earlier, it is no surprise to observe that absolutely **none of the promises of the free marketers has been fulfilled**. If productivity has increased for some deregulated sectors, it is not certain that this has not been at the expense of a deterioration in quality with widening inequality between categories of consumers and, especially, between businesses and households. Furthermore, most national oligopolies have been destabilized and new boundaries established between new emergent sectors, but ultimately a new form of oligopolistic competition has emerged at national or international level. As the trajectory of the airline industries shows, free entry and pure competition between two oligopolistic configurations are only transitory phenomena. More generally, these sectoral deregulations have not yet triggered a new wave of productivity increases spreading from one sector to another, even though there are some problems in measuring this.

Even the most conservative governments continue to suffer from significant public deficit and the external imbalances between the United States, Japan and Europe have persisted for more than a decade. The flexibilization of labour contracts and the decentralization of collective bargaining have curbed real wage increases, but they have not solved the mass unemployment problem, so crucial for Europe. Reforms and the rationalization of the welfare state in the direction of privatization have not overcome the basic discrepancy between increasing social needs in relation to pensions, benefits, health care for the elderly, adjustments to the educa-

Table 2.2: The free-marketers' promises and what they delivered

	PROMISE	OUTCOME
1. Capital labour relation	• Deregulation will allow full employment	• No clear impact
2. Forms of competition	• Deregulation will bring, more efficiency by the entry of new producers	• Re-regulation, less producers: from one national oligopolistic form of competition to another more internationalized
3. Monetary regime	• Control of monetary base is possible • It provides price stability, without departing from full employment	• Monetary innovation prevents this control • Price stability, but mass employment
4. State	• Minimal State will enhance growth and pro-ductivity	• Lack of public investment • Poor private productivity due to the lack of education and infrastructures
5. International regime	• Smooth currency and exchange rates adjustments • External disequilibria will not more exist • Complete autonomy of national economic policies	• Large ups and downs of exchange rates • Unprecedented and stable polarization of deficit and surplus countries • Stronger constraints upon the national degree of anatomy in economic policy choices.

tional system. The major disappointment concerns the outcome of financial deregulation: exchange rates are more volatile than ever, uncertainty has become radical and systemic, economic policies are more and more constrained by the daily assessment of the financial markets. Clearly, the only success is the drastic reduction of inflation rates but, contrary to the promises of the neoclassical economists, the costs have been particularly high in terms of employment and of foregone potential growth.

Of course, neoclassical fundamentalists reply that this disappointment stems from reforms in the direction of pure market mechanisms that were insufficiently thoroughgoing and unduly timid. They argue, for instance, that all European unemployment problems derive from what continues to be an excessive degree of regulation (Becker 1996). But there is no clearly established theory or modelling to show that a small dose of market mechanisms is bad for economic performance, but that a complete shift would completely reverse the outcome. Historical record in the tradition of Fernand Braudel (1979) or Karl Polanyi (1944) suggests, on the contrary, that there is an optimum degree in the use of market mechanisms, since the two extreme configurations (no market at all or the full "marketization" of society) generally deliver relatively poor outcomes.

Therefore, if a jury were to assess the merits and limits of the market, long-run historical evidence, modern general equilibrium formal models and the contemporary record of economic and financial liberalization do not argue in the least in favour of the thesis that the market is able to solve most, if not all, contemporary problems. **The reality of capitalism institutions seems strongly to contradict the ideology of free markets**. But since politicians are left without any doctrine to legitimize public intervention, they have preferred to issue strong public statements in favour of markets rather than admit that they had no alternative institutional proposal for solving the more urgent contemporary problems, such as unemployment in Europe and rising inequalities in North America. This could be an explanation of the second paradox, observed earlier, which is directly linked to the scheme of an alternative approach.

IV. Regulation Theory: An Analysis of Evolving Capitalism

When political economy was recognized as a separate discipline, only a few dedicated scholars were working on the understanding of emerging capitalism system. The theory was in its infancy, the concepts quite crude ... and errors in logic frequent, but the intent was to capture the essence and logic of the whole social and economic system. Nowadays, the configuration is the very reverse: the vast majority of economists are contemporary, they benefit from the conceptual breakthroughs of their predecessors, they discuss each notion and concept carefully and are trained to express their ideas in rigorous mathematical models.

In addition to these efforts, they use numerous and detailed statistics and econometric techniques to test the predictions of their models and are joining academic groups devoted to the collective improvement of a given discipline or subdiscipline. The principle of the division of labour, so important in the history of capitalist production and manufacturing, has been extended to intellectual life. One might therefore expect economic theories of unprecedented quality and an exceptional degree of accuracy in the predictions that can be derived from them. After all, to read some of the leading academic journals, one might have the impression that economics is becoming a normal science on the model of the physics or biology.

A third paradox is precisely that this extreme division of labour among economists no longer provides a coherent and satisfactory representation of the contemporary capitalist system. The numerous results obtained in each subdomain are either negative or partial (the theory explains a very minor fraction of the phenomenon observed) or are mutually contradictory since adding up all the competing theories produces no general result whatsoever.

1. The failed hopes of a "generalization" of General Equilibrium theory

A short summary of the achievements of the distinguished scholars who have tried to generalize General Equilibrium Theory by removing each of the specific and limiting hypotheses sheds light on the current state of economics as a discipline (Diagram 2.7). Back in the 1960s, the hope was progressively to generalize the highly idealized and unrealistic model and converge towards a **new formalization** which was to be simultaneously grounded in clear axioms and representative of economies that actually existed.

Unfortunately, two decades later, mathematical economists as a body had to acknowledge that the removal of each separate hypothesis opens up a new and highly specific economic world, which can no longer be compared with the path followed by other scholars. Each realistic hypothesis opens a whole spectrum of models, which are so complex and rich in terms of results, that they cannot be pooled into a renewed general equilibrium model.

Although the hypothesis of the Walrasian auctioneer (*secrétaire de marché*) has been rejected, new neoclassical theory has stressed the role of expectations in the sense that the rational calculus of each economic agent is trying to mimic what a pure Walrasian economy would deliver (Lucas 1983). But this elegant construct has been challenged, since it requires incredible informational and computational capability from each economic unit ... and assumes that the economies are in reality Walrasian, certainly a very doubtful hypothesis indeed. By contrast, new Keynesians argue that many of the results of General Equilibrium theory can be obtained within the same rational expectations theory (Weil 1989). But previously, other theoreticians had modelled temporary equilibrium, under the hypothesis that expectations are not necessarily rational, nor perfect, and that is the essence of Keynesian unemployment (Benassy 1982). Therefore diametrically opposed results have been obtained within the same research programme: convergence towards full employment on the one hand, persistent unemployment on the other.

It is easy to imagine still more divergent results if the asymmetry of information is the primary concern (Stiglitz 1987). Unemployment then derives from the inability of the managers to monitor the private information held by workers about the intensity of their effort, and unemployment is then used as a disciplinary device to elicit commitment from opportunistic workers. But a totally different economic world is opened up by overlapping generational models: economic dynamics stem from the imperfect transmission of wealth from one generation to another, and this is a response to the fact that contingent markets are quite rare, despite the numerous financial innovations involving derivatives. Yet another path is followed by the imperfect competition school which builds upon Chamberlin's intuition concerning the restrictive conditions required for pure competition. Then, unemployment is the joint consequence of oligopolistic pricing upon the good markets and some oligopolistic rents extracted by unions (Benassy 1992).

Diagram 2.7: From general equilibrium theory to game theory: Analyses by domains but not any theory for the complete economic system

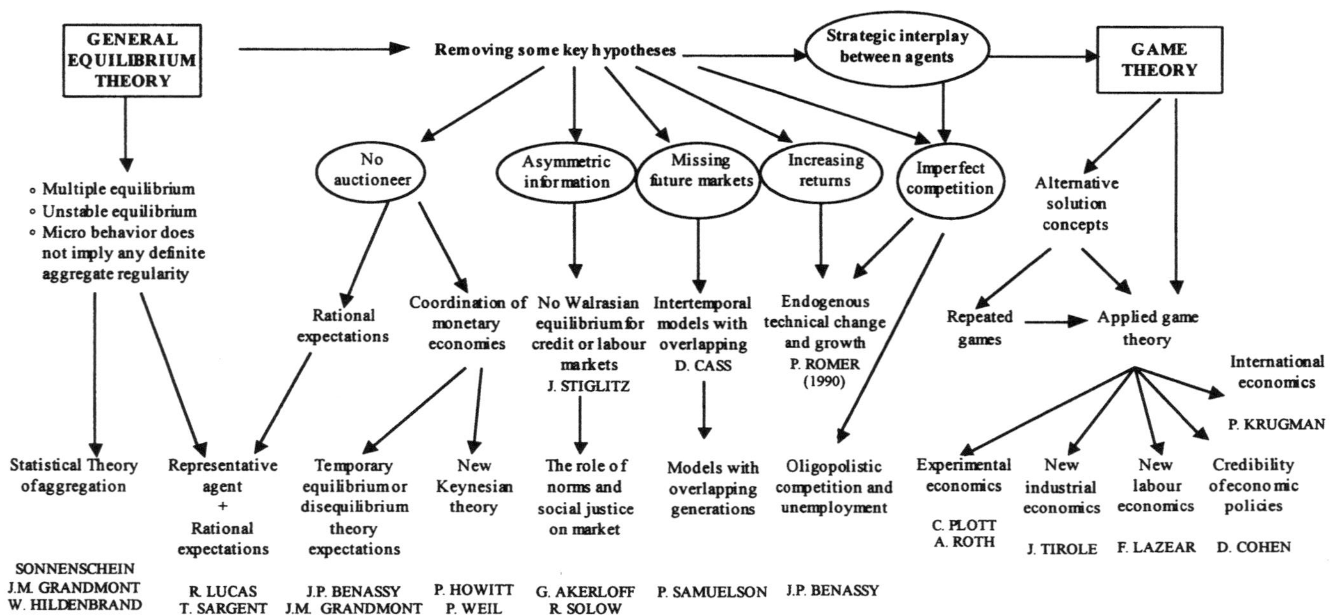

The role of econometric methods should be to test one theory against another and thus progressively work out a general theory which would encapsulate all the basic axioms and deliver accurate predictions for any past known experience. Unfortunately, there are serious methodological problems in testing one model against another and there are few examples where econometric techniques have delivered such a crystal clear message. Even if applied economists refer to Popperian principles about the falsification of their hypotheses, their everyday work is basically "confirmationist": "the data do not reject the hypothesis at a given confidence level". There is therefore no strong selection among theories concerning their respective explanatory power, and in many cases the core hypotheses of the model explain only a limited fraction of the variance, because many *ad hoc* technical hypotheses have to be added and these contribute to the global explanatory power. Therefore either axiomatic methods predominate and/or *ad hoc*, dreamed or imaginative models proliferate, either with no empirical control whatsoever, or controls so few and weak that they are insufficiently discriminating. In other words, despite the intensive use of mathematics and frequent statistical tests, economics is far from having attained the rigour and relevance of physics (Amable/Boyer/Lordon 1995).

Therefore, more than a harmonious division of labour among diverse economic specializations (theoreticians, statisticians, econometricians, economic policy analysts and so on), contemporary research displays **a form of balkanization** among totally contrasting approaches, tools and, still more, results. There is no better example than the current fashion for game theory: these rather attractive tools can deliver almost every required result, provided that the structure of the game, the flow of information and the sequencing are properly defined. The relevance of the conclusions is directly linked to the adequacy of the core hypotheses ... which are never tested but simply introduced by few subtle references to some features of the real economies. But why select one feature rather than another? Contemporary research is far removed from the ambition of General Equilibrium theory to provide a unified model, valid at any time and in every place. There is a huge number of models differing in their core hypotheses, the tools used, the subdomains investigated (industrial economics, labour markets, economics of technical change, growth theory, business cycle models, regional economics, international relations and so on) and the position of the economist (manager, consultant, international expert, civil servant, central banker, union adviser).

Contemporary economies are thus more complex than ever, and the economic profession has specialized accordingly ... but unfortunately its intellectual relevance lags far behind as far as the need for a minimal understanding of economic systems as a whole is concerned. At a period when financial markets interact with monetary policy, forms of competition, labour market adjustments and, of course, international trade, production and finance, it is highly detrimental to compartmentalize research in economics. Contemporary research understands many detailed mechanisms, but it does not grasp how these components form a more-or-less coherent picture. It is reminiscent of amateur palaeontologists who collected a lot of scattered bones, studied them carefully and even built wonderful models about each piece ... without even daring to imagine that they belonged to the same dinosaur, the concept of which would help structure the whole research programme and, of course, the relevant theory.

2. From the missing auctioneer to the five institutional forms

Is this present state of economic theory irreversibly fatal? It so happens that various researchers from different disciplines (history, sociology, law, economics and so on) have begun a careful investigation of the institutions necessary for a capitalist system to function and adapt. This new institutional economics is frequently accused of being constructed upon *ad hoc* hypotheses, by inductive methods rather than by a deductive or, still better, an axiomatic approach which are assumed to be the distinguishing characteristics of "true science". A first step in this direction can be made by starting from the old problem of political economy, which is still useful for understanding contemporary capitalism: why does a system built upon competition and conflict not result in chaos (Diagram 2.8)?

Basically, scholars are divided between two visions and two approaches. Starting from **Thomas Hobbes**, many philosophers and political scientists focus on the role of the sovereign, or in modern parlance the state, in imposing his authority and thus the rules governing interaction among individuals. On the other hand, however, **Adam Smith** starts from a totally different vision of mankind: man is not necessarily anybody's enemy but exhibits a propensity for exchange which is the basis for mutually advantageous bilateral exchanges. The invisible and anonymous hand of the market is replacing the hierarchical authority of a political order. Basically, many difficulties in understanding capitalism derive from its dual aspect,

Diagram 2.8: At the roots of political economy

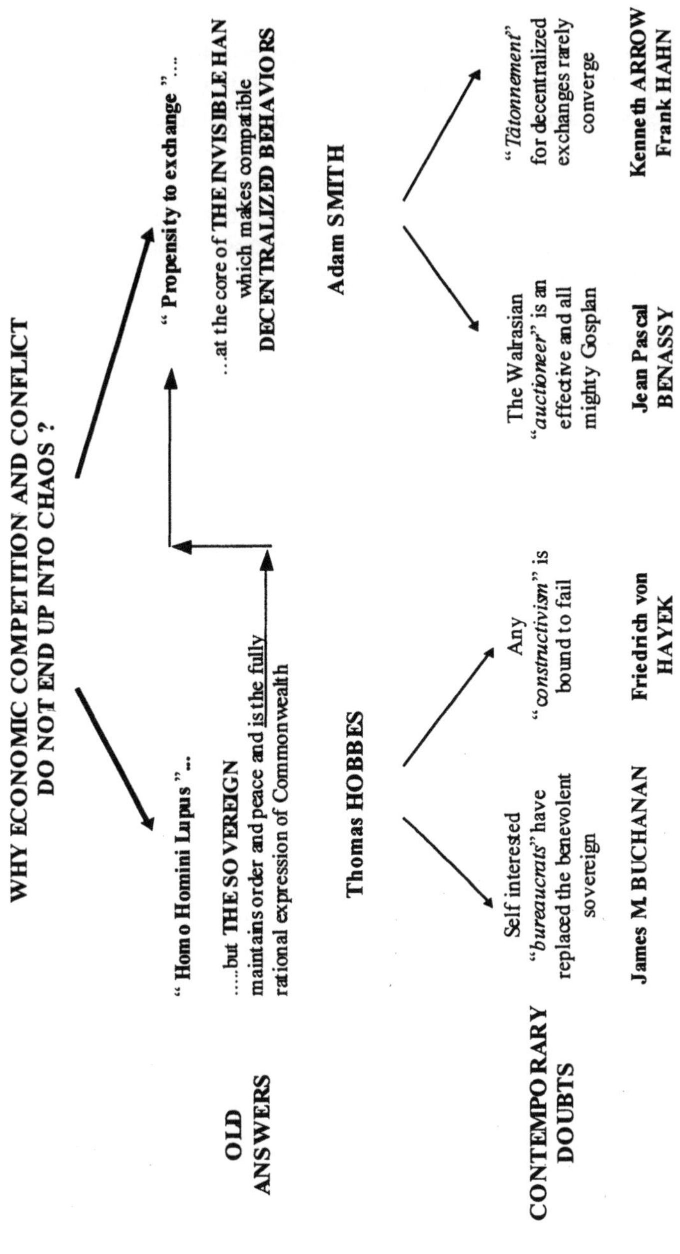

both political and economic, and academic specialization makes the two visions increasingly divergent.

For simplicity's sake, let us follow the trajectory of political economy and attempt to demonstrate that the body of last resort which warrants the existence of a market economy is the state. It has already been shown that the general vision of a peaceful adjustment via a series of market transactions is not substantiated by modern economic theory: as soon as the Walrasian auctioneer no longer exists – and Gosplan has after all been disbanded in contemporary Russia! – there is not necessarily a market equilibrium if transactions are totally decentralized and take place before the equilibrium price is reached (Arrow/Hahn 1971). Nevertheless, capitalist systems are not that chaotic and usually function rather well, which means that some key institutions must be playing the role of diffusing information and allocating goods and resources following some rather effective rules.

These institutions can in fact be built conceptually, starting from the requirements of decentralized exchanges themselves (Table 2.3). Clearly, a payment system has to replace the central planning agency, but this system manages credit and debt, and does not directly regulate the circulation of goods. As previously demonstrated (Diagram 2.4, above), this calls for a series of banks which themselves require a central bank in order to compensate for recurrent disequilibrium among commercial banks. Therefore, the first constitutive institution is the **credit and monetary regime**. Incidentally, this explains why the absence of a robust and coherent payment system prevents the very constitution of product and factor markets, as demonstrated by contemporary Russia, where barter exchanges are spreading because of the lack of any reliable system of payment. Then, if a stable credit regime of this kind exists, markets can be organized, provided that the more powerful actors have an interest in implementing them. At this point we encounter a second basic institution, **the form of competition**, which describes the conditions of entry, the number of existing firms, their strategic interactions, public regulations governing fair competition and so on.

One of the more misleading hypotheses of Walrasian economics has then to be removed: there is no such thing as a market in which the services of labour is exchanged. Labour is not a conventional commodity, even if wages seem to be set on the so-called labour market. Basically, labour underpins a social relationship: wage-earners accept the authority of the firm in exchange for remuneration, which means that strategic interactions between workers and managers necessarily take place during the wage bargaining process, but also afterwards in the dense organization of work in everyday production. The third institution to replace the service of labour is the **wage–labour nexus,** defined as the precise set of conditions and rules which affect the productive use of labour and the income formation of wage-earners.

Table 2.3: When the Walrasian auctioneer no longer exists, five institutional forms are required

What does the auctioneer do ? ⇨	**It is coherent and relevant ?** ⇨	**The role of institutional forms**
1. Money is only a numeraire	Money is a medium of exchange and reserve of value	The need for rules of money creation and destruction
The auctioneer centralizes all the transactions	This not a market economy but Gosplan	A credit/**monetary regime** decentralizes transactions
2. All agents are price takers	Strategic behavior is frequent	**Form of competition** differs from pure and perfect ideal
3. The service of labor is exchanged on a typical market	The dual nature of labor : present wage against future effort	The labor contract is embedded into a **wage labour nexus**, which implies a social relation
4. No State	A required authority for sustaining property, contracts, monetary exchange, public goods	The configuration **state/economy**
5. No Nation-State	Any given State is ruling upon a well defined territory	The insertion into the **international regime**

Could all these three institutions be self-implementing? Yes, most economists usually reply, but other social scientists immediately argue that the political and legal order is necessarily involved in the emergence and implementation of these forms. Until now, the only viable monetary regimes have been controlled and endorsed by a public authority. Similarly, forms of competition necessarily imply the imposition by the political process of some constraints or rules to be enforced, merely to prevent collusion which would be detrimental to the rest of the society. Finally, even a competitive wage-labour nexus calls for some laws which, for example, forbid any coalition between workers or firms. In this context, one cannot conceive a capitalist economy without an explicit **role for the state**. This is a fresh paradox, of course. On the one hand, capitalism emerges when economic activity is separated from the direct control of the sovereign. But on the other, the very founding institutions of this new economic system call for original forms of state intervention. The nature of the relationship between the political order and the economic institutions therefore defines a fourth major institution.

But the legitimacy and coercive power of a state is limited to a given territory and a fifth and complementary institution must accordingly be introduced. The contemporary nation-state is simultaneously defined by the *internal political process* of constitution of a domestic constitutional order and by the *external recognition or imposition* of a style in the relationship with other nation-states. The modalities according to which a nation-state organizes its relationship with an **international regime or configuration** thus define the last general institution necessary for the analysis of a capitalist economy. We are then far removed from the atemporal and aterritorial Walrasian model, and the five institutions now organize interactions among economic units, both in terms of information flows and allocation of resources and products. The economist has now to assess the viability or the incompatibility of an institutional architecture brought into being by past political processes and economic specialization.

3. Accumulation regimes and regulation modes: The diverse forms of capitalism

One might recognize the core concepts of Regulation theory. But it should be stressed they were derived first, not from the limits of General Equilibrium theory but from a critical appraisal of Marxian theory. Observation of some methodological errors in the original works of Karl Marx persuaded the majority of economists to abandon the analysis of capitalist dynamics. But they mistakenly attributed to the global and systemic character of Marxian theory, errors largely relating to methodological approximations and inadequate empirical hypotheses. Regulation theory was developed in response to these criticisms, while preserving a macroeconomic approach to capitalist accumulation, growth and crisis (Diagram 2.9). This theory is built upon three guiding principles:

- Contrary to Marx's hypothesis, the conjunction of a market relation and a capital–labour relation, defined at the most general level, is insufficient to define a single and unique accumulation regime, which would for instance induce a tendency for profit to fall. Even if accumulation is the coercive law of capitalism, **several regimes of accumulation** may exist, from both a theoretical and a historical viewpoint.

- Similarly, the very form of institutionalization of the capital–labour relation (in other words, the wage–labour nexus), of the market relation (that is, the monetary regime and the form of competition) may be creating new processes of socialization and economic dynamics. **The precise nature of institutionalization matters for capitalism**, which can therefore vary over time (it is the old question of the stages of capitalism mentioned in Introduction) and space (this theme was more neglected, even if touched upon by the theory of imperialism).

- In opposition to methodological individualism, or more specifically the new classical school, economic actors do not have to have perfect knowledge of, and internalize, the regularities governing the macro level of capitalist reproduction. By definition, **institutional forms summarize the rules of the game**, reduce the relevant information to be assembled and analysed, restrict strategic interactions to a very limited set and monitor the solutions given to possible and recurring conflicts and disequilibria. This is why **the regulation mode**, which is actually a guide-line for individuals and groups, is to be distinguished from the accumulation regime, which is an abstraction created by an external analyst in order to capture the inner structure of a given capitalist system.

It is not the purpose of this chapter to present a survey of the main conclusions obtained by this research programme, since this was done a few years ago (Boyer/Saillard 1995) and is periodically updated (*Année de la Régulation* 1997; 1998; 1999). The most recent advances deal with the issue of emerging supranational rules, with a special emphasis upon the issue of European monetary integration (Dehove 1997), the significance of the South Asian crisis (Contamin/Lacu 1998), the emerging regulation modes for the twenty-first century (Petit 1998) or the role of the polity in the transformation

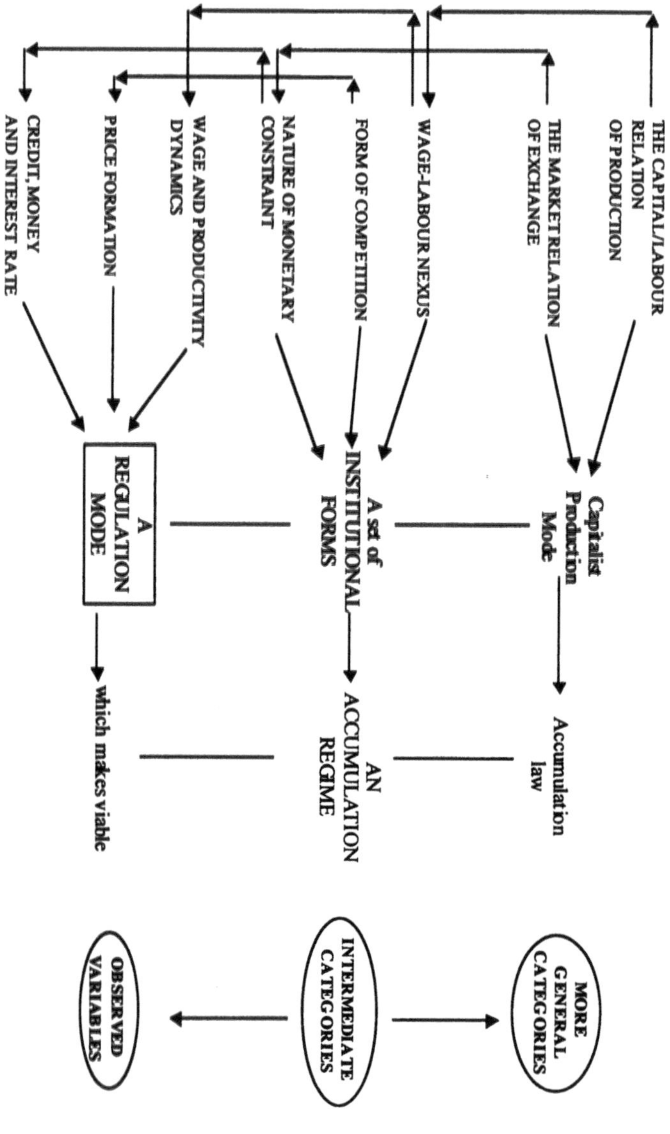

Diagram 2.9: From General Equilibrium theory to Game theory: Analyses by domain but no longer any theory for the whole economic system

of institutional forms and accumulation regimes (Boyer 1999). For the time being, it is more interesting to note a surprising convergence with some recent approaches in institutional economics which have had some impact.

4. Regulation theory as one ingredient in the melting-pot of institutional economy

Any alternative to conventional neoclassical economics has to develop a set of common founding concepts, rigorous methods, which are shared by a large community of researchers, whose advances can be used by other members involved into the same research programme. This is particularly the case if the objective is to understand the dynamics of a capitalist system, which by definition is supposed to form an entity, meriting an integrated approach. It would be dismal if the new wave of institutional analyses were to produce a series of disconnected results organizing from competing, or (worse) fragmented, communities. Therefore, this section proposes a provisional common basis for promoting such an understanding of capitalist systems, and overcoming the paradox that neoclassical theory is currently facing. The taxonomy builds upon recent proposals from economic historians (North 1990) and political scientists (Sabel 1997) and is fairly compatible with a logical extension of the theory of Regulation.

Too often, any coordinating mechanism that is an alternative to markets is equally labelled as "constitution", "institution", "organization", "convention", "collective" or "individual routines" as if these terms were synonymous, whereas they have to be properly defined – and then they are not at all equivalent (Table 4). The definitions proposed indicate a clear hierarchy according to the general character of the related coordinating mechanism. In this respect, it should be agreed that the **constitutional order** is the higher principle, directing a solution for possible conflicts between rules at lower levels. The principle of action derives from the legitimacy of a past formative political event and the solutions given proceed from deliberation, which is a principle quite distinct from the economic principle of exit. We are a long way from the Walrasian auctioneer, since the issue is about the compatibility of a complete set of potentially conflicting legal rules; clearly, compatibility cannot be achieved by organizing ... a market. It is an open question, and a different one, to investigate whether or not the constitutional judges maximize an economic criterion such as social welfare (Posner 1982).

Table 2.4: A tentative taxonomy for some key components of an institutional economy

NATURE / COMPONENT	DEFINITION	PRINCIPLE OF ACTION	FACTORS OF CHANGE
CONSTITUTIONAL ORDER	A set of **general rules** to set lower level conflicts among institutions, organizations, individuals	**Legitimacy** via deliberation	• Large inertia in democratic states • Role of **political** process in the redesign
INSTITUTION	An immaterial method for **structuring interactions** among organizations and eventually individuals	Reduces or removes the **uncertainty** associated to strategic behaviour	• Structural **crises** • Low efficiency is not a **sufficient** reason for change
ORGANIZATION	**A structure of power and a series of routines** to overcome coordination failures among agents or their opportunistic behaviours	**Carrot and stick** (i.e. the pay system and control) are related to external institutions or conventions	• Poor outcomes in the **competition** with other organizations • Major crises trigger redesign
CONVENTION	**A selfenforcing set of shared expectations and behaviours**, emerging from decentralized interactions.	**Lost memory** of the origins of the convention which seems "natural"	• **General crisis, invasion, translation** • Efficiency is rarely a selection criteria
HABITUS	A set of **embodied patterns of behaviour**, forged during the socialization process of an individual	**Adaptation** to a given field, possible disequilibria out of this field	• Shift to a new field of an habitus forged into another • New **learning**, even if quite difficult

The different definitions will not be detailed, but they are organized from the more overarching coordinating mechanisms in the direction of those that are more individualized and context-specific. **An institution** takes into account the constraints and opportunities brought by a constitutional order and structures interaction among organizations by purely non-material methods. **An organization** is built upon a structure of power and a series of routines in order to overcome coordination failures among more-or-less opportunistic agents and behaviours. **A convention** is totally different, since it is a self-enforcing set of shared cross-expectations and behaviours, emerging without any third party enforcement. Finally, **a habitus** is a set of embedded patterns of behaviours, forged during the socialization process of an individual, whereas **a routine** can be shared by a group and externalized, without implying any hierarchical authority to monitor it.

According to this vision, the principles of action are very different indeed and *a priori* the more difficult the change in these coordinating mechanisms, the higher their place in the hierarchy in Table 2.4. Of course, this hypothesis can be challenged, especially from the point of view of methodological and ontological individualism: is not the habitus the longer-lasting element, featuring the larger path and past dependency? In fact, one could argue first that the pattern of behaviour is highly sensitive to the institutional and general context – the modernization of European and Japanese economies during the Golden Age is a good example of such malleability of so-called "cultural values". Secondly and conversely, it has been observed that quite contrasting national trajectories for capitalism are closely related to the permanence of a constitutional order (Berger/Dore 1995; Sabel 1997; Amable/Barré/Boyer 1997) and that it is difficult to shift from one constitutional style to another, since all institutional forms, organizations and even conventions have to evolve in accordance with the new constitution (Diagram 2.10). This is a matter of a generation rather than of years or quarters.

This does not mean that in the very long run, an institutional architecture cannot change, quite on the contrary. During normal times – when a well-established regulation mode is operating – the logic of interaction between a constitution, institutions, organizations, conventions and habitus is **top-down**. But by the very definition of a capitalist and democratic system, some actors always have an interest in innovating and the very generalization of a given economic order simultaneously generates new and unprecedented imbalances, which challenge the higher-level coordinating mechanisms, according to a process which is now **bottom-up**. It is not only institutions

Diagram 2.10: A suggested map for the relationship between constitution, institutions, organizations and conventions

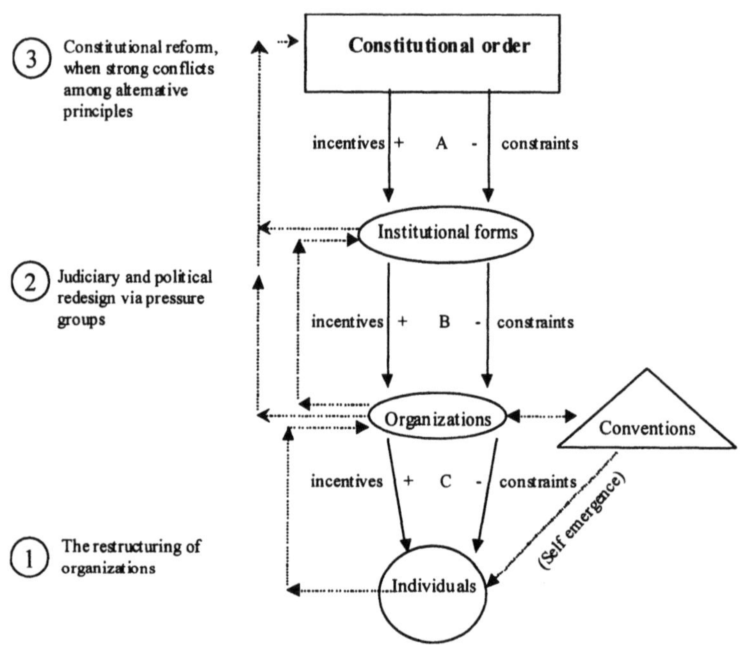

Top-down: From the constitution to individuals: a clear hierarchy

A → B → C

Bottom-up: Emerging disequilibria and conflicts call for a revision of upper level rules of the game

1 → 2 → 3

Degree of persistence:
Constitutional order > Institutional forms > Organizations > Individual behaviours

Source: Freely adapted from Douglas NORTH (1991), Charles SABEL (1997)

which play a role in this, but they are changing over time while remaining quite distinct across nations. There are major phenomena to be explained by any satisfactory theory of capitalism.

The balkanization of conventional economic theory thus gives rise to the third paradox already mentioned: many detailed analyses and much specific knowledge but no general understanding of the more pressing contemporary issues which are systemic and related to a precise configuration of capitalist economies. But this is not necessarily inevitable, since a **new generation of institutionalist thinking** is reinvigorating the concern for a more encompassing theory. It is to be hoped that the regulation approaches are part of such an ambitious process.

V. An Agenda for the Twenty-First Century

This chapter has argued that modern economic theory, in spite of its prolific development and technical achievements, is far from delivering an intelligible analysis of the transformations of contemporary capitalism. The inability to analyse a system as a whole is very detrimental to economic relevance. This explains some of the discrepancies between the development of theory and contemporary requirements.

But it is not remotely inevitable, since the first steps towards a more relevant theory of modern capitalism and its transformations have already been made by a variety of scholars, whether historians, sociologists, political scientists, heterodox economists or philosophers. Economic issues are too serious to remain the exclusive responsibility of neoclassical economists. Regulation theory is immersed in this melting-pot and research agenda and aims to galvanize once again the ambition of the old political economy.

Might this interdisciplinary research programme find a voice when disenchanted politicians realize the limitations of current policies, before or after a major world financial crisis ... or some unexpected but devastating domestic social unrest?

References

AGLIETTA, M.: *Regulation and Crisis of Capitalism*, New York (Monthly Review Press) 1982. Original: Régulation et crise du capitalisme. L'expérience des Etats-Unis, Paris (Calmann-Levy) 1976.

AMABLE, B., BARRÉ, R., BOYER, R.: *A la recherche des systèmes d'innovation et production: Théorie, configurations et évolutions*, Paris (Economica) 1997.

AMABLE, B., BOYER, R., LORDON, F.: "L'ad hoc en économie: la paille et la poutre", in: A. D'AUTUME, J. CARTELIER (Eds.): *L'Economie devient-elle une science dure?*, Paris (Economica) 1995, pp. 267-290.

ARROW, K. J., HAHN, F.: *General Competitive Analysis*, San Francisco (Holden Day) 1971.

BARRO, R. J., SALA-MARTIN, X.: *Economic Growth*, Stratford (Lucille H. Sutton and Scott D.) 1995.

BECKER, G.: "Pourquoi le chômage est-il si élevé en Europe?", *Le Monde*, 14 May 1996.

BENASSY, J. P.: *The Economics of Market Disequilibrium*, Boston (Academic Press) 1982.

BENASSY, J. P.: "Imperfect competition and unemployment", *Couverture Orange CEPREMAP*, 1992.

BERGER, S., DORE, R.: *Globalization and national diversity*, Ithaca (Cornell University Press) 1996.

BOWLES, S., GORDON, D. M., WEISSKOPF, T. E.: *Beyond the Waste Land*, Garden City NJ (Double Day) 1983.

BOYER, R.: *The Regulation School. A Critical Introduction*, New York (Columbia University Press) 1990.

BOYER, R.: "Markets: History, Theory and Policy" in: R. HOLLINGSWORTH, R. BOYER (Eds.): *Contemporary Capitalism, The Embeddedness of Institutions*, Cambridge (Cambridge University Press) 1997.

BOYER, R.: "Le politique à l'ère de la mondialisation et de la finance: le point sur quelques recherches régulationnistes", *L'Année de la régulation*, 3 (1999), pp. 13-75.

BOYER, R., DRACHE, D. (Eds.): *States Against Markets. The Limits of Globalization*, London (Routledge) 1996.

BOYER, R., SAILLARD, Y. (Eds.): *Théorie de la régulation. L'état des savoirs*, Paris (La Découverte) 1995.

BRAUDEL, F.: *Civilisation matérielle, économie et capitalisme XV-XVIIIe siècles*, Paris (Armand Colin) 1979.

CONTAMIN, R., LACU, C.: "Origines et dynamiques de la crise asiatique", *L'Année de la Régulation*, 2 1998, pp. 11-63.

DEBREU, G.: *Theory and Value: An Axiomatic Analysis of Economic Equilibrium*, New York (John Wiley) 1959 [= Cowles Foundations Monograph No. 17].

DEHOVE, M.: "L'Union Européenne inaugure-t-elle un nouveau grand régime d'organisation des pouvoirs publics et de la société internationale ?", *L'Année de la régulation*, 1 (1997), pp. 11-55.

DELORME, R., ANDRÉ, C.: *L'Etat et l'économie*, Paris (Seuil) 1983.

DEWATRIPONT, M., ROLAND, G.: "Transition as a Process of Large-Scale Institutional Change", *Economics of Transition*, 4/1 (1996), pp. 1-30.

GARCIA, M. F.: "La construction sociale d'un marché parfait: le marché au cadran de Fontaines-en-Sologne", *Actes de la Recherche en Sciences Sociales*, 65 (1986), pp. 2-13.

GLYN, A.: "Contradictions of Capitalism", *New Palgrave Dictionary of Economics*, London (MacMillan) 1988, pp. 638-640.

GRANOVETTER, M., SWEDBERG, R.: *The Sociology of Economic Life*, San Francisco CA (Westview Press, Boulder) 1992.

HEILBRONER, R. L.: "Capitalism", *New Palgrave Dictionary of Economics*, London (MacMillan) 1988, pp. 347-353.

HOLLINGSWORTH, R., BOYER, R. (Eds.): *The Embeddedness of Capitalist Institutions*, New York (Oxford University Press) 1997.

HOLLINGSWORTH, R., SCHMITTER, P., STREECK, W. (Eds.): *Governing Capitalist Economies: Performance and Control of Economic Sectors*, New York (Oxford University Press) 1993.

INGRAO, B., ISRAEL, G.: *The Invisible Hand. Economic Equilibrium in the History of Science*, Cambridge MA (The MIT Press) 1990.

LORDON, F.: "Endogenous Structural Change and Crisis in a Multiple Time-Scales Growth Model: A Stylized Formalization of the Exhaustion and Crisis of the Fordist Growth Regime", *Couverture Orange CEPREMAP*, 9324 (1993).

LUCAS, R. E.: *Studies in Business Cycle Theory*, Cambridge MA (The MIT Press), 1983.

LUCAS, R. E.: "On the Mechanisms of Economic Development", *Journal of Monetary Economics*, 72, July (1988), pp. 3-42.

NORTH, D.: *Institutions, Institutional Change and Economic Performance*, Cambridge/New York (Cambridge University Press) 1990.

PETIT, P.: "Formes structurelles et régimes de croissance de l'après fordisme", *L'Année de la Régulation*, 2 (1998), pp. 177-206.

POLANYI, K.: *La grande transformation*, Paris (Gallimard) 1983. Original: *The Great Transformation* (1944).

POSNER, R. A.: *The Economics of Justice*, Cambridge MA (Harvard University Press) 1981.

POWELL, W., DI MAGGIO, P. J. (Eds.): *The New Institutionalism in Organizational Analysis*, Chicago (The University of Chicago Press) 1991.

ROMER, P.: "Increasing Returns and Long-Run Growth", *Journal of Political Economy*, 94 (1986), pp. 1002-1037.

SABEL, C. F.: "Constitutional Ordering in Historical Context", in: F. W. SCHARPF (Ed.): *Games in Hierarchies and Networks*, 1997.

SAPIR, J.: *Le Chaos Russe*, Paris (La Découverte) 1996.

SHONFIELD, A.: *Le capitalisme d'aujourd'hui. L'Etat et l'entreprise*, Paris (Gallimard) 1967.

STARK, D.: "Privatization in Hungary: From Plan to Market or From Plan to Clan?", *East European Politics and Societies*, 4/3 (1990), pp. 351-392.

STARK, D.: "Path Dependency and Privatization Strategies in East-Central Europe", *East European Politics and Societies*, 6/1 (1992).

STIGLITZ, J. E.: "The Causes and Consequences of the Dependence of Quality on Price", *Journal of Economic Literature*, 25 (1987), pp. 1-48.

WEIL, P.: "Increasing returns and animal spirits", *American Economic Review*, 4 (1989), pp. 889-894.

WOLF, C. JR.: Markets or Governments: Choosing Between Imperfect Alternatives, Cambridge MA /London (The MIT Press) 1990.

Chapter 3

German Contemporary Analyses of Economic Order: Standard Ordnungstheorie, Ordoliberalism and Ordnungsökonomik in Perspective

SYLVAIN BROYER

Characterizing the processual orderliness of economic structures is a strong tradition of German economic discourse (Tribe 1995 pp. 1-8). The oldest contributions came from the cameralist sciences which, by defining the role of the state established a link between welfare objectives, the social as well as the legal order, and consensus. Friedrich List, for his part, defines a national and historical system of the economy that is opposed to the "cosmopolitan" view of British paleo-liberalism. For List, the structure of an economy can only be understood with reference to the political interests of nations which themselves depend on the stage of economic development reached. Subsequently, the German Historical schools developed appropriate tools for analysing economic structures and evolution. By introducing a revised method of analysis that leads to the definition of the concept of economic order, the Freiburg school finally emerged as the contemporary synthesis of the ideas relating to this issue.

Since the Freiburg school links this ancient issue with the dominant economic theories, its *Ordnungstheorie* is often considered as the final stage in the evolution of this thought. For that reason, *Ordnungstheorie* usually serves as a standard for comparisons with other theoretical approaches. However, this focus on the work of the Freiburg school should not hide the new developments dealing with the same topic on the basis of criticisms concerning the Freiburg theory.

Hence, this chapter proposes to distinguish between this and the redevelopments by the Freiburg school which are most directly related to standard *Ordnungstheorie*, but by criticizing it moves away from this paradigm in order to create a new economics of order (*Ordnungsökonomik*).

The first part of this presentation exposes the contemporary treatment of the above issue - which is understood as standard *Ordnungstheorie* - stressing the way in which it conceives of the topic of economic order, its definition, as well as the tools and methods of its analysis. The second part describes the political principles that the Freiburg school deduces from the former analysis. Ordoliberalism is thus defined as a political action and a neoliberal theory of competition and the state. The third part underlines the criticism that can be addressed to standard *Ordnungstheorie* and Ordoliberalism in order to present and identify recent fruitful redevelopments, in particular *Ordnungsökonomik* as that which is most complete.

I. Standard Ordnungstheorie

Ordnungstheorie, on the one hand, represents a way of thinking economics and its relations to the other spheres of society that leads to the concept of economic order and its interdependence with other orders of society. On the other hand, *Ordnungstheorie* is a method of analysing this order based on a synthesis of the historical and theoretical approaches.

This exposition requires consideration of Walter Eucken as the founder of standard *Ordnungstheorie*. By reconstructing the subject and the method of economics in his *Grundlagen der Nationalökonomie*, Eucken is led to think in terms of the combination of elements that constitute an "order", on which the efficiency of factor allocation depends. His second major work - the *Grundsätze der Wirtschaftspolitik* published after his death in 1952 - then develops the economic policy program that makes it possible to realize this optimal order and to ensure its continuation. As Lenel puts it, *Ordnungstheorie* indeed evolves in the perspective of the analysis and implementation of economic orders (Lenel 1989 pp. 3-4). Eucken himself steadily underlined this orientation (Eucken 1989 p. IX).

1. "Denken in Ordnungen" as the theoretical starting-point

Two statements constitute the basis of this theoretical construction. First, the subject it refers to exhibits the coordination of economic actions as the central problem of economics. Secondly, since economic action is only seen as a part of human activity in a society, economic order – as a pattern of coordination – must not be viewed in an isolated manner but the interdependence of different existing orders must be taken into account.

a) The problem of coordinating economic activity

Coordinating economic actions has become the main concern of modern societies based on the general division of labour. "The problem of order" leads to the demonstration that the process of efficient resource allocation in order to overcome scarcity has been subject to tremendous transformation and complexification due to the evolution of a technoeconomic environment (Eucken 1948 p. 1). In his view, economic activity is recognized as a global process in which individual decisions result partially from those of others individuals, but without anybody having complete knowledge of this process. Coordinating these decisions assures unity of the "economic process". Hence, *Ordnungstheorie* centres around the question of how coordination functions, that is to say which mode it takes: how do supply and demand actually fit together on real markets without being managed by anybody? Which phenomenon makes it possible to gain an impression of "order" despite the "anarchy" of decision-taking, asks Röpke (Röpke 1994 p. 17).

The mode of coordination referred to is not entirely characterized by means of a Walrasian *tâtonnement* in search of an equilibrium price, even if the matching of supply and demand through a price vector – the equilibrium conditions – of course forms part of the analysis. The coordination of all individual actions, taking into account their interdependence, arises within the framework of a general equilibrium approach. However, as will be shown, for several reasons this conception differs from the neoclassical one, especially as far as the processual dimension is concerned, that is, the way in which the equilibrium is reached. First, time is considered as non-fictitious and consequently it can be derived from this that agents' expectations are adaptive. Next, even if Eucken does not make himself clear, the achieved equilibrium does not necessarily represent a maximum – something that is in particular true for the labour market (Eucken 1990 p. 323). Even if this equilibrium is fully realized, a social policy is needed to correct the outcome. However, it is still an equilibrium because it possesses an internal stability, despite the fact that the same equilibrated state cannot be reached twice and that the very first time it is achieved it needs a strong institutional framework. This conception of equilibrium is nevertheless close to the neoclassical one in its instrumental dimension, considering that it requires an institutional framework that guarantees the functioning of the price mechanism as the instrument of coordination.

Five aspects of coordination are more generally distinguished: the allocation of factors of production and their spatial distribution, the distribution of income resulting from the production process, the coordination of needs in a time perspective and finally the choice of the production technique (Eucken 1989 pp. 2-5). These five components all depend on each other for satisfying needs under the constraint of limited resources or, to use the Euckenian terminology, for "overcoming scarcity". Moreover, when Eucken treats the issue of coordination, he links it to the different types of decision makers, distinguishing between an individual, a household, a firm, and so on, with regard to their capacities and means to satisfy their needs. Standard *Ordnungstheorie* thus formulates the problem of coordination in a broader sense (Streit 1992 pp. 678-680), including: competence aspects of decision-taking (who decides on what, referring to which value system, respecting which constraints?), coordination in a narrow sense (how are decisions coordinated, through which information system?) and the process of controlling or evaluating decision-taking (how are decision-makers controlled, by means of which sanctions or incentives?).

Coordination in its broad sense, as the starting-point of standard *Ordnungstheorie*, can in a certain way also be viewed as the conclusion of Eucken's earlier works on capital formation. The argument which leads to that interpretation is related to the different aspects of coordination mentioned above. In particular, the evaluation and control function can be conceived of as an expression of adaptive expectations through the introduction of a time variable. This makes it possible to extract the decision-taking process from the theoretical framework of a complete market system: "Wirtschaften ist anpassen", states Eucken with respect to the evolution of agents' plans considering *ex post* mismatched *ex ante* planning data, such as investment and saving (Eucken 1948 p. 60). These dimensions are already present in his theory of capital formation, the *Kapitalistische Untersuchungen* published in 1934, which can be related to that of Böhm-Bawerk. In the latter, time is an element of the value of capital goods, as this value rises in relation to the detour of production and the time it takes to produce consumption goods. Besides, technical progress is reflected in a decrease in production time for a consumption good of equal value (Fehl 1989 p. 73). Hence, the introduction of time and space variables when addressing the issue of coordination in the context of standard *Ordnungstheorie*, in other words refraining from applying atemporal theories, must clearly be seen as a result of this former work.

b) Interdependence of orders

Standard *Ordnungstheorie* is based on a second element directly derived from the first one. The economic process of coordination changes according to the historical framework in which it is situated:

"The way in which everyday economic life proceeds depends on the nature of the country, the race, culture, and beliefs of the inhabitants, on the political institutions and structure of the state, in fact on the entire historical environment" (Eucken 1952 p. 35).

As far as the decisions of individual economic agents are concerned, they no longer interact only with those of other agents but also with the whole social and political organization of the society. The institutional framework in standard *Ordnungstheorie* has to be understood in a broader sense than the neoclassical circumstances of general equilibrium. The former refers not only to market structures but also includes exogenous variables which do not belong to the latter conception of economic science. The coordination of economic actions as a whole depends on the other orders of society.

At the same time, the evolution of society (for instance, the emergence of new social classes, the decline of professions, the rise and fall of certain corporations, family organization, and so on), the evolution of the individual as well as the state is dependent on the evolution of economic structure. The role that industrialization, early capitalistic organization of production and monopolies play in shaping political and social orders is particularly considered (Eucken 1948 pp. 70-71).

On the basis of these mutual dependencies, standard *Ordnungstheorie* reaches the concept of the "interdependence of orders". It should then be understood as a general equilibrium approach in which circumstances are contextualized in political, social and legal organization. Hence, the economist must make himself aware that the coordination process of individual decisions – the equilibrium conditions – takes place in an order which is itself connected to non-economic orders. This view, called "Denken in Ordnungen", represents a specific way of economic thinking that causes Eucken to criticize existing theories and to shape a method of economic analysis.

c) The economic order: subject of analysis

From this double starting-point, economic activity is conceived as a complex phenomenon, specific to a particular point in time and a certain situation, the complexity of its constitutive elements not being recognizable through direct observation. A complex economic phenomenon simultaneously constitutes an "individual historical problem" as well as a "general theoretical problem" (Eucken 1989 pp. 15-18). Consequently, neither the historical method alone nor the purely theoretical method of analysing economic reality can explain at once to what extent an economic phenomenon has particularities as a result of its context, which elements it consists of and which relations between these elements lead to the observed reality. Eucken's aim is to develop a synthesis of the two approaches (Schefold 1995a p. 14; Herrmann-Pillath 1987 p. 37), the economic order is defined as the only concept that makes it possible to overcome their antinomy. It is at once ideal-type, meaning here that it can be subject to theoretical reasoning, and real-type, since at all times and places economic action takes place in a specific economic order.[1] The shape of the order, unlike the style, is a function of the range of decisions it wants to outline and explain. Thus, for a national economy "the economic order of a country corresponds to the totality of existing forms in which the economic process daily takes place",[2] or, to put it slightly differently, the economic order is "the totality of forms in which the management of the daily economic process *in concreto*, here and there, in the present and the past, ran – and still run – their course".[3] In his later writings, Eucken sticks to this second definition (Eucken 1948 p. 62; 1990 p. 180).

The forms have to be understood both as the institutions that influence the coordination process and as the function that these institutions fulfil within the order, being interdependent with other institutions. The term "institution" refers to both the institutions designed according to an economic constitution and to the actual institutions which develop according to historic evolution. The distinction between factual and designed elements of economic orders is reminiscent of Menger's opposition between, on the one hand, institutions created on the basis of a deliberate calculation and a collective will, called "pragmatic" institutions, and, on the other hand, "organic" institutions conceived of as the non-planned result of specific individual actions by the members of a society (see Garrouste 1994).

---

1 Note that the real- and ideal-types do not have the same meaning in Eucken as in Max Weber (CASSEL 1968 pp. 146-158; MEYER 1989 pp. 32-44).

2 EUCKEN 1989 p. 51: "Die Wirtschaftsordnung eines Landes besteht in der Gesamtheit der jeweils realisierten Formen, in denen der Wirtschaftsprozeß alltäglich abläuft" (my translation).

3 Ibid. p. 167: "Die Wirtschaftsordnung ist die Gesamtheit der Formen, in denen die Lenkung des alltäglichen Wirtschaftsprozesses *in concreto* – hier und dort, in Gegenwart und Vergangenheit – erfolgte und erfolgt" (my translation).

The coherence of the two kinds of institutions in an order is not certain, but all of them taken together define the unity in which coordination functions. This means that economic orders are reflected in the combination of institutions and their functions. If there are so many combinations that the set of possible orders is a priori infinite, the set of possible forms that make up economic orders is closed and defined.

2. The analytical tools

An important point of standard *Ordnungstheorie* consists in orientating the analysis – from the beginning to end – towards the facts and not the concept. Yet the focus on the concrete is useless without analytical concepts. In other words, the theory has thus to define the tools and the way they are to be used in order to resolve the antinomy.

However, the instruments must not already be concepts, they have rather to exist outside themselves in economic reality. Hence, a link is needed between the definition of tools and this reality. It consists in a variable that is both found in every single economic fact and liable to theoretical interpretation. The analytical tools will be characterized as functions of this variable. Finding the latter refers back to the problem of coordination of economic agents' actions, that is to say their decision-taking. In fact, Eucken states that decisions have always been, and are still, taken in accordance with a plan.

"At all times and places man's economic life consists of forming and carrying out economic plans. All economic action rests on plans" (Eucken 1952 p. 118). This variable is the central element of all economic action since, as the starting- and finishing-point of decision-taking, its definition includes the three functions of coordination in a broad sense: who makes the plan, how, and with which result. The variable "economic planning" constitutes the interface that allows us to define theoretical tools of analysis with regard to economic reality. It is at one and the same time useful for describing real-types and analysing ideal-types.

The economic orders, instruments of analysis, are characterized on the basis of the above variable as a combination of "pure basic forms". The latter definitively describe each past or present economic order and they can be expressed by means of theoretical relations. It follows that the analytical tools which standard *Ordnungstheorie* develops are "pure" economic orders, meaning ideal-types or "economic systems".

a) The plan as defining variable

Planning is based on a set of data and on rules of experience. The first piece of data relates to the identification of needs. Their satisfaction can be viewed as the normative element of the variable. To meet the norm, the agent must take the productive organization into account, which is described by three pieces of data: the stocks of raw materials and other natural resources, the labour force available, and also the stocks of existing production goods ready to be used. In addition to this purely economic calculation, the planning agent must consider two further pieces of data, namely the level of technical knowledge acquired, as well as the legal and social organization. The latter is understood in a broader sense, including not only pragmatic but also organic institutions, such as traditions, customs and the "spirit" that guides economic agents:

"It not only includes traditions, laws and customs, but the spirit in which men live and keep to 'rules of the game'"(Eucken 1952 p. 186). By integrating the "economic spirit" at this stage as a piece of economic planning data, Eucken relinquishes the ambition of the Historical school to explain the ethical dimension, as attempted by the concept of economic style. Economics is embedded in a cultural context, the characterization of the second by no means being the object of analysis of the first. Standard *Ordnungstheorie* represents a theory of systems (Schefold, 1995a p. 15)

In addition to this set of data, the economic planning agent has three rules of experience at his disposal. These rules are innate and manifest themselves unconsciously during the calculation process. Consequently, they are routines which determine individual rationality and which, characterized as laws, cannot logically be broken. The first rule to which every single economic action conforms is Gossen's first law. This confirms the diminishing satisfaction of needs with each additional unit of satisfaction until a satisfaction-point is reached. The second rule of experience concerns comprehension of the possibility of decreasing labour productivity. The third rule finally states a positive relation between the "productiveness" (*Ergiebigkeit*) of both labour and capital and the intensity of production detours, that is, the time the production of consumption goods takes and the time that one has to wait (Eucken 1989 pp. 135-137).

Provided that the rules of experience are common to all economic agents, the nature of the plan modifies consideration of the presented data. For the central authority of a planned economy, the relevant data correspond to the whole set described above. In this case, planning concerns the entire economic process and the data refer to the whole economy. The time-scale to which the planning relates also modifies the composition of the data set. Short term planning can view the two last-mentioned pieces of data – the state of the art as well as the legal and social organization – as fixed. In the case of individual agents, planning does not focus on the functioning of the global economic process. They have no complete vision of it and for them only partial processes – in which their actions take place – are relevant. As a consequence, the important data are different from the preceding set. It now contains prices of consumption and production goods that are part of the decision-taking framework. Differences between individual and global data consists in both the nature and the range of information, as well as the mode of information transmission.

In standard *Ordnungstheorie*, the time-horizon, planning level and the nature of the economic agent explain – on the basis of different relevant data sets – planning processes that are too different to be compared. Eucken confirms in his "Theory of the centrally directed economy" that the system of this type and that of a "market" economy should be subject to distinct theories in so far as the underlying economic processes do not function in the same way (Eucken 1990 pp. 62 ff.). Leading to the "duality" of the theoretical construction, this statement of standard *Ordnungstheorie* has often been criticized in the literature (see e.g. Herder-Dorneich 1991 p. 354).

b) The economic systems and their elements

The two ideal-types opposed here differ from one another with respect to their forms of coordination in a broad sense, that is, with respect to the three dimensions of competence, the coordination pattern narrowly defined, and control over decision-taking. These two systems themselves take various forms according to the latter three features.[4]

The first is the centrally directed economy. Here, competence and decision-taking control are in the hands of a central authority. Coordination in a narrow sense does not take place via a price system but via a value system belonging to this authority.

Two forms of the centrally directed economy can be distinguished with regard to the size of the plan: an elementary form, representing a self-sufficient economy of the "Robinson type", in which no transaction occurs: *Eigenwirtschaft*. The second system corresponds to a centrally administered economy: *Zentralverwaltungswirtschaft*. Its plan determines the utilization of productive forces, the set of products and the distribution of consumption goods for the entire economy. The plan consists here in the "artificial" reconstruction of the economic process through resource allocation by means of opposing transactions.

4 The original presentation of the two economic systems is to be found in EUCKEN 1989, pp. 78-127.

These three elements can be planned to different degrees, leading to the definition of three types of the central administered economy, which differ in the degree of centralization of the plan, or in the inverse degree of freedom of individual planning (see Graph 1 for details).

The second ideal-type system is that in which each economic agent works out a plan that is realized by the help of transactions. In this case, each individual plan constitutes only a tiny part of the global economic process and is coordinated with the plans of other economic agents. Coordination in a narrow sense is guaranteed by the existence of a value measure, money, and of an exchange-place, the market. It takes different forms according to the prevailing market form and monetary system.

These forms, of course, refer to the possible supply and demand configurations. The distinction between competition, partial or complete oligopoly and monopoly relates to the way in which individuals do their planning, to the expected reactions from other sellers and buyers, and to the functioning of the price mechanism.[5] The market form is also a function of the degree of freedom that the economic agent has concerning his planning activities. A market is open, competitive, if the agents do not know what influence their decision has had, and will have, on the market, and if prices are not fixed by a certain economic agent or by political powers. A market is closed, as in the monopoly case, if an agent possesses a superior degree of planning freedom compared to other market participants, for example, if his or her decisions can force the planning of others into a direction that suits him or her and if he or she is able to impose a market price. According to this typology, a market can take twenty-five different forms, bearing in mind that five can be applied to each side of the market.

Three possible forms of the monetary system have to be added to the market forms. The criterion for distinguishing between the former is the degree of tangibility of money. The first form consists in a barter system, where the money used was originally ordinary material goods which could be used again as goods at any time. In the second system, money has a persistent form and is systematically used for payment of goods and services. Finally, money can, in a third system, be created by credit and disappear with its repayment.

It should be noted that the typology of monetary systems has not been subject to Eucken's own developments. He takes over the existing typology developed by Hildebrand (quoted in Piettre/Redslob 1986 p. 229) with the difference that these three systems do not correspond to chronological stages

---

5 This "Datengestaltung der Wirtschaftspläne" serves as a differentiating criterion between competition and monopoly (EUCKEN 1989 p. 100).

of development, but to possible forms that can coexist in the same economic order.

The difference between the treatment of market forms and money systems seems surprising. It might be asked why Eucken does not consider the money system with respect to the market forms it can take, examining for instance the credit supply and demand sides according to their degree of openness. Such an approach appears reasonable, especially considering Germany's pre-war monetary history and particularly the power of finance capital over the industrial cartellization during this period (see e.g. Hardach 1980 chaps. 2, 3; Wehler 1974 pp. 36-57). Why the question of market forms of monetary systems is absent from Eucken's analysis, including his later works, is hard to comprehend - above all in view of the objective of extracting from historic reality the pure basic forms of economic orders. However, this lack seems to indicate a concept of money that was still largely exogenous to the functioning of the rest of the economic process. This interpretation is verified by Eucken's later propositions for a quantitativist monetary policy.

3. The methods: Isolating abstraction and morphological analysis

Using economic systems as tools for the analysis of concrete economic orders necessitates a particular method, derived by Eucken from that which von Thünen applies in his book, *Der isolierte Staat*. It consists in extracting the pure forms that manifest themselves in economic phenomena. Eucken then differentiates two extraction or abstraction methods. The first implies the extraction of generalized forms from many economic phenomena. "Generalizing" abstraction, employed among others by the Historical school, presents, however, the risks of identifying either only real-types that are not useful for a theoretical analysis or ideal-types that will never occur in reality. Hence, Eucken prefers the "isolating" abstraction method. Here, just one phenomenon is examined at a time, from which the different realized forms of economic systems are extracted. A concrete problem is thus broken down into "categories", in other words, the set of elements that compose the ideal-types and can be found by examining one single economic phenomenon.

Using this first method, the complete cartography of the economic reality gives access to its "morphology". The complexity of an economic phenomenon is then ordered into simple forms, which leads to a clear understanding of the economic process. This double step method resolves definitively the antinomy inherent in economic phenomena. The discovery of their morphology makes it possible to discern their historical identity by categorizing without dissolving it. Each economic order reveals an individuality through its combination of forms. Hence, it is understood as a part of the global historical environment. In a further step, the analyst is able to decide which theoretical relations are suitable to the analysis of the categorized forms of the phenomenon. This approach elucidates why economic theories are not in all cases verifiable. They are always correct, yet, unlike those of the natural sciences, they cannot be applied at each period and point of the economic process, but only to an economic order defined by a specific combination of elements. The modification of this composition does not challenge the theory itself but only its relevance for analysing a contextualized phenomenon, namely its topicality.

Economic evolution is the factor explaining why theories can lose relevance. Two different kinds exist: the evolution of the order and that of the process. The former consists in the modification of the composition of prevailing forms, institutional changes because of legal and political reforms. Evolution of process is due to the non-identical reproduction of the coordination pattern – referring to the manner and level of the satisfaction of needs, the use of productive forces, the production technique and place. Economic evolution as a whole represents the result of data variations that alter the economic plans. Each change in economic order modifies the economic process by altering the data set. On the other hand, a modification of the process does not influence the combination of forms that make up the economic order. In order to describe economic evolution it is therefore sufficient to measure the variations in planning data. The method consists in characterizing two records of the economic process. Then one looks at which pieces of data or forms were modified and tries to explain the fluctuations. This static comparative approach examines two moments of economic reality which are not necessarily two equilibrium points in the process.

Morphological analysis – based on isolating abstraction with the intention of characterizing the economic order and process – and the data variation method – aiming to measure the gap between two realizations of the economic order and of the process at different points in time – form a unity. As can be seen in German and East European writing on this subject, this methodological set is applied to the study of the transition process from socialist to market economies, as well as to comparison of Germany's monetary reforms in 1948 and 1990 (see e.g. Willgerodt 1994). Hence, standard *Ordnungstheorie* – despite its extensive methodological definition – has not only reached the objectives fixed by Eucken but also, fifty years after being developed, its relevance is indisputable. However, well before these quite recent applications of standard *Ordnungstheorie*, another theoretical development of the Freiburg thought played an important role: certain reflections on, as well as direct references to, Ordoliberalism served to guide Germany's post-war economic policy until the Keynesian approach grew in influence (Allen 1989 pp. 267-289).

The argumentation of this first part is summed up in the graph 3.1.

II. Ordoliberalism

Ordoliberalism has to be located in the conceptual framework of the Freiburg school. However, it goes beyond the strict boundaries of the school in the sense that it combines with its scientific definition a political action which directly and consciously refers to the theoretical approach. Ordoliberalism can also be considered as an extension of standard *Ordnungstheorie*, since it developed an *Ordnungspolitik* primarily aiming at the reintegration of Germany into the group of western economies after the Second World War. Ordoliberalism as a theory refers to Eucken's analysis of the economic order, to a neoclassical theory of competition and to a particular theory of the state.

1. Political action

Political activity was particularly strong after the Second World War because of Walter Eucken's responsibilities as a member of the scientific council of the economic administration of the Bizone, Franz Böhm's position in the cartel office and later as CDU-deputy in the new German Parliament, and above all with Ludwig Erhard's role first as head of the economic administration, then Minister for economic affairs during Konrad Adenauer's chancellorship, and ultimately as Chancellor between 1963 and 1966. Their most significant contributions to political decision-taking in their country resulted in the emergence of a market economy order – thanks to scientific councils' debates, as well as active support for the monetary reform of 1948, the liberalization of the law of prices and production and later of wages[6].

6 This contribution will not examine in greater detail the political role of ordoliberals during the period of FRG emergence. For a complete history, see NICHOLLS 1994, especially chaps. 4-10.

Graph 3.1: Standard *Ordnungstheorie*

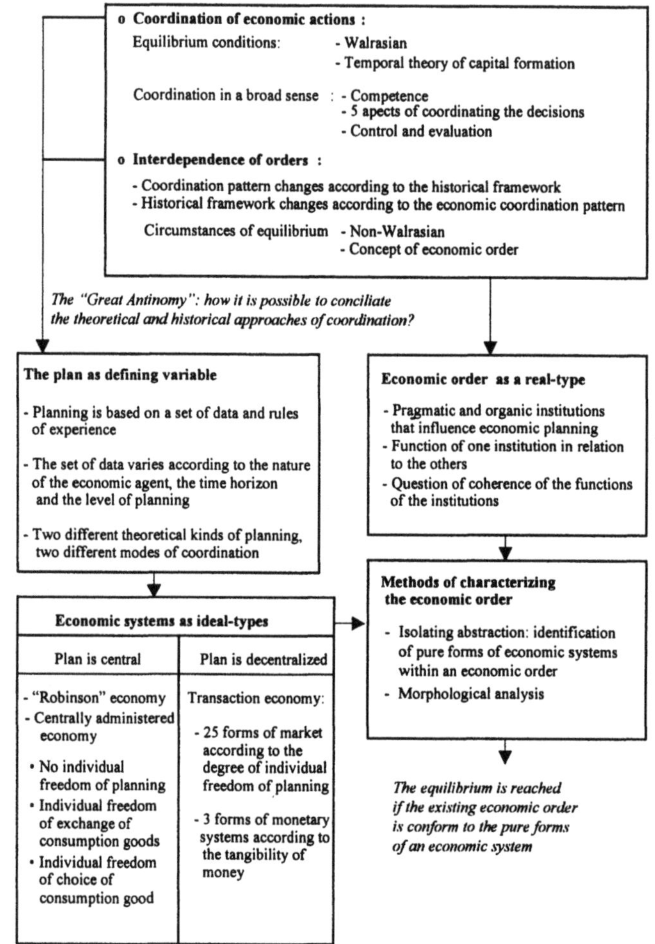

Furthermore, part of the development of the concept, as well as the diffusion of these ideas was accomplished outside Freiburg. Despite their political exile, Alexander Rüstow and Wilhelm Röpke took an active part in the scientific development of liberal thought in Germany (Starbatty 1994 p. 240).[7] Political action was supported by a large initiative to disseminate these ideas, for example by Röpke's writings,[8] Erhard's articles in the *Neue Zeitung*, and those by Müller-Armack addressed to the German industrial and business world, as well as the creation of the *Ordo* journal by Böhm and Eucken in 1948.

Ordoliberalism does not represent the only liberal current in German universities at the end of the war. However, the intellectual dictatorship imposed by the Nazi regime had not allowed much development of this thought (Rieter/Schmolz 1993 pp. 88-91). For the second quarter of this century, three liberal groups can be identified in Germany (Tuchtfeldt/Willgerodt 1994 pp. 369-371). The first and oldest one was considerably influenced by the Austrian marginalist school, and was most threatened by the Nazi dictatorship. Alexander Rüstow and Wilhelm Röpke belonged to it. They were both forced into exile. The geographical distance just allowed them to participate in the after-war debate on liberal ideas from afar. Another group, often called the Cologne school, engaged belatedly – compared with the Freiburg school – in work on the liberal paradigm. They developed the concept of the Social Market Economy, drew upon the work of the third group, that of the Freiburg school (Müller-Armack 1974 pp. 244-251). Thus the latter constituted one of the very few organized places that was in a position to support the promotion of liberal ideas in Germany at that period. This fact certainly helps to explain why the term "Ordoliberalism"

7 It is questionable whether Alfred Müller-Armack should be associated with this work, as in Starbatty, at least as an early advocate of economic liberalism. Before 1936 his theories ran rather counter to liberalism (NICHOLLS 1994 pp. 71-73). Besides, he did not really participate in the definition of Ordoliberalism, but only used the ordoliberal ideas on competition and the state in his late theory of the Social Market Economy (MÜLLER-ARMACK 1974 p. 8). His role in diffusing liberal ideas during the war and in post-war Germany, on the other hand, is indisputable.

8 The publication of *Die deutsche Frage* in 1945 before the Postdam Conference is especially noteworthy. Nicholls writes: "By opposing collectivism and Prussian hegemony, by accepting the Elbe as Germany's frontier – at least for the time being – and by concentrating on rapid economic rehabilitation of a Western-oriented economy, Röpke was writing as a prophet of the Federal Republic of Germany" (NICHOLLS 1994 p. 134).

alone is employed with reference to German neoliberalism, despite the existence of different currents of thought.

2. A neoliberal theory

Ordoliberalism proposes a legal organization of market competition determined by the action of the state – the implementation ensuring an optimal economic coordination pattern that is coherent with the political and social imperatives of a free society and that concurrently supports them. By emphasizing the institutional framework of an economy, the functions of competition as pillars of this framework, and the role of the state in securing these functions, Ordoliberalism breaks away from paleo-liberalism as a neoliberal theory.

a) A natural order of economics?

This theory proceeds from the distinction between the order that results from the totality of decentralized decisions taken within the economic process ("gewachsene Ordnung") and a positive, designed order of the economy ("gesetzte Ordnung") (Eucken 1990 p. 373). This difference goes back to the Physiocratic opposition between *ordre positif* and *ordre naturel*. The role that Ordoliberalism attributes to the state as the actor of economic policy, is inspired by that which the Physiocrats attributed to it: the state has to intervene until the desired order is reached.

However, German Ordoliberalism differs considerably from French Physiocracy on two points. First, for the latter, economic policy intervenes in the economic process – for instance by means of a trade control policy aiming at the stabilization of prices (Schmidt 1994 pp. 56-57) – the ordoliberal state, in contrast, directs its interventions principally to the institutional framework of the economy. Secondly, the Physiocrats expected the state to respect the natural order which is rather stable (Gonnard 1930, book 3, chap. 4). The desired order – *Ordo* – on the contrary, is not natural in this sense since it has no chance of spontaneous realization. This apparent contradiction with Physiocracy is also emphasized by Grossekettler, when he examines the etymological significance of *Ordo* (Grossekettler 1989 p. 43).

In order to reach a better understanding of Eucken's thought, it should be added that his *Ordo*, being a system of "complete competition",[9] is natural in

9 "Complete competition" ("vollständige Konkurrenz") is a key phrase in Eucken's theory. It can be defined as the market form that allows the price system entirely to direct the economic process (EUCKEN 1990 p. 249). This concept is often criticized, since at first sight it seems indistinguishable from

the sense that it possesses an internal stability thanks to the price mechanism, since its functioning in combination with an adequate monetary system always brings the economic process back to equilibrium. Precisely, this means that the conditions of equilibrium are natural in the sense of being "self-enforcing" while the circumstances of equilibrium have to be enforced. The "natural" character of the order conceived as internal stability can only be ensured by a set of institutions.

b) "Complete competition" as a socially and economically efficient order

Another characteristic of Ordoliberalism resides in the content of *Ordo*. It describes an "efficient and fair" equilibrium that connects the economic with the other values of the society (Eucken 1990 p. 166). *Ordo* is achieved by a particular competition order, that is, market forms and monetary system, to which different functions are attributed, based on classical justifications of efficient resource allocation, coordination of plans and income distribution, but also on a justification in terms of political and social organization that respects the Smithian conception of competition (Myers 1976 pp. 563-567).

Six functions of competition can be isolated in Eucken's work (Lenel 1975 pp. 49-62). First, in the case of "complete competition", the corresponding coordination pattern liberates the individuals by allowing them to develop their strengths – thanks to freedom of planning and decision-taking, as well as the freedom of contracting – and by minimizing illegitimate power positions. Moreover, competition in its complete form also brings about a tendency towards harmonization of private and general interests, this argument referring back to Smith's concept of the invisible hand as the best mode of coordination in a system characterized by the general division of labour. Coordination is optimal in the sense that its mechanism, the price system, measures the relative scarcity of goods. Fourthly, competition possesses an internal tendency towards stability, that is, towards equilibrium. The latter has a function of fairness, that is to say a function of income distribution according to production contributions. Finally, competition embedded in the economic order positively influences the other orders of society, in particular respect for the law and the general interest, free movement and free choice of profession, as well as private ownership as the guarantee for freedom in the private individual sphere. Besides, on the one hand, private property rights are a condition necessary

---

neoclassical perfect competition based on the atomicity of agents on both sides of the market. The only difference lies in the rejection of the neoclassical hypothesis of homogeneity of goods and agents.

for competition, on the other, competition ensures that these rights are not violated.

The employment of market competition as a social regulation pattern represents an important feature of ordoliberal thought. This pattern is based on the will to minimize the powers that can arise in free societies. Eucken - by reference to his early academic writings as well as thanks to his work with Franz Böhm at Freiburg - puts it like this: "Economic politics, like all politics, faces the problem of power" (Eucken 1990 p. 169).[10] The positive order of competition protects the individual from both public and private economic powers by ensuring the greatest degree of equality concerning starting positions (*Startgleichheit*). Thus, ordoliberal competition contains a specific ethic which defines a welfare function based on a Pareto optimum. For Röpke, only competition provides a pattern of "ethic neutral relations" between one individual and the others (Röpke 1994 p. 39). The process that drives competition - which Eucken conceives as the "overcoming of scarcity" - is in Röpke's opinion the "social form of struggle against deficiency". The neutrality of social relations is due to Pareto optimality, since in the economic world individuals get in touch with each other only to increase their welfare without reducing that of others.

By stressing this ethical role of economic relations, ordoliberal thought in a final dimension again meets the Physiocratic utopia (Streit 1992 p. 681). In fact, the *Ordo* is not only desired because of its economic efficiency and social regulation functions that are both the best conceivable solutions. But also, these functions correspond to the essence of all things and mankind, what is rather postulated than demonstrated. As a consequence, the implementation of *Ordo* is justified as the only and permanent objective of economic policy.

3. The achievement of the competition order: Normative principle of a positive economic policy

The achievement of the competition order constitutes the prime principle that all economic policy must serve and conform to. In other words, the

10 In 1914 Eucken published his doctoral thesis on the formation of interest groups in the shipping industry (*Die Verbandsbildung in der Seeschiffahrt*), submitted the same year at the University of Bonn to Professors Hermann Schumacher and Heinrich Dietzel. He wrote his *Habilitation* about the world nitrogen supply (*Die Stickstoffversorgung der Welt*). In this second book the aim was also to isolate power structures and regulations that affect the world markets.

competition order has to be regarded as an immovable norm. In this context, the action of the state can be differentiated according to two groups of positive principles that are defined in a very rigorous way by Eucken in the *Grundsätze der Wirtschaftspolitik*. Economic policy is divided into a legal type of action of creating the institutions of the competition order and a regulative type of action with respect to these institutions. The former enforces the functioning of the market system. The latter aims principally at correcting the organic institutions of the existing economic order in order to render them conformable to the set of pragmatic institutions that define the competition order. In this sense, the second action consists in an institutional regulation policy.[11] Thirdly, particular principles could be underlined, which guide the scope and scale of state action as far as decision-taking in economic policy is concerned.

a) Constitutive principles

In order to achieve an efficient competition system, the realization of the following six constitutive principles is required.

[1] Priority is given to monetary policy. Its aim consists in ensuring stability so that prices always remain a reliable indicator of scarcity and so that individual planning is as little as possible distorted by monetary phenomena, such as the use of money for political objectives. The difficulty in realizing such a policy lies in the fact that money is based on a credit system, and thus is intrinsically unstable. Hence, to protect individual planning from all monetary instability, the liquid assets of commercial banks must entirely cover their credit operations. At the same time, a rational regulation mechanism of the value of money must be set up by an agency independent of political powers. The objective of monetary stability attained by a quantitative control of the money supply and ensured by an independent central bank should be compared with Milton Friedman's theory and the monetarist policy of the 1980s.[12]

[2] In order to avoid the formation of monopolies, a systematic policy of open markets must be pursued. Here, all kinds of market entry barriers are taken into account: legal, technical or tariff restraints.

[3] Property rights, as a legal institution, have to be defined in their forms and functions, so that the individual's power and freedom to dispose of his or her property are combined with the impossibility of preventing others from

11 This particularity of Ordoliberalism is never stressed enough. The second type of political action by no means consists in an investment policy of a Keynesian type, even if a redistribution policy is envisaged.

12 For further details on the "monetary constitution" of Ordoliberalism, see DEHAY 1995.

using their rights. This definition makes it possible for the pursuance of private interests not to hurt the general interest.

[4] Parallel to the definition of property rights, the freedom to contract must be guaranteed. However, this right must not permit the conclusion of contracts that could endanger this freedom, for example price cartels. Finally, freedom to contract can only be institutionalized where competition already exists, so that it cannot lead to the reinforcement of a power position.

[5] Constitutional right must also address the issue of responsibility for economic actions. Eucken refers to the tendency towards increasing separation between property owners and decision takers in modern capitalism, stated by Schumpeter. The aim should be both to guarantee a certain morality of economic actions and to brake the concentration process.

[6] Lastly, economic policy must not distort the agents' planning, so that their plans always refer to price cost relations and so that they can also correct market fluctuations. As a consequence, economic policy has to be stable. In this respect, Ordoliberalism is again close to Friedman's conception in the way that the stability principle refers to an individual rationality based on adaptive expectations. The stability effort of economic policy concerns the stability of planning data. *Ex ante*, the agents anticipate the relevant data set on the basis of that corresponding to the previous period and possible forecast errors they made. They adapt themselves. If the state disturbs this process, an equilibrium can never be reached.

It should be noted that only the combined realization of all these principles ensures the fulfilment of each of their functions necessary for the achievement of the competition order. These principles are thus constitutive in the sense that they form the pillars of an economic constitution.

b) Regulative principles

In order to render the prevailing economic order conformable to the above constitution, the non-conformable organic institutions have to be dismissed and market failures have to be eliminated. Four principles make it possible to pass from the existing order to the positive order.

[1] An independent competition office must monitor signs of economic market powers in order to avoid the emergence of new monopolies and to force the existing ones to act in conformity with competition.

[2] A progressive direct fiscal policy must redistribute income, but without affecting the propensity to invest by inducing a reduction in saving. The market failures that Eucken considers here are, however, not linked to the income distribution function of the market but to its resource allocation function. Taking into account the fact that the income distribution based on individual contributions to production leads to disparity in purchasing power,

a spillover effect from the production of basic products to producing luxury goods takes place, with the need for basic goods not being satisfied. Nevertheless, the mechanism of this spillover effect seems unclear. Eucken's argumentation presupposes the existence of a higher profit rate in the luxury goods industry. Yet, if the demand for basic goods is not met, the price must increase until the profit rates in these two industries are equalized. Consequently, capital must return to the production of basic goods. Failure to take this dynamic into account leads to the assumption that his statement is based on the historical observation of the shortage of basic goods and the simultaneous surplus of convenience goods (such as electric lamps, ashtrays, and so on) in Germany immediately after the war. In this case inefficient resource allocation was not due to market failures, but rather to the occupation legislation that included a list of all German products to be requisitioned: ashtrays and electric lamps were not on the list.

[3] The third regulative principle concerns negative externalities. They cannot only occur in a monopoly situation, but also in the case of complete competition, since the individuals – when making their plans – cannot take into account all the elements influenced by their actions. Despite this modern conception, Eucken does not envisage internalizing or to "merchandizing" the external effects. The solution – considered from too global a point of view to be discussed here in greater detail – simply affirms the necessity for limiting freedom of planning in some spheres.

[4] According to the last regulative principle, the labour market must be monitored in order to observe – as far as wage-fixing is concerned – the reactions of the supply side to demand variations or to productivity gains induced by technical progress. Corrections – as for instance the creation of a minimum wage – must be made in the marginal cases where the competition process does not restore an equilibrium.

c) Principles of political action

The competition order – understood as a norm translated into reality by a positive economic policy – requires that the action of the state is guided by principles.

Following Eucken, the modern economic state is characterized by the loss of authority due to quantitative and qualitative increases in its action, as well as to the behaviour of private interest groups. This behaviour is reflected in the pressures that such groups exert on the definition of economic policy and in the legitimacy that the state itself gives to these groups, for example in trade negotiations. In such a society, rent-seeking actions lead to the reduction in consumer surplus, since the producer includes the costs of such behaviour in his cost function (Krueger 1980). In Eucken's view, these activities not only lead to a non-Pareto-optimal resource allocation and social waste. But also, according to the interdependence of the orders of society, the danger of such phenomena principally consists in a political crisis with the lack of respect for the constitutional state.

In order to resolve and to avoid these problems of authority, two principles define the role of the ordoliberal state. It has to eliminate the economic pressure groups or constrain their functions, and to focus its action on the institutional enforcement of the competition norm (Eucken 1990 pp. 334-336). These principles show that the ordoliberal state is not guided by quantitative results. *Ordnungspolitik* – which consists in implementing the competition order by means of the constitutive principles – as well as *Prozeßpolitik* – which corrects the existing order by means of the regulative principles – are both rules-orientated and not output-orientated (Vanberg 1988 p 22). Moreover, the ordoliberal state is not at all a welfare state. In the case that rules-orientated measures are inefficient, the actions of the state have thus to be "market-conformable". This conformity is, however, not a very well-defined concept which can be applied to measures that modify the planning data set as little as possible.

The ideas of the second part are summarized in the graph 3.2.

III. New Ordnungstheorie

The process of departing from standard theory can be separated into three stages. Dialectically, the first consists in criticizing the old theory without rejecting its core construction. The second attempts to bring the standard conception close to other theories, and in particular to the New Institutionalist school in order to show synergies and differences. Next, on the basis of the identified inadequacies of standard *Ordnungstheorie* in the face of the modernity of the new theories, a third synthetic stage aims at modifying its theme in an appropriate way.

1. External criticisms

Several criticisms can be made concerning Eucken's theoretical work. Three groups of remarks concern the standard conception of *Ordnungstheorie*. Another group refers to the theory of the ordoliberal state, and a last the political manifestations inspired by Ordoliberalism.

Graph 3.2: Ordoliberalism

o Objective

Passing from the existing economic order to the system of the transaction economy that maximizes the degree of individual freedom of planning with respect to the political constitution: enforcement of the competition order

o Why ?

- The competition order is socially the most efficient order
  - Coherent with and complementary to the social and political goals of a free society (respects and promotes the freedom of movement, free choice of profession, private ownership)
  - Ethic neutrality of social relations
  - Function of fairness
- The competition order is economically the most efficient order
  - Circumstances which are necessary for equilibrium
  - Optimal resource allocation
  - Tendency towards harmonization of private and general interests
- But, the existing order of the economy never corresponds to this competition order
  - Non coherence of institutions
  - Existence of economic powers

o How ?

Economic policy: legal organization of competition through action of the state in order to create an economic constitution

Constitutive principles Creating the institutions of the competition order	**Regulative principles** Modification of the existing institutions in order to render them conform to the constitutive principles of the competition order	**Role of the state**
• Monetary stability • Open market policy • Definition of property rights • Freedom of contract • Responsability of economic agents for their actions • Stability of economic policy	• Minimizing economic powers (Cartel office) • Fiscal policy • Legislation on property rights in order to avoid negative externalities • Monitoring the labour market	• Confomity to the economic constitution • Rules-oriented economic policy

First, the concept of order can be criticized on the grounds that the connections between the different orders are never analysed. Indeed, the interdependence of relations is never demonstrated. Even if the examples undoubtedly reflect the idea of mutual dependence of orders, no causal relation or dynamic process is shown. The concept of interdependence of orders remains an empirical statement which has rather to be seen as a postulate (Streit 1992 p. 692). At the same time, the one-way path – which links the evolution of the economic process to that of the order – seems as deterministic as the Structure-Conduct-Performance paradigm.

With respect to Eucken's criticism of the concepts of the Historical school, his integration of historical particularities into the pattern of coordination appears rather poor, since the economy is embedded in the ethical dimension of a society. This conception results in the latter being understood as an exogenous variable, while it is in fact an endogenous one in the historicist concept of economic style.

When employing standard concepts of economic theory, Eucken only refers vaguely to the general equilibrium approach, but also his departure from the latter does not really seem convincing, since the conditions and the circumstances of equilibrium are definitively neoclassical. The concept of "complete competition" is barely differentiated from that of perfect competition, because both the former and the latter use the size, the number of agents and the investment and price policy of the market as criteria for the degree of market openness. Eucken's concept of individual rationality can be criticized for being quite a simplistic concept of adaptive expectations and for not integrating the risk factor as endogenous variable.

As far as Ordoliberalism is concerned, the utopian version of a political state free from any "lobbyist" behaviour in a parliamentary democracy should particularly be discussed. Furthermore, in view of the possible evolution of the economic order, it seems problematic to define permanent goals of economic policy. Finally, the principles of the action of the state are not defined in such a way that they can be applied in reality.

Another set of criticisms can be addressed to the political realization of German economic policy between 1948 and 1966, which was during the period inspired by Ordoliberalism. The exogenous character of money led to overlooking the economic power which commercial banks – *via* the market forms – have over the competition order. This political omission directly induced the reconcentration of the German banking system in December 1956. Moreover, the fact that competition law was still incomplete in 1957, in the ordoliberal view, together with Ludwig Erhard's political resignation in 1966, leads one to question whether the minimization of economic powers is a feasible objective for competition policy and as a social mode of regulation.

The criticisms about standard *Ordnungstheorie* and Ordoliberalism are numerous and the brief listing presented here is certainly not exhaustive. However, it should be noted that they could be formulated thanks to the progress made by economic science since Eucken's theorization. The criticisms primarily concern the adequacy of the instruments Eucken used to construct his subject and method of analysis: the conception and the functions of competition, the rationality of the economic agents, the exogeneity of money, the lack of any dynamic analysis of institutions. Moreover, the last group of criticisms of the political successes of Ordoliberalism must be relativized. As far as the macroeconomic results of Erhard's policy are concerned, it should be observed that the evaluations directly compare the theoretical principles with their partial realization, the latter being the expression of a political game rather than of theoretical incoherence. In addition, the conception of the role and the nature of the ordoliberal state is now assessed by means of theories that benefit from new experiences of parliamentary democracies. To sum up, all these criticisms mainly affect the satellite hypotheses of Eucken's work.

By stressing the contextualized mode of economic coordination – which manifests itself in institutions that pass beyond the only market system; by underlining the interdependence of institutions according to their functions; and in particular by emphasizing the political influence of economic powers, the Freiburg school initiates research on the concept of order and economic systems that remains modern and relevant today. For this reason, several contributions try to overcome the external criticism and work again on the topic by integrating new economic tools. The foundation of a new *Ordnungstheorie* should thus be put in perspective with the former standard definition. These new developments can exclusively be found in Germany and several of them are to be differentiated.

Two major distinct works contribute to the renewal of the issue. The first way represents an extension of standard *Ordnungstheorie*. It rejects the static conception of orders and introduces the analytical tools developed by Hayek and the New Institutionalist school (Streit 1995 pp. 3-5). The criticism of the second contribution on standard theory goes further. This work only adopts the topic of order and frees it from all its neoclassical mechanisms. In this point, the break with the standard theory is thus too large to be understood here as a simple extension of it. On the one hand, this latter contribution demonstrates the complementarity between economic systems and styles in order to reintroduce the ethic as an endogenous variable of order (Schefold 1995a pp. 16 ff.). On the other hand, this work reexamines the income distribution function with the help of the English Cambridge theories (Schefold 1995b pp. 9-38). That leads to rejection of the conception of *Ordo* as the best economic and social regulation pattern and results in a radically different *Ordnungspolitik*. In what follows, only the first development will be treated, considering that it is closest to the old Freiburg conception and more unified.

2. Parallels with the New Institutionalist school

During the 1980s, at the time of the expansion of the New Institutionalist school, several contributions tried to replace the Freiburg school in the topical debates. Putting it into perspective with the New Institutionalist school is justified by the fact that they share a common methodological basis (Vanberg 1988 pp. 17-18) which Hutchison considers as a Smithian mode in contrast to a Ricardian one (Hutchison 1981 p. 162). The latter describes a strictly economic method of free market policies which seeks to find an equilibrium or an utopian competition optimum. The Smithian mode also refers to a competitive market economy in a broader sense which includes the social and political order and especially the foundations and the legal structures of the economic order. This approach is based on a more realistic view of individual rationality and accepts the lack of complete information. Consequently, the goals of economic policy are defined more modestly and the rules of the law are given the role of reaching a certain economic efficiency.

Besides, the two theories can be brought together, since their axiomatics are similar (Tietzel 1990 pp. 10-26). They both tend to extend neoclassical theories. The methodological individualism, confirmed by New Institutionalism, is not incompatible with the German concept of order. Considering the exchange economy, efficiency is based on the coordination of individual planning. The rationality hypothesis – understood as maximization by means of an individual utility function – is enlarged in the two theories through the introduction of a non-fictitious time variable. The question of rules becomes a central problem of coordinating agents and in addition, the two approaches propose a similar integration of the ethical dimension. The economy is embedded in society in the same way, which justifies the moral dimension as an exogenous variable.

Taking into account these similarities, Tietzel, however, rejects any kinship between standard *Ordnungstheorie* and New Institutionalism. Nevertheless, this conclusion does not mark the end of all attempts to establish parallels between the theories in question. Other later works have tried to compare more precisely the instruments of the two analyses.

For example, Eucken's conception of the coordination pattern is examined on the basis of the property rights theory. The role of this element as a constitutive principle of *Ordo*, as well as the interdependence that exists between the economy and legal rules makes one anticipate a fruitful comparison. Yet, Eucken's view quickly proves to be the larger, in the sense that – apart from the property of production goods – standard *Ordnungstheorie* also takes into consideration the fact that the coordination pattern results from the distribution of individual planning competence (Schüller 1987 p. 77). The essential feature of the pattern does not consist in the property rights, but in the degree of planning freedom, the property rights constituting only a determinant of the latter (Schmidtchen 1984 pp. 61-62).

At the same time, the necessity of a sound theoretical instrument –which Eucken required for the analysis of institutions – leads to a comparison with Williamson's transaction costs theory. This approach certainly chooses the same starting-point, but it differs from standard *Ordnungstheorie* in two respects. First, the level of measuring coordination is not the same (Grossekettler 1989 p. 65). The transaction costs theory focuses mainly on microeconomic relations whereas Eucken considers only macroeconomic ones, but both believe in a relative continuity between micro- and macro-phenomena.[13] The really important point that distinguishes the two theories resides in the relation between "market and hierarchy". For Williamson, there is continuity due to transaction costs between the two forms of coordination (Williamson 1994). The agent can choose the mode of coordination – market or hierarchy – that minimizes the costs of his transaction. In Eucken's opinion, on the contrary, there is no continuity between market coordination and centralized planning. As it has been shown, the two modes of coordination cannot be compared, since they do not apply the same method of evaluating "transactions". Standard *Ordnungstheorie* perhaps implies a notion of transaction costs, as Schüller affirms, but these costs are limited to the case of market coordination.

Finally, it is possible to draw parallels between the principles of economic policy of the Freiburg school and *Constitutional Economics* (Leipold 1987; Vanberg 1988; Streit 1992). Vanberg even states that the German school is a precursor of the latter theory. Both approaches consider the economic order as an order of rules rather than an order of actions. This order must be positively achieved by means of an economic constitution. In that respect, the two theories share the same constructivist conception of institutional

13 Eucken refuses to compartmentalize the disciplines and affirms the continuity between "Betriebswirtschaftslehre" and "Nationalökonomie" (EUCKEN 1989 p. 249).

structure and do not aim at quantitative output efficiency. Lastly, the concept of a constitutional order necessitates in any case the definition of normative evaluation criteria in order to assess the institutional framework.

All these contributions show that the Freiburg school is still relevant today. Standard *Ordnungstheorie* – as a field and a method of analysing the institutional framework of an economy – and Ordoliberalism – as a constitutional theory of competition and of the state – prefigure in several points the work of the New Institutionalist school. However, any kinship between them must be denied, since the thought of the latter remains unaffected by the earlier German work (Hutchison 1981 p. 167). The conclusions reached are taken into account in the second stage, which redevelops the German theories by taking advantage of this proximity to the New Institutionalist school.

3. Ordnungsökonomik as an attempted synthesis

The contribution of Manfred Streit lies at centre of *Ordnungsökonomik*. On the basis of preceding works – to which the above developments belong – he defines a revised "economics of order". Four departures from standard *Ordnungstheorie* give this approach a theoretical autonomy.

a) The subject of analysis extended to the dynamics of orders

Ordnungsökonomik considers the order as the central issue of its analysis. But the order characterizes at once the set of rules in which all economic action takes place and the result of the totality of human actions given particular rules. Hence, order means both order of rules and order of actions. This difference between *Ordnungsökonomik* and standard *Ordnungstheorie* leads to redefinition of the subject of analysis and examination of the emergence, imposition and efficiency of orders (Streit 1995 p. 3). With regard to this first point, *Ordnungsökonomik* is close to the New Institutionalist school. By devoting its analysis to identifying the possibilities and limits of the enforcement of orders according to given objectives, however, it accepts the continuity between *Ordnungstheorie* and *Ordnungspolitik*. In this respect, *Ordnungsökonomik* comes closer to the Freiburg school.

Moreover, treating the order of actions at first sight does not seem to render *Ordnungsökonomik* really different from standard theory. Indeed, Eucken distinguishes between the constructed institutional order and that which results from the decentralized coordination of individual actions (Schmidtchen 1989 p. 164). However, the two conceptions differ in the way that they deal with these two types of orders. In Eucken's view, the "order of actions" (*gewachsene Ordnung*) only constitutes a suboptimal order of rules which has to be corrected. The analysis of the former is not justified except in order to explain the mismatch between the existing and the positive order. In Hayek's theory, the two types of orders are radically different, since the process of emergence is not the same. Their analysis is thus necessary to understand how competition works. In new *Ordnungsökonomik*, understanding the order of action changes Eucken's existing order from an explicative variable into an explained variable, in other words, from an exogenous into an endogenous variable, and leads to the extension of the subject of analysis to the study of the dynamics of economic order.

b) Revised foundations of the competition order

Like standard *Ordnungstheorie*, *Ordnungsökonomik* opposes two types of order according to the nature of the plan. The system of a centralized economy is characterized by a legally planned order, the system of a market economy by a spontaneous order. In accordance with standard *Ordnungstheorie*, the latter order implies the interdependence of actions which cannot be perceived by the individual agent. Similarly, the price system ensures market coordination. But unlike Eucken – in whose eyes the two economic systems were only opposed as a result of their coordination pattern – here, the centralized economy and the market economy are also opposite as far as normative foundations are concerned. Streit refers back to the Hayekian distinction between "right and wrong individualism" (Hayek 1948) and states that the two systems lead to diametrically opposed conceptions of the political society. The discontinuity between market economy and centrally planned economy is much more underlined than in standard theory. This difference is due to a deeper analysis of the issue.

A second departure from standard *Ordnungstheorie* concerns the cognitive foundations of orders. This aspect – which is not considered by Eucken – once more relates to Hayek's work. The individual choice in decision-taking is "open" in the sense that it is always possible to discover alternative actions in the time between planning the decision and carrying out the transaction. The decentralized coordination pattern, that is to say the competitive system, is hence no more an order of rules but an order of actions. Size and number of market agents are no longer the relevant criteria for competition policy. The cognitive foundations of competition give it a dynamic dimension that renders it different from the neoclassical theories. The critical view of competition as an information-discovery process strongly diminishes the possibilities of achieving an ideal order based on a competitive system.

c) A less utopian conception of the state

The process of discovering knowledge supposes a limited stock of currently available information and leads to the hypothesis of bounded individual rationality. *Ordnungsökonomik* consequently envisages the human ability to design a normative economic order in a sceptical way. Hence, this theory moves away from Ordoliberalism. The latter holds a very optimistic view of rationality and considers it to shape the external institutions of a normative order, as well as to evaluate its internal institutions.

According to the new approach, the institutional framework is limited to a set of formal rules which restrains the behaviour of agents, that is to say their economic relations to other agents. All these rules are comprised in an economic constitution (Streit 1995 p. 24). The latter consists only in external institutions that are expressed as universal rules. Legally, universality means that the rules are known by everybody, that they apply to everybody and in all circumstances. Economically, this triple characteristic of the universality of rules guarantees stability of the agents' expectations and so diminishes their transaction costs. The economic constitution ensures that the competition order remains an open process of information discovery, since it authorizes the choice of new alternatives as long as they are not sanctioned by a new set of rules. The economic constitution is thus restricted to defining and to ensuring property rights, that is, the rights and responsibilities of the owner concerning the utilization of his property, as well as the rights and responsibilities of the user of a property. The constitution also defines and guarantees the agents' autonomy in the sense of the equality of their legal status and it protects individual economic freedom from public power.

The role of the state in economic policy is not only revised with respect to the modification of the rationality hypothesis but also on the basis of an analysis of the political decision-taking process in a parliamentary democracy. In the ordoliberal approach, the political sphere is not subject to such an analysis. The politician is never understood as an "entrepreneur" but only as the executive power of the economic constitution, to which he strictly has to conform his action. *Ordnungsökonomik*, taking account of more than fifty years of parliamentary democracy proposes an analysis based on the interaction between politicians, voters, and lobbies, as it is developed in the "Rent-Seeking Society" (for a lecture on this issue, see Buchanan/Tollison/Tullock 1980, especially chap. 4).

d) An analysis of the interdependence of orders

The economic constitution in large part depends on the political constitution which guarantees the former. Hence, in contrast to Eucken – for whom each order is positioned on the same level – *Ordnungsökonomik* introduces a hierarchy between the orders of society. Indeed, the economic order is subordinated to the political one. But this must not be understood as the predominance of the state over the economy. On the contrary, the political constitution embodies the legal order, to which the state is also subordinated. The legal order is at the top of the hierarchy and consequently the rules are the variable that explains the interdependence of orders. This "relay" allows one to understand how economic powers not only exert an influence on the citizens within the economic sphere, but also on the political order. *Ordnungsökonomik* provides the analytical element of the interdependence of orders which was simply postulated in standard *Ordnungstheorie*.

The graph 3.3 depicts the new theoretical developments from Manfred Streit on the one hand and Bertram Schefold on the other hand concerning the issue of economic order with regard to their criticisms of the core concepts of the Freiburg school.

IV. Conclusion

This chapter aimed to give precise definitions of, and discern the differences between, *Ordnungstheorie* and Ordoliberalism as the two core concepts of the Freiburg school. *Ordnungstheorie* centres on the question of the coordination pattern of economic agents' decision-taking. It is conceived as a general equilibrium approach, the outcome depending less on the conditions of the equilibrium than on its circumstances, that is, on the economic order. This order is composed of pragmatic and organic institutions which go beyond the framework of the market system. The economic order that materializes in the real world can be compared with different economic systems, or ideal-types defined on a theoretical basis according to the existing degree of individual freedom of planning. The plan is considered as the central variable of the theory and allows the opposition of the centrally directed economy and the exchange economy as two dual economic systems. The efficiency of the mode of coordination is thus measured by the degree of coherence of the materialized order with the ideal economic system towards which that prevailing tends.

Ordoliberalism only refers to the system of a exchange economy. It proposes principles of economic policy to transform the actual economic order into this ideal-type system. These principles serve to implement a legal organization of market competition by state action in order to ensure an optimal mode of economic coordination that is concurrently coherent with, and complementary, to the political and social objectives of a free society. Moreover, thanks to a broad initiative to diffuse their ideas and by guiding political decision-taking in order to create a functioning market economy in post-war Germany, Ordoliberalism overcame the boundaries of the Freiburg school.

The presentation of the individual concepts also stresses the fact that German contemporary analyses of economic order are still not to be found in the Freiburg school. New theoretical developments are based on standard *Ordnungstheorie*, showing at once the topicality and the limits of this latter theory. In particular, the present contribution examined *Ordnungsökonomik*, which extends the subject studied by standard *Ordnungstheorie* to the analysis of the dynamics of order and tries to integrate an analysis of the interdependence of orders, of revised foundations of the competition order and a less utopian conception of the state than its forerunner.

Graph 3.3: *New Ordnungstheorie*

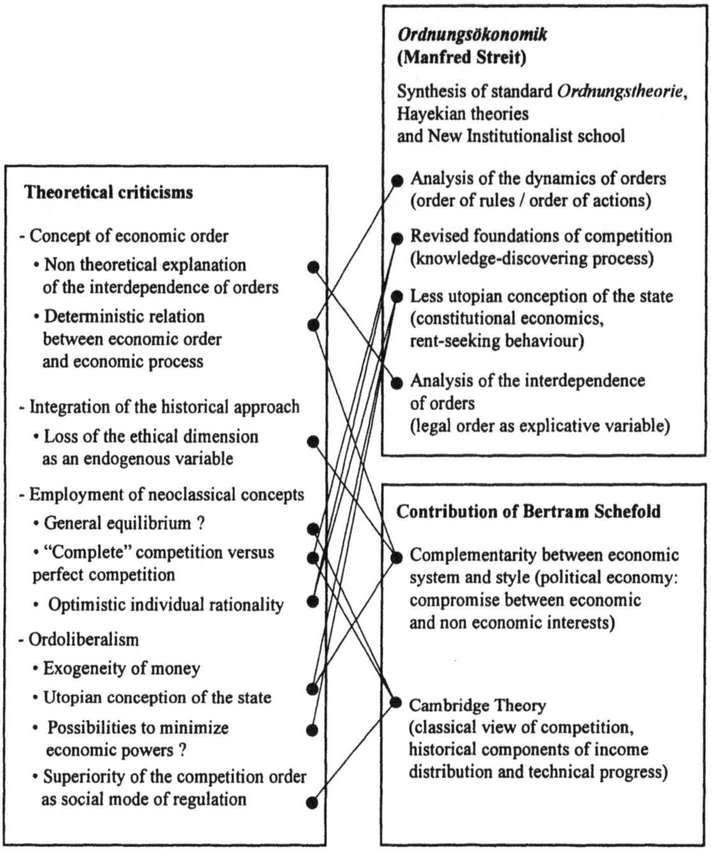

Theoretical criticisms

- Concept of economic order
  - Non theoretical explanation of the interdependence of orders
  - Deterministic relation between economic order and economic process
- Integration of the historical approach
  - Loss of the ethical dimension as an endogenous variable
- Employment of neoclassical concepts
  - General equilibrium ?
  - "Complete" competition versus perfect competition
  - Optimistic individual rationality
- Ordoliberalism
  - Exogeneity of money
  - Utopian conception of the state
  - Possibilities to minimize economic powers ?
  - Superiority of the competition order as social mode of regulation

***Ordnungsökonomik* (Manfred Streit)**

Synthesis of standard *Ordnungstheorie*, Hayekian theories and New Institutionalist school

- Analysis of the dynamics of orders (order of rules / order of actions)
- Revised foundations of competition (knowledge-discovering process)
- Less utopian conception of the state (constitutional economics, rent-seeking behaviour)
- Analysis of the interdependence of orders (legal order as explicative variable)

Contribution of Bertram Schefold

- Complementarity between economic system and style (political economy: compromise between economic and non economic interests)
- Cambridge Theory (classical view of competition, historical components of income distribution and technical progress)

References

ALLEN, C. S.: "The Under Development of Keynesianism in the Federal Republic of Germany", in: P. A. HALL (Ed.): *The Political Power of Economic Ideas - Keynesianism across Nations*, Princeton (Princeton University Press) 1989, pp. 263-289.

BUCHANAN, J., TOLLISON, R. D., TULLOCK, G. (Eds.): *Toward a Theory of the Rent-Seeking Society*, Texas (University Press) 1980.

CASSEL, D.: 1968, *Methodologische Systeme der Wirtschaftswissenschaft*, Dissertation, Marburg (Philipps Universität) 1968.

DEHAY, E.: "La justification ordolibérale de l'indépendance des banques centrales", *Revue française d'Economie*, 10/1 (1995), pp. 27-53.

EUCKEN, W.: "Das Ordnungspolitische Problem", *Ordo-Jahrbuch für die Ordnung der Wirtschaft und Gesellschaft*, 1 (1948), pp. 56-90.

EUCKEN, W.: *Kapitaltheoretische Untersuchungen*, Tübingen/Zürich (J.B.C. Mohr) 1954. First edition: 1934.

EUCKEN, W.: *Grundlagen der Nationalökonomie*, Berlin (Springer Verlag) 1989. First edition: 1939.

EUCKEN, W.: *Grundsätze der Wirtschaftspolitik*, Tübingen (J.B.C. Mohr) 1990. First edition: 1952.

EUCKEN, W.: *The Foundations of Economics*, London (William Hodge and Company Limited) 1950. Sixth German edition translated by T.W. Hutchinson.

FEHL, U.: "Zu Walter Euckens kapitaltheoretischen Überlegungen", *Ordo-Jahrbuch für die Ordnung der Wirtschaft und Gesellschaft*, 40 (1989), pp. 71-83.

GARROUSTE, P.: *C. Menger et F. Hayek à propos des institutions: continuités et ruptures*, Université Lumière Lyon 2 (ECT Discussion paper) 1994.

GONNARD, R.: *Histoire des doctrines économiques*, Paris (Librairie Valois) 1930.

GROSSEKETTLER, H.G.: "On Designing an Economic Order. The Contributions of the Freiburg School", in: D. A. WALKER (Ed.): *Perspectives on the History of Economic Thought*, Vol. 2, Upleadon (Edward Elgar) 1989, pp. 38-84.

HARDACH, K.: *The Political Economy of Germany in the Twentieth Century*, Berkeley (University of California Press) 1980.

HAYEK, F. A.: "Wahrer und falscher Individualismus", *Ordo-Jahrbuch für die Ordnung der Wirtschaft und Gesellschaft*, 1 (1948), pp. 19-55.

HERDER-DORNEICH, P.: "Ist eine Dynamisierung der Ordnungstheorie möglich? Der Beitrag der neuen Politischen Ökonomie zur Weiterentwicklung der Ordnungstheorie", *Ordo-Jachbuch für die Ordnung der Wirtschaft und Gesellschaft*, 42 (1991), pp. 353-361.

HERRMANN-PILLATH, C.: "Kritischer Rationalismus, Strukturalismus, und die methodologischen Prinzipien von Eucken/Hensel", in: MARBURG PHILIPPS UNIVERSITÄT (Ed.): *Arbeitsberichte zum Systemvergleich*, 11 (1987), pp. 32-73.

HEUß, E.: "'Die Grundlagen der Nationalökonomie' vor 50 Jahren und heute", *Ordo-Jahrbuch für die Ordnung der Wirtschaft und Gesellschaft*, 40 (1989), pp. 21-30.

HUTCHISON, T. W.: "Walter Eucken and the German Social-Market Economy", in: T. W. HUTCHISON (Ed.): *The Politics and Philosophy of Economics - Marxians, Keynesians and Austrians*, Oxford (Blackwell) 1981, pp. 155-175.

KRUEGER, A. O.: "The Political Economy of the Rent-Seeking Society", in: J. BUCHANAN *et al.* (Eds.): *Toward a Theory of the Rent-Seeking Society*, Texas (University Press) 1980, pp. 51-70.

LEIPOLD, H.: "Constitutional Economics als Ordnungstheorie", in: MARBURG PHILIPPS UNIVERSITÄT (Ed.): *Arbeitsberichte zum Systemvergleich*, 11 (1987), pp. 101-134.

LENEL, H. O.: "Walter Euckens ordnungspolitische Konzeption, die wirtschaftspolitische Lehre in der Bundesrepublik und die Wettbewerbstheorie von heute", *Ordo-Jahrbuch für die Ordnung der Wirtschaft und Gesellschaft*, 26 (1975), pp. 22-76.

LENEL, H. O.: "Walter Euckens 'Grundlagen der Nationalökonomie'", *Ordo-Jahrbuch für die Ordnung der Wirtschaft und Gesellschaft*, 40 (1989), pp. 3-20.

MEYER, W.: "Geschichte und Nationalökonomie: Historische Einbettung und allgemeine Theorien", *Ordo-Jachbuch für die Ordnung der Wirtschaft und Gesellschaft*, 40 (1989), pp. 31-54.

MYERS, M. L.: "Adam Smith's Concept of Equilibrium", *Journal of Economic Issues*, 10 (1976), pp. 560-573.

MÜLLER-ARMACK, A.: *Genealogie der Sozialen Marktwirtschaft – Frühschriften und weiterführende Konzepte*, Bern (Paul Haupt Verlag) 1974.

NICHOLLS, A. J.: *Freedom with Responsability – The Social Market Economy in Germany, 1918 – 1963*, Oxford (Clarendon Press) 1994.

PIETTRE, A., REDSLOB, A.: *Histoire de la pensée économique et analyse des théories contemporaines*, Paris (Dalloz) 1986.

RIETER, H., SCHMOLZ, M.: "The Ideas of German Ordoliberalism 1938-45: Pointing the Way to a New Economic Order", *European Journal of the History of Economic Thought*, 1/1 (Autumn 1993), pp. 87-114.

RÖPKE, W.: *Die Lehre der Wirtschaft*, Bern (Paul Haupt Verlag) 1994. First edition:1937.

SCHEFOLD, B. (1995a): "Theoretische Ansätze für den Vergleich von Wirtschaftssystemen aus historischer Perspektive", in: B. SCHEFOLD *et al.* (Eds.): *Wandlungsprozesse in den Wirtschafts-systemen Westeuropas*, Marburg (Metropolis Verlag) 1995, pp. 9-40.

SCHEFOLD, B. (1995b): *Wirtschaftsstile - Band 2: Studien zur ökonomischen Theorie und zur Zukunft der Technick*, Frankfurt (Fischer Taschenbuch Verlag) 1995.

SCHMIDT, K. H.: "Merkantilismus, Kameralismus, Physiokratie", in: O. ISSING, (Ed.): *Geschichte der Nationalökonomie*, Munich (Verlag Vahlen) 1994, pp. 37-62.

SCHMIDTCHEN, D.: "German 'Ordnungspolitik' as Institutional Choice", *Zeitschrift für die gesamte Staatswissenschaft (Journal of Institutional and Theoretical Economics)*, 140 (1984), pp. 54-70.

SCHMIDTCHEN, D.: "Evolutorische Ordnungstheorie oder: die Transaktionskosten und das Unternehmertum", *Ordo-Jachbuch für die Ordnung der Wirtschaft und Gesellschaft*, 40 (1989), pp 161-182.

SCHÜLLER, A.: "Ordnungstheorie - Theoretischer Institutionalismus, ein Vergleich", in: MARBURG PHILIPPS UNIVERSITÄT (Ed.): *Arbeitsberichte zum Systemvergleich*, 11 (1987), pp. 74-100.

STARBATTY, J.: "Ordoliberalismus", in: O. Issing: *Geschichte der Nationalökonomie*, Munich (Verlag Vahlen) 1994, pp. 239-254.

STREIT, M. E.: "Economic Order, Private Law and Public Policy, the Freiburg School of Law and Economics in Perspective", *Journal of Institutional and Theoretical Economics*, 148 (1992), pp. 675-704.

STREIT, M. E.: *Ordnungsökonomik - Versuch einer Standortsbestimmung*, Max-Planck-Institut zur Erforschung von Wirtschaftsystemen (Ed.): Diskussions-beitrag 04-95 (1995).

TIETZEL, M.: "Der neue Institutionalismus auf dem Hintergrund der alten Ordnungstheorie", UNIVERSITÄT DUISBURG (Ed.): *Diskussionsbeiträge des Fachbereichs Wirtschaftswissenschaft*, 135 (1990).

TRIBE, K.: *Strategies of Economic Order – German Economic Discourse, 1750-1950*, Cambridge (Cambridge University Press) 1995.

TUCHTFELDT, E., WILLGERODT, H.: "Wilhelm Röpke – Leben und Werke", in: W. RÖPKE: *Die Lehre der Wirtschaft*, Bern (Paul Haupt Verlag) 1994, pp. 340-371.

VANBERG, V.: "'Ordnungstheorie' as Constitutional Economics - The German Conception of a 'Social Market Economy'", *Ordo-Jahrbuch für die Ordnung der Wirtschaft und Gesellschaft*, 39 (1988), pp. 17-31.

WILLGERODT, H.: "1948 und 1990: Zwei deutsche Wirtschaftsreformen im Vergleich", in: C. HERRMANN-PILLATH, O. SCHLECHE, H. F. WÜNSCHE (Eds.): *Grundtexte zur Sozialen Marktwirtschaft - Marktwirtschaft als Aufgabe - Wirtschaft und Gesellschaft vom Plan zum Markt*, Vol. 3, Stuttgart (Gustav Fischer Verlag) 1994, pp. 65-78.

WILLIAMSON, O. E.: *Les institutions de l'économie*, Paris (InterEditions) 1994. Original: *The Economic Institutions of Capitalism*, London (Free Press) 1985.

WEHLER, H. U "Der Aufstieg des Organisierten Kapitalismus und Interventions-taates in Deutschland", in: H. A. WINCKLER: *Organisierter Kapitalismus - Voraussetzungen und Anfänge*, Göttingen (Vandenhoeck and Ruprecht) 1974, pp. 36-57.

Chapter 4

A Systemic Perception of Eucken's Foundations of Economics

JEAN-DANIEL WEISZ

I. Introduction: A Paradoxical Success

Husserl's remark, "tradition is omission of the origins",[1] does not apply only to philosophy. It is valuable in the economic field as well, where the scientific and public achievements of a research programme often appear strangely proportional to the distance travelled by the latest heirs from the founding message. With the help of some often-undetected reductions, if not deliberate falsifications, the research takes a different course from that initially indicated by the intuitions of the founders. What, for example, would Leon Walras say today of the studies conducted by the neoclassical *mainstream* by those who claim, even indirectly, his theoretical economics heritage?

It is often difficult to escape the motives of this differentiation process. The need to make the theory operational leads to a reduction in its initial scope, which was more sensitive and receptive to the complexity of the world

1 Quoted by H. G. GADAMER, "The Science of the Life-World", in: A. T. TYMIENIECKA (Ed.), *The Later Husserl and The Idea of Phenomenology*, Dordrecht Holland (D. Reidel Publishing Company) 1972, p. 173.

under consideration. Sometimes, too, the sociological dynamic of the research within the paradigm, as well as developments in formal tools oblige the disciples, if they are to preserve conformity with their scientific environment, to stand some distance away from the intuitions on which their own research programme was initially based. The initial message is then set aside, to be called up only in times of emergency for the paradigm.

The ordoliberal paradigm, if it still merits this appellation, seems to follow this description. Its public achievements cannot be contested: closely related to the model of the "social market economy" (*soziale Marktwirtschaft*), considered as its theoretical matrix, its major historical theoreticians in Germany enjoy a fame extending far beyond merely academic circles. However, this public fecundity has been accompanied by some theoretical marginalization. What is claimed to be the normative character of Ordoliberalism and its lack of formalization (see Delorme, below, chap. 8) are seen as serious handicaps for research in this field today. Ordoliberalism seems to have been definitively formulated by the radical criticism to which it was exposed (Riese 1972; Kirchgässner 1988). This is particularly noticeable as far as Walter Eucken, the father of Ordoliberalism, is concerned.

"Eucken is dead!" exclaimed Hajo Riese as early as 1972, believing the ordoliberal paradigm to be closed, unproductive and focused on itself. The closest heirs of Ordoliberalism, the *Ordnungsökonomie* theoreticians today lean clearly towards concepts such as the New Institutional Economy, transaction costs or constitutional economy (see Wohlgemuth, below, chap. 7). If Ordoliberalism seems to have drifted away from its initial intuitions, there is perhaps a historical irony that Walter Eucken had formerly stigmatized this kind of evolution in his treatment of the historical and theoretical schools. Walter Eucken's initial ambition, as set out in his book, *Die Grundlagen der Nationalökonomie (The Foundations of Economics),* was to rebuild economics on solid ground and to overcome the deep antinomy between the historical German school and the theoretical Austrian school. Therefore, Ordoliberalism was born from a genuine questioning as to the very bases of the scientific reality of economics, proposing an original answer to this question. The concepts of economic policy later to be widely discussed were only developed in a second phase.

A few noticeable exceptions aside (see Herrmann-Pillath 1987; 1991; 1994), this initial aspect of Ordoliberalism has been systematically hidden from view in later researches by both critics and defendants. This has resulted in a progressively reductive slide from an open theory to a closed paradigm, neglecting the specificity of the initial project. Was this reduction necessary and might it be fatal? Starting from Walter Eucken's *Grundlagen der Nationalökonomie,* I shall attempt in the first part of this chapter to underline what seems specific to the initial ordoliberal research programme. Then I shall try to show, following Carsten Herrmann-Pillath, that the book's main originality apparently owes much to transcendental phenomenology. The third part will attempt an analysis of this reduction process, which half-way along its course, forgot Eucken's initial ambition to consider the complexity of the economic world.

II. The Scientific Specificity of Ordoliberalism

The Ordoliberalism of the founding fathers, those of the 1936 ordoliberal manifesto, is neither limited to the distinction between two antagonistic ideal types – the centrally directed economy (*zentralgeleitete Wirtschaft*) and the exchange economy (*Verkehrswirtschaft*) – nor to the sole justification of the moral and functional superiority of the competition order (*Wettbewerbsordnung*), both roles in which Ordoliberalism is today generally confined. This is a very narrow and crippling approach to Ordoliberalism. Beyond such a standardized vision[2] Ordoliberalism proceeds from a radical interrogation concerning the scientific character of economics and proposes an original answer to the problem of its scientific foundations.

1. A radical interrogation concerning the scientific character of economics

The Ordoliberalism I am referring to is the specifically German doctrine, also known as the "Freiburg school" born in the thirties around a research programme involving economists and jurists. At that time, their diagnosis

2 Beyond this distinction between ideal-types used in describing the economy – the centrally administered economy corresponding more-or-less to the socialist model of planned economy and the exchange economy describing the market economy – this standardized version includes the demonstration of the "systemic" superiority of the latter and of its real forms, as well as the choice made of the competition order and of all the basic and regulating principles of this order as enumerated by Walter Eucken in his *Grundsätze der Wirtschaftpolitik* (I shall not dwell upon this particular point, see Broyer, above, chap. 3, for further discussion).

was one of a **crisis affecting both economy and economics.** A quick survey of the context of the period helps to understand this attitude, caused by an acute crisis in the discipline of economics in both its "scientific" (that is, scholarly and intellectual) and its practical dimensions.

A good example is the Bad Pyrmont conference, which was organized by the Friedrich List Society in 1928 under the direction of the economist Edgar Salin. To discuss the Young plan and study the problem of German reparations, it gathered together the top names in German economic thought, as well as a number of high-level economic participants from the institutional and industrial sectors.[3] In the face of these expectations, the economists were unable to sustain a common language. This conference was a failure for economics, which had the gravest consequences for the future of the academic group, later known as the "ordoliberals".[4] The economists were unable to build a continuous line of defence in response to the practical questions asked by the practitioners, something rendered still crueller by the great crisis of 1929. This perception of a crisis in economics as an intellectual discipline greatly heightened awareness of the problem facing for the economic order. A number of jurists and economists are now focusing on the original form of Ordoliberalism and confronting the question of private and public economic powers: the economic chaos of the Weimar Republic could well have been partially generated by the random power of increasing influential interest groups, lobbies, trusts and big business concerns. The ordoliberal thought process starts from a critique of "organized capitalism". It tries to identify the economic and juridical conditions of the economic order, in other words, of a careful balance between public and private power compatible with a certain conception of liberalism, to answer the pressing question: how can we limit those powers without generating compensatory mechanisms destructive of freedom?[5] This is nothing new and already lay at the core of thought on "organized capitalism".[6]

---

3 "Ordoliberals-to-be", such as Walter Eucken, Wilhelm Röpke, and Alexander Rüstow, were present, and economists such as Werner Sombart, Erwin von Beckerath and Friedrich von Gottl-Ottlilienfield, as well as a number of high-level actors, e.g. the Weimar Minister of Finance, Rudolf Hilferding, and Hjalmar Schacht.

4 On this point, see H. JANSSEN, *Nationalökonomie und Nationalsozialismus. Die deutsche Volkwirtschaftlehre in den dreißiger Jahren*, Marburg (Metropolis-Verlag) 1998, p. 20.

5 This aspect was in particular underlined by Michel Foucault in his talk, "La naissance de la biopolitique" (The Birth of Biopolitics), given at the Collège de

The famous jurist Franz Böhm has contributed much towards the development of the juridical aspects of the ordoliberal construction in this direction. I shall therefore concentrate rather upon the contribution of Walter Eucken, considered as the spiritual father of this theory, and on his efforts to try to resolve the crisis of economic scientificity.

Eucken's starting-point is that economics, lodged between the Historical school, exemplified by the emblematic figure of Gustav von Schmoller, and the Austrian school characterized by the works of Carl Menger, do not share common scientific fundamentals. Some reconstruction is therefore required in order to restore a **strong theoretical basis.**[7] His reflection is a radical query as to the scientific character of economics, such as Descartes or Husserl undertook for philosophy: "Considering the present state of economics, this problem must be posed in the most radical possible way".[8] The originality of Eucken's approach is to place himself midstream between the general theory of knowledge and economic theory, within a gap that neither the historical school, awaiting some coherence from multiple partial surveys, nor the theoretical school, built upon a hypothetical-deductive dynamic, could fill. Eucken considers that neither the expectant scientificity of the historic school nor that imported from the hard sciences by the theoretical school can establish economics as an autonomous science. What he is really looking for, as is apparent from the very first pages of the *Grundlagen,* are the conditions for an **autonomous economic scientificity,**

France 24 January 1979: "For both the German liberals of the Freiburg school of *c.*1927-1930 and the present-day American liberals (the 'libertarians'), in one way or other, the analytical starting-point, to which they nailed their problem, is: some intervention mechanisms were established in the economy to avoid the reduction of liberty that would result from the transition to a Socialist, Fascist, or National Socialist regime. But surely these very mechanisms induce processes which themselves become as dangerous for liberty as the highly political visible and manifest forms that we want to escape?" (my translation).

6 Rudolf Hilferding, the theoretician of organized capitalism is credited with having tried essentially to solve the problem of how to organize the power controls so as to avoid the creation of new power concentrations, themselves a threat to liberty. On this point, see H.-A. WINCKLER, "Zur Hilferdings Theorie des Organisierten Kapitalismus", in: H.-A. WINCKLER, (Ed.), *Organisierter Kapitalismus*, Göttingen (Vandenhoeck & Ruprecht) 1974 (= Kritische Studien zur Geschichtswissenschaft, Vol. 9), p. 16.

7 For him economics is torn between two schools, "two different sciences with two different intellectual objectives and two different ways of thinking" in: W. EUCKEN, *Foundations of Economics*, p. 324.

8 Walter Eucken, *Grundlagen der Nationalökonomie*, p. XII.

freed from predominant influences in the hard sciences. As Herrmann-Pillath observes, Eucken considers that the interactivity between the theory of knowledge and the methodological reflection applied to the sphere of science establishes the former as the regulator of the research processes.[9]

2. The foundations of the ordoliberal approach

Nobody can remain indifferent when reading the *Grundlagen*, particularly if, following the author's advice, one reads the book from the first page to the last.[10] This book brings something new, quite rare in economics, and here I shall underline some original aspects of the Euckenian[11] approach, while briefly describing it.

The economist's *cogito*, his first certitude, comes from observation. Besides any question about the essence of the events that we perceive – that is the philosopher's task – we can ask ourselves about the economic order we can daily observe. This is what Walter Eucken does about an economic process organized in such a way that the apparent disorder resulting from the social division of labour creates a perceived order, where the object fulfills its task here and now. How can a need be satisfied by the economic system? This is no new question *per se*: it is the prelude to all answers using the invisible hand. But Walter Eucken's answer is original in its own construction.

Eucken's theory is built upon an iterative dialogue between a theoretical grid, the so-called "morphology of pure forms" and the observation of economic orders, past and present. This choice implies a particular meaning of the expression "do science", since the definition of the scientific work in hand is based on an object, an attitude and a target.

The object here is economic reality such as everyone can daily observe. The scientific attitude consists in raising oneself above the common perception of one's own economic environment limited by partiality so as to

9 C. HERMANN-PILLATH, "Kritischer Rationalismus, Strukturalismus und die methodologischen Prinzipien von Eucken/Hensel", in: Arbeitsberichte zum Systemvergleich, No. 11, Marburg, 1987, p. 37.

10 "This book forms an entity. Someone who reads only some pages will not even be able to understand what they mean, as they exist as a whole", Walter Eucken *Grundlagen der Nationalökonomie*, p. XII (my translation).

11 For a more thoroughgoing presentation, see Heinrich von Stackelberg's discussion of the *Grundlagen* of 1941.

adopt a general view of the economic system. Although each individual is able to understand his or her immediate economic environment, the scientist is the individual who can rise above this perception and understand the economic world as a whole. With such an attitude, it is possible to ask the pertinent questions.[12] The scientific method is then defined as a rigorous survey of individual facts.[13]

"The scientific approach to economics is from the start distinguished from the pre-scientific by the profundity of the questions asked and the analytical penetration with which the facts are approached".[14]

The objective is to reconcile theory and empiricism. Eucken insists that a final judgement on his conception of science can be made only after his system has been fully developed:

"This may seem paradoxical at first, to try to formulate generalised problems for theoretical analysis by studying detailed facts. But we shall see where this leads us".[15]

A conception of "doing science" leads him to refuse *a priori* anything that does not come from observation.[16] Whence his refusal to use concepts that have not first been defined in the course of his demonstration.[17]

To begin with, we can learn three things from observation: first, that each individual, inserted in an economic structure, acts in accordance with a plan; then that each economic act takes place within an economic form (a family or a company, for instance) itself ultimately part of a given economic system. This is why he tries to identify the different and ideal-typical forms that make it possible to organize the coordination of these various plans.

---

12 "A science and its earlier doctrines can only remain alive by formulating direct questions about the real world" in: W. EUCKEN, *Foundations of Economics*, p. 25.

13 "Theoretical analysis is thus simply the full use of our power of thought" in: W. EUCKEN, *Foundations of Economics*, p. 40.

14 W. EUCKEN, *Foundations of Economics*, p. 105.

15 *Ibid.*

16 Such as the concepts: "Because everyday concepts cannot be defined immediately in a scientifically satisfactory way, economists must use them as they are used in ordinary life, that is, without definitions. In this way they get on with a direct analysis of their subject-matter. The results of their investigations are then summarised in definitions and these definitions can be used in further study" in: W. EUCKEN, *Foundations of Economics*, p. 25.

17 "Models constructed a priori and imprecise 'blanket' concepts like 'capitalism', 'socialism' and the like can be of little help in the investigation of economic reality" in: W. EUCKEN, "On the Theory of the Centrally Administered Economy", p. 193.

To do this, Walter Eucken uses the technique of the isolating abstraction.[18] It is a way to find intuitively the elementary elements existing in every economic order and at all levels. The observation of a particular form, a farm for example, allows him to define a hierarchy of power (organizing the production), combined with aspects of exchange economy (organizing the market product supply). It is then possible, through past and present examples of economic organization, to distinguish pure types, invariables, which are elementary power-organizing structures, making coordination between the plans possible.[19] The isolating abstraction method uses the observation of isolated cases in which the pure forms are revealed in their diversity and combination. Within this process of discover priority is given to the coordination mode of the plans, therefore to economic power. Eucken's analysis of selected present and historical examples is based on the survey of the institutional background comprising production, exchange and consumption activities, starting from the analysis of power relations. Whether he is studying a monastery, a family farm, a lordship, the Inca Empire or 1940s Germany, Eucken always details the institutionalized power relations which constitute the frame of the considered economic order.[20]

All these elementary forms filtered through historical study constitute the economic morphology or morphology of the pure forms. It is shaped and refined through the survey of past and present economic orders – of the way economic plans are organized – by means of isolating abstraction. The two pure leading types are the centrally directed economy, where plans are centrally coordinated and ruled by a central plan, and the exchange economy, where a multitude of plans confront each other. This apparently logical and factually confirmed distinction is completed by a historical survey that shows

---

18 The method of the isolating abstraction is defined by Eucken as follow: "We examined the individual phenomena, the farms, estates, households, etc., studied them from every standpoint, and in the course of our analysis extracted one by one the forms realised in each case (the varieties of centrally directed economy, the market forms and monetary systems)" in: W. EUCKEN, *Foundations of Economics*, p. 229.

19 "Nevertheless, in the immense number and variety of economic systems under which men carry on, and have carried on, their economic existences, certain recurring elemental forms are to be found" in: W. EUCKEN, *Foundations of Economics*, p. 116.

20 H. von Stackelberg in particular agrees on this point: "The way Eucken inserts the economic power problem in his system is remarkable. His elementary forms are at the same time the conditions for the manifestation of power, and the resources it uses" in: VON STACKELBERG: "Die Grundlagen der Nationalkökonomie", p. 280 (my translation).

both types can display an important number of variants. The variants of the centrally directed economy are the totally centrally directed economy comprising *Eigenwirtschaft* (literally the small autarkical or self-sufficient economy), the centrally directed economy including a free market for consumer goods (not a consumer-choice situation, but the possibility of exchanging goods produced and centrally distributed) and the centrally directed economy with free consumer choice.[21]

In the elementary type of centrally directed economy (*Zentralgeleitete Wirtschaft*), the *Eigenwirtschaft* describes a type of familial economy under patriarchal rule. The shift to the more highly developed type of totally centrally administrated economy (*Zentralverwaltungswirtschaft*) results from the multiplication of the actors involved and the resultant higher complexity of the coordination process. This first hierarchic type is in fact defined by the power exerted from the top to the bottom, from the central to individual plans. At the other end of the spectrum, the forms of the market and the organization of the monetary system characterize the ideal-type of the exchange economy. The different types of market differ in their supply-side and demand characteristics,[22] and Eucken also identifies several monetary systems. Furthermore, each individual takes six facts into account when elaborating his plan: the needs, the stocks of goods, nature, labour, technical knowledge, and social and juridical organization. In the exchange economy, these six factual elements are completed by a seventh: prices, which are considered as independent (on the *Daten,* see C. Herrmann-Pillath 1994).

When Eucken had completed his identification work on pure forms, he inserts the morphology comprised of all the forms to be found in the real economy, mixed together at different levels. He uses an analogy with the alphabet: it is possible to build an immense number of words with the alphabet. All economic systems can be understood through morphology as well[23].

---

21 A fourth version was added in the first four editions of the *Grundlagen*, namely, a centrally directed economy with free choice of labour.

22 They are competition, partial oligopoly, oligopoly, partial monopoly, monopoly. Demand and supply can be open or closed, according to new operators entering the market or not. With these five characteristics of demand and supply and the open/closed variable, a given market can assume as many as 100 different facets.

23 The comparison with chemistry (the Mendeleiev classification) that Eucken evoked to some extent has no relevance for two reasons: first, it refers to a natural order; secondly, it is always possible for a new atom to be added to this classification.

But this intellectual construction creates a number of problems. First, Eucken considers observation as the unique starting criterion for science. Once the central character of the plan has been affirmed, he starts researching pure forms through an abstract reasoning, focusing on some individual facts. When, however, can economic morphology be considered as complete? Eucken believed that he has listed all the pure forms necessary for his morphology. Built once and for all, it is not supposed to undergo further evolution other than for reasons of simplification (as in his *Grundsätze*).[24] He does not even think that any other pure form could be discovered, or that the present forms could be made obsolete by the evolution of the economic world.[25] To him, the multiplication of examples confirms the existence of a limited number of pure forms. Herrmann-Pillath gives one answer: "Like structuralism today, Eucken fundamentally starts from a dynamic comprehension of the theory: an apparently static ideal-type shows through its conceptual evolution its own capacity to produce more and more differentiated worlds".[26]

Finally, a strong hypothesis is advanced concerning the status of history. It uses some of the examples chosen by Eucken to find his pure forms or to verify their relevance. Placed therefore on the same level as observation, it does not take into account the bias introduced by the historian's work. Does, however, the extent of the investigation into the scientificity of economics permit one to bypass the same question about the scientificity of history? All these problems are left to one side by Eucken and make the *Grundlagen* generally difficult to understand. For that reason, I should like now to underline some analogies between the Euckenian system and Husserl's philosophy. I do not believe that this system can be considered coherent and comprehensible unless it acknowledges that a number of its elements stem from the system and the method of transcendental phenomenology.

---

24 For instance, he refused to use the term, "controlled market economy" (*gelenkte Marktwirtschaft*), introduced, among others, by A. Müller-Armack. See W. EUCKEN, *The Foundations of Economics*, p. 226.

25 That a constant of theory is to be atemporal is one of Eucken's main objections to the historical school. The historical school identifying layers in economic history cannot lead to theory because these concepts concern particular places and times.

26 C. HERMANN-PILLATH, "Kritischer Rationalismus, Strukturalismus und die methodologischen Prinzipien von Eucken/Hensel", p. 64.

III. The Sources of an Original Path

Generally, it is "thinking in orders" that is underlined to characterize the novel approach introduced by Eucken. Stress is also laid on the systemic aspects of the work: the economy seen as a whole, the accentuation of the interdependence of social and economic orders, the importance of their interactions. I aim to show that this scientific specificity leads further than this first degree of systemicity. Eucken follows a systemic conjunction pattern. This undoubtedly has its roots in the influence of phenomenology upon him.[27] As far as I know, the contemporary economist, C. Herrmann-Pillath, was the first to consider Eucken as a phenomenologist[28]. I shall return to this point and make sustained use of the ideas advanced by Herrmann-Pillath.

1. The contribution of phenomenology

As the son of Rudolf Eucken, the neo-Kantian philosopher and winner of the Nobel Prize for Literature, Walter Eucken had the opportunity for sustaining a close relation with phenomenology. He also met its figurehead, the great philosopher Husserl, who taught at the end of his life in the same University of Freiburg im Breisgau.

We know that Husserl's objective was to bring back to life Descartes's philosophical radicalism, in order to "free philosophy from all possible preconception, to give to it the status of a truly autonomous science, built on utmost obviousness drawn from the subject itself, finding there its absolute justification".[29] The cause of this undertaking was the deep crisis of philosophy in the first decades of the twentieth century. Its purpose was to make philosophy a **rigorous science,** according to the title of one of his books,[30] through rebuilding "the edifice that might correspond to the concept of philosophy, conceived as a universal scientific unity, raised upon a

27 For this brief consideration of the links between phenomenology and Euckenian theory, I have relied mainly on the *Méditations cartésiennes*, published in 1931, the product of a series of conferences given by Husserl at the Sorbonne in 1929.

28 C. HERRMANN-PILLATH, "Der Vergleich von Wirtschafts- und Gesellschafts-systemen: Wissenschaftsphilosophische und methodologische Betrachtungen zur Zukunft eines ordnungstheoretischen Forschungs-programmes", *Ordo Jahrbuch*, 42, (1991), p. 18.

29 E. HUSSERL, *Méditations cartésiennes*, p. 24 (my translation).

30 The book's title is *Die Philosophie als strenge Wissenschaft [Philosophy as a Rigorous Science].*

foundation absolute in character".[31] To make connections more easily with Eucken's theory, I shall make the common distinction between a system and a method of phenomenology.[32] The phenomenological system refers to transcendental phenomenology. The term "transcendental" "points to the original motives ... of the reactive question as to the ultimate source of all development in knowledge".[33] The primary goal of phenomenology is to find the final principle of all kind of reality. Is phenomenology then a "general revolution" *via* "a radical return to the pure *ego cogito*"? By so doing, transcendental phenomenology becomes an eidetic science.[34]

"The uncertainties of science, already noticeable in humanistic sciences, but finally reaching those which even looked like their model, *physics and mathematics*, have their origin in a blinding experimental concern. Before studying physics, one must first study the physical fact, its essence; and this applies to all disciplines. It will then be possible to draw from the definition of *eidos*, as captured by the original intuition, the methodological conclusions on which the empirical research can be oriented."[35]

It is difficult not to draw a parallel with the approach used by Walter Eucken in his *Grundlagen*. Like Husserl, he is confronted with a crisis in his discipline. Although scientific, philosophical and economic writings abounded, they lacked any absolute foundation.[36] The phenomenological system attempted to overcome the mistakes of the Cartesian system and of empiricism. Does not it, too, face a great antinomy? Husserl criticizes Descartes for whom it was normal that: "universal science should be a deductive system, whose edifice would lie *ordine geometrico* on a founding axiom used as an absolute base for deduction".[37] On the other hand, phenomenology tries to go back to "things for themselves": "The ultimate source of right for any rational affirmation is contained in 'seeing' in general,

---

31 E. HUSSERL, *Méditations cartésiennes*, p. 18 (my translation).

32 A. LALANDE, article "Phénoménologie", in *Vocabulaire technique et critique de la philosophie* (my translation).

33 E. HUSSERL, *La crise des sciences européennes*, p. 113 (my translation).

34 i.e. "a science whose object is to consider the relations between ideal forms, as do logic or geometry": A. Lalande, *Vocabulaire technique et critique de la philosophie*, s.v. "eidétique"(my translation).

35 LYOTARD, J.-F., *La phénoménologie*, p. 13 (my translation).

36 "In place of a philosophy that is unique and alive, what do we have? An ever-growing number of philosophic works, but without any internal link". E. HUSSERL, *Méditations cartésiennes*, p. 22 (my translation).

37 *Ibid.* p. 26.

that is to say in the original *donating* conscience".[38] It goes beyond empiricism which "confused the imperatives to go back to things for themselves and to base all knowledge on experience".[39] Now, this reconsideration of the Cartesian method can also be found in Eucken.

It was no coincidence that the *Grundlagen* began with a nod to René Descartes. Looking, like Descartes at the beginning of his *Discours de la méthode*, at the stove in his room, Eucken does not try to explain how it works by taking it apart: his first approach is not analytical. He tries instead to explain the existence of this stove by analysing backwards, step by step, the different economic acts "caused by, or at the origin of" that lead to the satisfaction of his need for heat. By going abstractly backwards in time and space, he becomes conscious of the multitude of economic acts that were necessary so that he might have this stove in his room, here and now.

In another book, he takes a different bird's-eye view.[40] He simultaneously observes both some regularity and a slow and discontinuous evolution in human activities. Guided by observation, as by intuition of the economic world, he asks again the fundamental economic question: "how can this huge mechanism be directed that functions on Earth and on which the existence of every human being depends?" The particular element of the system (the stove) or the whole system in itself (the bird's focus) lead him to admit one fact: there is an economic reality composed of observable events, and one of the purposes of economics is to recognize this reality in its evolution and diversity. First, it is necessary to refuse anything *a priori*, and even to reject some terms or words already bearing some explanatory elements. Here he comes face to face with the very purpose of phenomenology: to study events, reality, while abstaining from drawing hypothesis. The emphasis placed by Eucken on the role of observation meets this intuition of the world (original, "naïve", trying to explore available data). His refusal to introduce initial definitions (see his critique of concepts and of what he calls *Begriffsökonomen*), his quest for impartial, non-ideological observation are comparable to the phenomenological position.[41] It really looks as if Walter

38 J.-F. LYOTARD, *La phénoménologie*, p. 12.

39 Idem.

40 W. EUCKEN, *Nationalökonomie, wozu?* p. 13.

41 "Eucken moved, to formulate it in contemporary terms, from a 'statement-view' to a 'non-statement view'; simply because Husserl's *Logische Untersuchungen* which he called upon for help viewed the analysis of concept as an essential element of a 'Wissenschaftslehre'", in: C. HERMANN-PILLATH, "Kritischer

Eucken, to answer the basic antinomy with which he was confronted, attempted to call upon the Husserlian research process consisting in "basic truths that must and can be the foundations of all universal science".[42]

The equivalent of the philosophical obviousness on which the economist can rely to begin this work of refoundation and to overcome the great antinomy is economic reality.[43] By this, we mean the acute perception, by an alert mind, aware of the influence of lobbies and ideologies, of the economic order characterizing the activities of production, distribution and consumption. Eucken's initial thought process was therefore inspired by transcendental phenomenology. But Husserl's method of phenomenological reduction using bracketing (*epoche*) can also be found in Eucken under the appellation, "pin-pointing and isolating abstraction". Unlike Cartesian philosophy, phenomenology is not built upon deduction, starting from a unique concept. It consists of "looking for concepts in the way that mathematicians do, as ideal, fixed and juxtaposed terms, free from the flux of experience and with no concern for their genesis".[44] Emile Bréhier makes a good summary of the bases of that method: "The most naïve and usual intuition of the world gives us a mix of events, and fixed terms, that appear or disappear, but are always immutable [...]; this has nothing to do with what we call abstract or general ideas, built by combination and assembly, but it concerns immutable essences like the Platonic ideas, to which we accede through **a particular intuition called intuition of essence** [*Wesenschau*]; this intuition is a priori and independent of experience, but it can only be obtained through this phenomenological analysis".[45] The isolating abstraction process plays the part of this particular intuition, and makes it possible to determine the essences through the variations observed. Historical economic orders then become, beyond their apparent diversity, combinations of a certain number of intangible forms or essences, disclosed

---

Rationalismus, Strukturalismus und die methodologischen Prinzipien von Eucken/Hensel", p. 62.

42 E. HUSSERL, *Méditations cartésiennes*, p. 36.

43 "In the great debate between Menger and Schmoller both parties were wrong, nor was the truth somewhere in the middle between the two. Neither Menger's dualism, of which Schmoller perceived the danger, nor Schmoller's pure empiricism, the failure of which Menger foresaw, does justice to economic reality" in: W. EUCKEN, *Foundations of Economics*, p. 324.

44 E. BREHIER, *Histoire de la Philosophie*, Vol. 3, p. 973.

45 Idem.

by isolating abstraction. For Eucken, the survey of historical variety plays the same role as phenomenological variation.[46]

Can we say, then, that the Euckenian scheme is an extension of phenomenology to economics? More precisely, if an eidetic science corresponds to each empirical science, related to the local *eidos* of the studied objects, can we assert that Eucken tried to define the local *eidos* of economics? C. Herrmann-Pillath considers Eucken's methodology as an attempt at applied phenomenology.[47] In reality, Walter Eucken comes back to research practices belonging implicitly to economics. "The classical economists took their conception of the form of the competitive market from direct observation of reality" in the same way that "Cournot, when developing his theory of monopoly got his law of demand partly from existing economic theory, partly from observation".[48]

However, phenomenology certainly inspired his scientific radicalism and justified his method. For his intellectual construction to hold, the intellectual projections of phenomenology must be considered as given. It is no wonder references to Husserl are sparing in the *Grundlagen*, given the context of the period.[49] The analysis of this experience, which we can consider as unique in economics, merits completion by means of a comparison with the work of other authors, in particular Max Scheler on values.[50] This might help answer the above question more precisely. This is not my aim here. I should like instead to stress another aspect of the theory: its subsequent relations with works on economics.

46 "The imaginary variation process shows us the essence itself, the being of the object ...The essence or *eidos* of the object is composed of the invariant, which stays identical through variations" in: J.-F. LYOTARD, *La phénoménologie*, p. 12.

47 C. HERRMANN-PILLATH, "Der Vergleich von Wirtschafts- und Gesellschaftsystemen: Wissenschaftsphilosophische und methodologische Betrachtungen zur Zukunft eines ordnungstheoretischen Forschungs-programmes", *Ordo Jahrbuch*, 42, (1991), p. 18.

48 H. VON STACKELBERG, "Die Grundlagen der Nationalkökonomie", p. 258.

49 Husserl was of Jewish origin and from the advent of the National Socialists encountered difficulties in staying at the University of Freiburg, before he was actually excluded from it.

50 Scheler finds the numerical identity character in values through the diversity of events, which for Husserl is the identifying characteristic of an object and an essence.

2. A systemic modelization of economic orders

So far, it has, I think been demonstrated that the theoretical block erected by Eucken has achieved consistency with the help of phenomenology. Once economic morphology has been implemented, we have at our disposal a suite of pure forms that are essences identical to those of reality. The Euckenian theory draws its systemic character from this approach (see Robert Delorme in this volume). There is no doubt that, in the *Grundlagen*, Walter Eucken argues in terms of systems, the word being used with its primary meaning of an organized whole. The five facets of the economic problem as enumerated by Eucken (the direction of production to satisfy needs, the planning of production, distribution, establishing technologies, and economic planning in town and country) come jointly under one main problem, the scarcity of goods. He criticizes all disjunction of these elements considered as facets of the same problem.[51] Eucken's approach is not Cartesian in the sense that he denies the initial disjunction, the process that consists of dividing up an event into explicable parts, then recomposing it as preliminary to analysis. On the contrary, having gathered his invariants or ideal-types by means of the isolating abstraction, he uses them in a **systemic modelization** process of the observed economic orders: "This method of approach, that is, the application of our morphological system to actual cases, enables us to understand the details and character of each economic system. *Individual detail does not disappear, but is fitted into a system and thus becomes properly comprehensible*".[52]

Eucken does not employ a process of dissection to study the way power is organized in a given economic configuration by means of his pure forms. Rather, he composes an isomorphic model of the order under consideration through a conjunction process of pure forms. It is a process of double synthesis:

"In the process of applying one theoretical scheme, we make a twofold synthesis. First a number of purely formal elements are combined together

51 "These five questions, arising as they do from the direct contemplation of our immediate surroundings are not independent. They simply express different aspects of one and the same problem, looked at from five different viewpoints" in: W. EUCKEN, *Foundations of Economics*, p. 22.

52 W. EUCKEN, *Foundations of Economics*, p. 224 (*Grundlagen* p. 165).

into one economic system, and secondly the economic system is fitted into its geographical, intellectual, political and social surroundings".[53]

The science of economics can then be put into practice from our theoretical knowledge of the pure forms and of some of their related forms. On this particular point, Eucken acknowledges no precursors, with the possible exception of Heinrich von Thünen, whom he quotes to illustrate what he means by ideal-type.[54] For him, von Thünen's "*isolierte Staat*" is an ideal-type that acts as a model permitting theorization. This is the move that Eucken also attempts, in a more systemic manner, passing from pure forms to economic theorization. Assembled from observation, the tool of morphology makes it possible to survey real economic orders and, through the combination of pure forms, to propose a form of representation. This representation of order and process provides a grid onto which economic theorization can be grafted: "The pure forms or ideal types are rather to be used as `models' on the basis of which theoretical propositions can be worked out".[55]

What does this imply in concrete terms? Eucken gives a precise example of his method in the *Grundlagen*: his subject is the Silesian economy and the situation of the weavers in the mid-nineteenth century.[56] Equipped with his morphology and theoretical propositions, he turns to real historical facts: the Prussian state at this date, the religious and social concepts, the economic and political power of individuals and groups. He then sketches in the indigenous agricultural industry within which elements of the centrally directed economy were to be found complementing aspects of the exchange economy (such as the monopoly of demand on the labour market).[57]

"[...] Certain ideal types were relevant for Silesia at that time, by means of which a unified account of the Silesian economic system could be given

53 W. EUCKEN, *Foundations of Economics*, p. 229 (*Grundlagen* p. 169).

54 On this point, see the important note 66 about the ideal types in the *Foundation of Economics* page 347 (*Grundlagen* p. 268)

55 W. EUCKEN, *Foundations of Economics*, p. 233 (*Grundlagen* p. 172).

56 "A Simple Case" in W. EUCKEN, *Foundations of Economics*, p. 241 (*Grundlagen* p. 178).

57 He proposed another application of his method in two successive articles published in the journal, *Economica* (1948): "On the Theory of the Centrally Administered Economy: An Analysis of the German Experiment".

and its place in the historical situation of the country as the whole explained".[58]

The theoretical device can be used only as a second stage, thus completing the grid of non-economic data by consideration of wages and by insertion into the regional, national and international economy. Indeed, for this purpose, Eucken insists upon the differences of level. Now, the composition of an economic order starts bottom-up, from the elementary units of economy upwards to the aggregated level. The work of conjunction is done by means of the composition of elementary forms at various levels. The representation of an economic system (household, firm, nation) of any size must combine and join together with the elementary forms from the microeconomic to the macroeconomic level.

One cannot therefore say of an economy that it is either exchange-based or centrally directed. His sole conclusion is: it is an economy where the form of centralization or that of exchange dominates. In a socialist-type economy, the centrally directed form will prevail, but the logic of the *Verkehrswirtschaft* will still guide numerous areas. Study of the economic system of the directed economy in Germany from 1938 to the post-war years must inevitably focus on the level of the central authorities. However, it must also take into account the subsidiary plans of businesses.[59]

Thus Walter Eucken builds a coherent construct that binds philosophy to economic theory from top to bottom. With the help of a consistent methodology, his construction permits reconsideration of all the theoretical contributions of economics: "This implies no contempt for the great achievements of the past. On the contrary, just because we are turning away from established doctrines to the actual subject-matter we shall arrive in the course of our studies at a true relationship with the scientific achievements of the past."[60] Moreover, Herrmann-Pillath has observed that all the recent approaches in economics could be integrated within Eucken's

58 W. EUCKEN, *Foundations of Economics*, p. 242 (*Grundlagen* p. 179).

59 "Certainly the procedure in private firms was completely overshadowed by the plans in the central administration. But the firms had their own subsidiary plans, and to understand German economic life in this period, it is necessary to take into account this subsidiary private planning" in: WALTER EUCKEN, "On the Theory of the Centrally Administered Economy: An Analysis of the German Economy", p. 98.

60 W. EUCKEN, *Foundations of Economics*, p. 101 (*Grundlagen* p. 68).

methodological approach[61]. The genius of the Euckenian method seems to consist in integrating the contributions of phenomenology in order once more to erect the construct required to continue economic research in the wake of the classics.

IV. A Crippling Reduction

Having established the bases of his theoretical programme in *Grundlagen der Nationalökonomie*, Walter Eucken left the task of giving it operational status to others. He indubitably established the systemic superiority of the market economy model using the lessons drawn from the *Grundlagen*. However, his death in 1950 prevented him from completing that book, the edition of which was left to his wife and Karl Paul Hensel (cf. *Die Grundsätze der Wirtschaftspolitik*).

The radical innovation of Eucken's work then experienced a paradoxical after-life. Ordoliberalism established itself as a major paradigm, but at the cost of a radical reduction in Eucken's originality. Of course, the seeds for this reduction had been sown by Eucken himself, who seemed to drop some aspects of the *Grundlagen*, when, after the Second World War, he was able to take stock of the breakthroughs accomplished by economics, in particular in the English-speaking world. He certainly responded to the need for simplification, as illustrated by the shift from the *Grundlagen* to the *Grundsätze*. However, the decisive reduction in the range of his morphology took place after his death. Then a false interpretation of Eucken began.[62] For Carsten Herrmann-Pillath, this was caused by the inability of critical rationalism, so prevalent in economics, to grasp Eucken's contribution. Instead of taking a place, as he did, at the level of debate between critical rationalism and structuralism, I shall pursue that interpretation, trying to analyse some aspects of the mechanism that might have lead to these misinterpretations of Eucken.

To me they seem essentially to stem from **an analytical and disjunctive reading** of a work that appeals very broadly to a systemic logic. The reduction of Eucken's theoretical message has two facets: one of a dualism of the ideal-types (the centrally directed economy versus the exchange

---

61 C. HERRMANN-PILLATH, "Methodological Aspects of Eucken's Work", *Journal of Economic Studies*, 21/4 (1994), p. 53.

62 C. HERRMANN-PILLATH (1987), "Kritischer Rationalismus, Strukturalismus und die methodologischen Prinzipien von Eucken/Hensel", p. 37.

economy); the other, the reinterpretation of his concepts through the neoclassical filter.

1. The dualism of ideal-types

The system developed by Eucken had been the target of a number of criticisms.[63] Nevertheless, the main critical attack on the *Grundlagen* seems to have been the article about the economic and social orders published in 1955 by Norbert Kloten in the journal *Ordo*. This article is often cited as ending the Euckenian system. C. Herrmann-Pillath, for his part, is of the opinion that it marks the beginning of "the long sequence of erroneous interpretations of Eucken".[64]

Attempting to develop Eucken's "systematic", by establishing it on a "related but more realistic basis",[65] Norbert Kloten's fundamental contribution was to give the *Grundlagen* a more analytical character. For him, the morphological construction has to reflect the prevailing nature of the events to which it relates and be capable additionally of characterizing other sectors of the social system. He insists upon Eucken's inability to place the three types of centralized economy at the same level: "This is a kind of inconsequence which is comparable to a discrimination of the principle of creation of an ideal type."[66] For him, the leading principle of Eucken's morphology is indeed the number of plans.[67] Now, subcategories of the centrally directed economy permit the existence of a multitude of plans. Therefore they belong, even when he emphasizes the underlying paradox, to the area of the exchange economy.

However, and for two essential reasons, this vision is a simplification of the Euckenian schema. On the one hand, Eucken never sets in opposition the ideal-types of the unique plan and of the multiple plans. This would ensure that, as soon as there were two plans there would also be a model of the exchange economy. It is clear for Eucken that all individuals make plans according to the data (*Daten*), and that the only difference is that, in a

63 See the list drawn up by E. Heuss at the beginning of his article on the *Grundlagen*.

64 C. HERRMANN-PILLATH (1987), p. 37.

65 N. KLOTEN, "Zur Typenlehre der Wirtschafts- und Gesellschaftsordnungen", *Ordo Jahrbuch*, 7 (1955), p. 124.

66 *Ibid.* p.125.

67 "The central question is: how many people make plans?" *Ibid.* p. 125.

centrally directed economy, one plan gives orders to the others. However, nothing prevents the individuals from enjoying some autonomy in their consumption or labour choices when they draw up their subordinate plans.[68]

On the other hand, pure forms are not arranged around a single logical principle that would be the basis for a typology. The three types of centrally directed economy are most certainly subordinated to the general form, but they are not constructed upon the same principle. The discrimination according to the types of centrally directed economy is based upon their growing complexity.

The elementary form of the *Eigenwirtschaft* is a system planned by one individual, allowed by the small size of the system to master and control it. In a further step, Eucken singles out through observation the three other forms we have already mentioned. For its part, the exchange economy model depends upon the market and upon monetary forms. Therefore, the pure forms cannot be differentiated according to one and the same principle. The morphological tool is much more a matter of "parcelling out the principles that add themselves together like ideal juxtaposed data",[69] which is a characteristic of Husserl's *Wesenschau*. In fact, with his pure forms Eucken does not propose a typology organized around clear and logical criteria. Instead, we find a catalogue of pure forms, sometimes intermingled at various levels, in the economic orders studied.

The efforts involved in reducing Eucken's systemic thinking tended to reintroduce the Cartesian principles of disjunction, as clearly shown in Kloten's article. However, we then progress from the morphology of pure forms to a typology of logical forms. The main contribution of the morphology, the conjunction tool, has been transformed into a dualism of ideal-standards, into opposite poles of a logical continuum.

The typology proposed by Norbert Kloten is still more at variance with Eucken's original thought. He singles out two initial principles: the dominance of a public or a private direction in the economy, and the organization of property. From those two principles he deducts two conflicting ideal-typical forms: the pure centrally directed economy and the

68 "We can say, in fact, that for the economic process as a whole, it was not the plans and actions of individual businesses or households that were decisive, but the plans and orders of the central authorities" in W. EUCKEN, "Theory of the Centrally Administered Economy", p. 80.

69 E. BREHIER, *Histoire de la Philosophie*, Vol. 3, p. 973.

pure exchange economy. In addition, he introduces a logical continuum between those two forms:

"If we make a theoretical transition from the pure centrally directed economy to a pure exchange economy, the subordination principle is replaced by the principle of coordination through the market."[70]

From this he infers three dominant economic systems: the exchange economy, the directed economy, and the centrally directed economy. In so doing, he substitutes a disjunctive for a conjunctive logic.

Yet Eucken specifically stipulated that his two ideal-types did not correspond to the opposition between capitalism, identified with the exchange economy, and socialism, associated with the centrally directed economy.[71] The two forms exist within each of the two systems. The centrally directed economy can therefore fit a simple economic household, such as a patriarchally run family farm. The production of this farm can be exchanged on a market where fierce competition is the rule. In this way any individual can act in an economic capacity as the recipient of two basic economic forms: "For Eucken, each economic order represents one individuality, with the result that his description as advocate of a duality of orders is clearly and totally wrong."[72] What appears to Kloten like moving a cursor between two opposite poles to define the relative degrees of exchange economy and of a mixed form of centrally directed economy is for Eucken a conjunction of pure forms. The reduction introduced by Kloten is therefore much more than mere simplification, but a real mutilation of Eucken's thought. With it the possibility of building systemic models of orders using conjunction and scale modifications disappears.

It is true, of course, that the *Grundsätze* set out a form of dualism, but Herrmann-Pillath shows its consistency with the Euckenian construct:

"It is true that Eucken argues in a dualistic manner in the '*Grundsätzen*' when he attempts to summarize pragmatically the historical period of the 'economic policy based on experience' (*Wirtschaftspolitik der Experimente*).

70 N. KLOTEN (1955), p. 127.

71 "A centrally administered economy is not to be confused with collective Ownership of Property" in W. EUCKEN, "Theory of the Centrally Administered Economy", p. 80.

72 C. HERRMANN-PILLATH (1987), "Kritischer Rationalismus, Strukturalismus und die methodologischen Prinzipien von Eucken/Hensel", p. 61.

Doing this and basing himself on experience, he can conclude that any attempt to introduce elements of centrally directed economy within market economies must have as its consequence a total transformation of the economic order. However, such a normative and pragmatic proposition has nothing to do with a dualism at the level of theory of economic orders"[73].

However, despite his constant concern not to misinterpret the importance of concepts, Eucken can be criticized for not always being entirely clear on the real meaning he gave to the terms used. In so doing, he has encouraged a neoclassical interpretation of his categories.

2. The neoclassical filter

Is Walter Eucken a neoclassical economist? This is a tricky question, particularly because of the uncertainty surrounding the accepted definition of the term. To give a proper answer, we have to specify the three levels of the Euckenian method. The first is the meta-level of thought on the scientificity of economics, which leads Eucken to define the economic morphological tool. Its essential goal is to allow modelization of existing economic orders by the conjunction of pure forms. By that means, the institutional specificity of each economic order is taken into account, and it is only thereafter that a second-order theory, such as the neoclassical, can be applied. If the first-order theory is unique, there are a variety of second-order theories relating to the periods or countries under study. They are more-or-less "actual". The third-order theories, like the classical economic theories, are used to explain the economy of a given historical and geographical context, defined by the second-order theory (economic morphology). It is not out of the question to refer to a neoclassical approach, but its explanatory power would be limited and lacking any claims to hegemony.

Eucken himself is quite an acerbic critic – after he had used this approach himself – of neoclassical works. He is quite clear on this point in the *Grundlagen*, when he refers to the so-called "new theoreticians": "The stimulus of concrete problems and the force of historical facts is no longer sensed by many theoretical economists. The increasingly mathematical formulation of economic theory has had the same effect, though there is no reason why it should have done so. The resulting propositions, in spite of their formal and logical correctness, often imply a deterioration of theoretical economic analysis itself which has been mastered by the classics more fully

73 *Idem.*

than by many modern economists."[74] In a note upon market forms, he elaborates this criticism: "Modern research has aimed at defining the forms of market precisely. (...) often a system of market forms is constructed *a priori* instead of being obtained from economic reality and found in it. Systems of market forms of this kind do not reproduce the forms in the actual economic world. For example, perfect competition is described as that form of market in which the influence of the individual is non-existent, which obviously is only the case when there is an infinite number of suppliers or demanders. Monopoly is defined as the opposite of this. (...) Both cases are unrealistic and the real world lies in between them."[75] How, in this light, can we explain the fact that Eucken was perceived as a neoclassical economist?

One of Hajo Riese's criticisms of Ordoliberalism was that it claimed to hold out the possibility of choice while in fact predetermining its content. In fact, distinguishing between a plurality of systems, and the emphasis laid on this diversity, should lead to the affirmation of a great number of possible choices in the construction of an economic order. However, Hajo Riese demonstrates that this is not the case. Enmeshed in the values it supposedly affirms, Ordoliberalism proposes a constrained choice in favour of the market economy model. The apparently free choice is in fact a stratagem bringing us back to the liberal conclusion that the market economy belonged to a natural order. The same circularity can be observed in Eucken's position on the neoclassical school.

He refers in the *Grundlagen*, in relation to both the exchange economy system and the directed, to a notion of general equilibrium, without giving any further details. In his 1948 article on the directed economy, Eucken notes that the equilibrium of plan-makers (the balancing of the budget of physical quantities in their section of the economy) is not an economic equilibrium and he asks himself whether such an equilibrium is possible in that type of economy:

"This question is difficult to answer, because the concept of equilibrium in an exchange economy is not immediately applicable to a centrally directed economy."[76]

---

74 W. EUCKEN, *Foundations of Economics*, p. 242 (*Grundlagen* p. 34).

75 W. EUCKEN, *Foundations of Economics*, p. 335 (*Grundlagen* p. 256).

76 W. EUCKEN, "On the Theory of the Centrally Administered Economy: An Analysis of the German Economy", p. 97.

None of the three kinds of equilibrium he singles out (the household or individual firm, partial equilibrium of the individual firm or of the markets, and general equilibrium) is applicable to that form of economy.

Eucken's relation to the concepts of neoclassicism may be explained in part by his intellectual isolation during the war years. When it was over, the need to "catch up" on the theoretical level, or even to export his ideas, might explain his use of neoclassical terms, but we can only guess at this.

Nevertheless, I cannot view Eucken as an advocate of a neoclassical and liberal vision, and his schema makes it possible to delimit the validity of this approach.

Nor is he advocating methodological individualism. At the end of the *Grundlagen*, he displays an interest in *homo economicus* by asking, "has economic behaviour always been the same?" He concludes by singling out five pairs of opposites which make it possible to describe that behaviour.[77] The first is the "economic principle": "everywhere and at all times man finds himself in the daily situation of having to adjust his needs to the means at his disposal for satisfying them and *vice versa*."[78]

He quickly moderates this principle of utility maximization by splitting it into two components: the objective and the subjective. An economic actor can adopt an economic behavior of below-optimal maximization, while believing he is maximizing his utility. Eucken takes the example of a peasant using traditional techniques which are inefficient, but the best for him, as he knows no others. The peasant is then acting on a subjective economic principle over and against the objective principle.

Once again, methodological individualism is a distinguishing characteristic of this economic science, although not of all science. If Walter Eucken is generally viewed as a neoclassical author,[79] then this would appear to be another further significant reduction of his contribution to economics.

---

77 I have highlighted here the first pair of opposites, *viz.* objective vs. subjective economic principle. The four others are: stable or unstable needs, optimum profitability or optimum needs satisfaction, short- or long-term objectives and, finally, the high or low level of acceptable risk in the face to imponderable situations.

78 W. EUCKEN, *Foundations of Economics*, p. 281 (*Grundlagen* p. 211).

79 "Eucken belongs – within the tradition of German-speaking economists at least – to the mainstream" in: C. HERRMANN-PILLATH, "Methodological Aspects of Eucken's Work", p. 54.

Because of the higher level of his contribution, and the fact that it was left to others to develop the second-level theories, a neoclassical Eucken would inevitably have been an inferior one.

V. Conclusion: Read Eucken again

Which of Eucken's contributions can be considered as decisive? From a theoretical perspective, the question of economic power is central to his construct. It is his starting-point – the economic situation during the Weimar Republic and the National Socialist period – and his finishing-line – competitive economy as the superior form. Can we not also consider this Ordoliberalism as an attempt to set the economic power in order according to positively stated normative principles? For the ultimate objective of this first version of Ordoliberalism was to provide instruments for thought and analysis to the restricted elite of the political and economic world responsible for institutional engineering?

By integrating power and theory in this way, new areas of action are opened up for those who set the rules of the economic game. Ordoliberalism is not a kind of liberalism aiming to suppress what is political. On the contrary, it tries to restore to it some capacity for action. However, while Keynes succeeded with his general theory in both theory and in practice, Eucken seems to have failed. The intellectual tradition that he initiated undoubtedly provides an "autochtonous alternative to rational criticism"[80] but, with this image, was unable to become sufficiently established to obtain international status.

Apart from all these considerations, Eucken's main contribution lies in his definition of the economic problem and in his answer to it by means of an enquiry into the scientificity of economics. His radical questioning concerning this problem still prevails today, and the debate over the integration of institutions in the models conceived by economists still has all the tension which, in the past, gave the great antinomy its vigour.

How, then, can we deny the intrinsic interest of a return to the *Grundlagen*? This is well illustrated by the H. von Stackelberg's final comments in his review of the 1940 work: "Eucken's book is of incalculable value for every researcher or specialist. However, it should also be put into

80 *Ibid.* p. 38.

the hands of all students of economics, who should read it at least when they begin and when they finish their studies: when they begin, to become aware of these problems, and when they finish, so that they can solve them."[81] For all the criticisms, Eucken's book can still act today as a spur to oblige economist to return to essential questions that the dynamic of the paradigms by which they identify themselves leads them to forsake. For this reason, he goes far beyond the narrow borders of current *Ordnungspolitik* and belongs to the personal inheritance of all economists.

References

BRÉHIER, E.: *Histoire de la Philosophie*, Vol. 3, Paris (P.U.F.) 1989. First published: 1964.

EUCKEN, W.: *Die Grundlagen der Nationalökonomie*, 9th edition, Berlin (SpringerVerlag) 1989. First published: 1939.

EUCKEN, W.: *The Foundations of Economics*, London (William Hodge and Company Limited) 1950. 6t German edition translated by T.W. Hutchinson.

EUCKEN, W.: *Nationalökonomie wozu?*, 3rd edition, Godesberg (Verlag Helmut Küpper) 1947.

EUCKEN, W.: "On the Theory of the Centrally Administered Economy. An Analysis of the German Experiment", *Economica*, 15-16 (1948).

EUCKEN, W.: *Die Grunsätze der Wirtschaftspolitik*, 6th edition, Tübingen (J. C. B. Mohr) 1990. First published: 1952.

EUCKEN, W., GROSZMANN-DOERTH, H., BÖHM, F.: "Die Ordnung der Wirtschaft als geschichtliche Aufgabe und rechtsschöpferische Leitung", in: *Ordnung der Wirtschaft*, 15-16 (1937).

HERRMANN-PILLATH, C: "Kritischer Rationalismus, Strukturalismus und die methodologischen Prinzipien von Eucken/Hensel", in: Arbeitsberichte zum Systemvergleich, No. 11, Marburg, 1987, pp. 32-73.

HERRMANN-PILLATH, C: "Der Vergleich von Wirtschafts- und Gesellschaftsystemen: Wissenschaftsphilosophische und methodologische Betrachtungen zur Zukunft eines ordnungstheoretischen Forschungsprogrammes", *Ordo Jahrbuch*, 42, (1991), pp. 15-68.

HERRMANN-PILLATH, C: "Methodological Aspects of Eucken's Work", *Journal of Economic Studies*, 21/4 (1994), pp. 46-60.

81 H. von STACKELBERG, "Die Grundlagen der Nationalökonomie (Bemerkungen zu dem gleichnamigen Buch von Walter Eucken)", p. 281.

HEUß E.: "Die Grundlagen der Nationalökonomie vor 50 Jahren und heute", *Ordo Jahrbuch*, 40 (1989), pp. 24-30.
HUSSERL, E.: *Philosophie als strenge Wissenschaft*, Frankfurt am Main (Vittorio Klostermann) 1965. First published: 1910.
HUSSERL, E.: *Méditations cartésiennes*, Paris (Librairie philosophique J. Vrin) 1992 . First published: 1931.
HUSSERL, E.: *La crise des sciences européennes et la phénoménologie transcendantale*, Paris (Gallimard) 1976. First published: 1954.
KLOTEN, N., "Zur Typenlehre der Wirtschafts- und Gesellschaftsordnungen", *Ordo Jahrbuch*, 7 (1955), pp. 123-143.
LENEL, H.-O.: "Walter Euckens Grundlagen der Nationalökonomie", *Ordo Jahrbuch*, 40 (1989), pp. 3-29.
LYOTARD, J.-F.: *La phénoménologie*, Paris (Presses Universitaires de France) 1995.
RIESE, H.: "Ordnunsgidee und Ordnungspolitik - Kritik einer wirtschaftspolitischen Konzeption", *Kyklos*, 25/1 (1972), pp. 24-48.
SALIN, E.: *Das Reparationsproblem, Teil I: Verhandlungen und Gutachten der Konferenz von Pyrmont*, Berlin (Reimar Hobbing) 1929.
STACKELBERG, H. (von): "Die Grundlagen der Nationalökonomie (Bemerkungen zu dem gleichnamigen Buch von Walter Eucken)", *Weltwirtschaftliches Archiv*, 51 (1941), pp. 245-286.
WINCKLER, H.-A.: "Zur Hilferdings Theorie des Organisierten Kapitalismus", in: WINCKLER, H.-A. (Ed.): *Organisierter Kapitalismus*, Göttingen (Vandenhoeck & Ruprecht) 1974 (= Kritische Studien zur Geschichtswissenschaft, Vol. 9).

Part B

Evolution and Present Relevance of Ordoliberalism

Chapter 5

The Evolution of Ordoliberalism in the Light of the Ordo Yearbook: A Bibliometric Analysis

FRANK BÖNKER, AGNÈS LABROUSSE, JEAN-DANIEL WEISZ*

I. Introduction

It is well known from the history and sociology of science that the rise of new schools and the institutionalization of disciplines, research areas and specialties are normally associated with the launching of learned journals (Weingart 1974 pp. 30-32; Hagemann 1991). Such journals serve as important forums for elaborating an approach and help maintain the "corporate identity" of a new school. Setting up a new journal also represents a way of overcoming resistance within the established academic community and of spreading the new gospel. Cases which highlight this important role of journals in the development of schools of thought abound.

* We would like to thank Hans-Jürgen Wagener and other conference participants for useful comments, Grit Laudel for a crash course in science studies and Karl Brandt for valuable information on his namesake. Additional thanks are due to Andreas Reinhardt and, in particular, Martin Kowalewski for their professional research assistance.

Among the most prominent are *Actes de la recherche en sciences sociales* (Bourdieu school), *L'Année sociologique* (Durkheim school) (Besnard 1979; Karady 1979), *Geschichte und Gesellschaft* (Bielefeld school of social science history) (Raphael 2000) and *Zeitschrift für Sozialforschung* (Frankfurt school of critical theory). In some cases, including *Annales*, *Public Choice* or *Tel Quel*, the labels by which schools are known even coincide with the journal titles.

One interesting difference between the ordoliberals and the regulation school concerns the existence of such a flagship journal. Unlike the ordoliberals, the Regulationists have not known such a journal until recently.[1] A handful of issues of the new *L'année de la régulation*, established in 1997, contrast with about 50 volumes of the famous *Ordo* yearbook.[2] Founded by Franz Böhm and Walter Eucken in 1948, *Ordo* is widely acknowledged as the central journal of the ordoliberal movement and the major forum for ordoliberal thinking. Prominent ordoliberals have featured as regular contributors and key ordoliberal texts have been published in the yearbook. For example, 13 out of the 16 texts in the quasi-official collection of ordoliberal highlights edited by Alan Peacock and Hans Willgerodt (Peacock/Willgerodt 1989) were originally published in the *Ordo* yearbook. Likewise, Gebhard Kirchgässner (1988 p. 54), in his well-known polemic against Ordoliberalism, treats *Ordo* articles as the main documents of ordoliberal thinking.

Given this prominent role of *Ordo*, examination of the volumes published should shed some light on Ordoliberalism. However, no systematic analysis of the *Ordo* yearbook has yet been undertaken. This is where this chapter comes in. It fashions the first ever "bibliometric" analysis of *Ordo*. Tracing trends in the composition of *Ordo* editors, contributors and articles between 1948 and 1998, it aims at complementing the existing literature on Ordoliberalism, with its more traditional focus on the reconstruction of individual biographies and intellectual doctrines (Grossekettler 1989; Haselbach 1991; Nicholls 1994).

Roughly speaking, Ordoliberalism can be understood either as a set of ideas or as a network of people. The first interpretation is captured by

---

1 Given the relative "newness" of the Regulation school, this is somehow logical.

2 In 1951 and 1960, no *Ordo* volumes were published. The 1965 volume featured as a double issue. This is why the 1998 volume which marks *Ordo*'s 50th anniversary bears the number 49 and is in fact the 48th volume.

concepts such as paradigm or research programme; the second is associated with concepts such as school, group, movement[3] or circle.[4] Of course, these interpretations are closely related. A shared approach normally rests on a broader set of social and cognitive ties. From a different perspective, it can even be regarded as part of the ties that constitute a group. In this chapter, we subscribe to the second way of understanding Ordoliberalism and treat it as a group of scholars held together by various social and cognitive ties, including some kind of a shared approach. In our analysis, we examine part of these ties and investigate how they have evolved over time.

The remainder of the chapter is divided as follows. The next part takes a look at the composition of *Ordo* editors and contributors. It examines the ties among editors and the concentration and the disciplinary affiliations of editors and contributors. The third part focuses on the content of *Ordo* articles. It documents trends in the coverage of countries and issues with a view to identifying thematic preoccupations and shifts in interests. The fourth part investigates the "ordoliberal content" of *Ordo* articles, that is, the extent to which they rely on key ordoliberal assumptions and concepts. The fifth and final part concludes by summarizing results and by sketching some lines of future research.

---

3 Peter Burke (1990 p. 2), e.g., speaks of the Annales historians as a movement rather than a school.

4 It goes without saying that the different concepts are highly controversial and contested. Moreover, there is a certain gap between science studies and most disciplinary accounts. Although science studies have found an audience among economists, notions such as "school", "paradigm" or "research programme" are often used rather loosely (see HAUSMAN 1994). This is partly due to the fact that science studies have traditionally focused on the study of natural sciences. See e.g. BROCHIER 1990 on the problems of applying Kuhn's concept of paradigm and Lakatos's notion of research programme to economics and the history of economic thought.

II. Strong Ties, Weak Diffusion: Observations on the Composition of Ordo Editors and Contributors

The composition of *Ordo* editors and contributors carries much interesting information: it points out key figures and documents the representation of different disciplines; it reveals the importance of professor–student relationships and other ties among editors; finally, it sheds light on the weight and the position of Ordoliberalism within German economics in general. In what follows, we concentrate on three issues: the ties among editors, the concentration of *Ordo* contributors and the disciplinary affiliations of *Ordo* editors and contributors.

1. Composition of editors

The founding of *Ordo* in 1948 can be viewed from different perspectives. It coincided with the setting up of other learned journals after the end of Nazism and the Second World War (Hagemann 1991 pp. 48 ff.). Given the explicit commitment to neoliberalism, the launching of *Ordo* was also part of a broader mushrooming of new political journals at that time. Finally, the setting up of *Ordo* represented an important step in the post-war reorganization of liberal forces and the formation of what became known as "neoliberalism" (Hayek 1951/67; 1983/92).

The first volume of *Ordo* mentions no fewer than ten collaborators. Franz Böhm and Walter Eucken featured as editors (*Herausgeber*), Fritz W. Meyer and Hans Otto Lenel as managing editors (*Schriftleiter*) and Karl Brandt, Constantin v. Dietze, Friedrich A. Hayek, Friedrich A. Lutz, Wilhelm Röpke and Alexander Rüstow as advisory board members (*Mitwirkende*).[5] The founding of *Ordo* thus brought together a group of diverse individuals. Some had been forced to spend the Nazi years abroad, others had stayed in Germany. In 1948, those associated with the journal were affiliated to different institutions and worked in different places. Despite these differences, however, they were bound together by a dense network of ties which dated back to the Weimar Republic. These ties were in part academic:

5 In the following, we do not distinguish between the different categories, but group them together as editors.

- Lenel, Lutz and Meyer were former students and assistants of Eucken. Lutz and Meyer wrote their doctoral theses under Eucken's auspices and submitted their post-doctoral theses (*Habilitationen*) in Freiburg. Lenel started studying economics in Freiburg in the winter of 1937/38, received his doctorate in 1942 and served as Eucken's assistant from 1945 to 1947.
- During the late 1920s and early 1930s, Eucken, Röpke and Rüstow cooperated as leading proponents of a group of younger economists known as the "Ricardians" (Krohn 1981 pp. 132-141; Nicholls 1994 chap. 2; Janssen 1998 chap. 2). Led by Rüstow, at that time head of the economics department of the Association of German Machine Builders (*Verein Deutscher Maschinenbau-Anstalten*), this group sought to challenge the grip of the Historical school on German economics. It comprised both liberal economists and those with leanings farther to the left. Eucken, Röpke and Rüstow, the leading members of the liberal wing, maintained close contact with Hayek (Hayek 1983/92 pp. 185-190). When the Nazis took over, Röpke and Rüstow were forced into exile. Until 1943/44, however, they were able to continue their exchange with Eucken by corresponding extensively (Röpke 1960/61; Lenel 1991). Hayek even met Eucken regularly until the late 1930s (Hayek 1983/92 p. 190).
- Böhm and Eucken had worked closely together ever since the early 1930s. When Böhm left the Imperial Ministry of Economics in late 1931 and moved to Freiburg to start working on a post-doctoral thesis in law, a fruitful cooperation with Eucken, who had been Professor of Economics in Freiburg since 1927, began. Eucken not only refereed and promoted Böhm's thesis: along with Hans Grossmann-Doerth, a recently appointed law professor, Böhm and Eucken also organized a series of joint seminars in commercial law and economic policy and coedited a series of books on "The Order of the Economy" (*Die Ordnung der Wirtschaft*) (Böhm 1957/60). Both the seminars and the book series, introduced by a programmatic statement known as the 1936 Ordo manifesto (Böhm *et al.* 1936/89), represent key steps in the development of Ordoliberalism.

These bonds were complemented by political ties, as academic cooperation was deeply interwoven with joint political activities:

- During the Weimar years, Eucken, Röpke and Rüstow not only cooperated in academic matters, they also joined forces in an attempt to influence economic policy by publishing newspaper articles, by playing an active part in various interest organizations and by giving policy advice (Krohn 1981 pp. 132-141; Nicholls 1994 chap. 2). For instance, all three were involved in the executive committee of the German League for Free

Economic Policy (*Deutscher Bund für Freie Wirtschaftspolitik*). Minor differences in political positions notwithstanding, they saw themselves, and were widely perceived as such, as advocates of the same cause and as members of a group.

- Under the Nazi regime, Böhm, Eucken and v. Dietze worked together in a series of working groups and discussion circles that debated issues of economic reform and had contacts with the Christian-conservative opposition to the regime (Blumenberg-Lampe 1973; Rieter/Schmolz 1993).[6] All three played a prominent role in drafting the 1942 Freiburg Manifesto (*Freiburger Denkschrift*), an outline for a post-war economic and political order commissioned by the opposition wing of the Evangelical Church (*Bekennende Kirche*), and advised Carl Goerdeler, one of the key figures of German anti-Nazi resistance, in economic matters.

- At the same time, Brandt, Hayek, Röpke and Rüstow were politically active outside Germany. They all participated in the famous Colloque Walter Lippmann in Paris in 1938, which brought together a number of liberal intellectuals from various countries and introduced the very notion of "neoliberalism". All four were also founding members of the Mont Pèlerin Society, an influential association of liberal intellectuals that was established in Switzerland in 1947 (Hartwell 1995). It was as a result of their influence that Eucken became one of the two non-emigrants from Germany invited to the founding conference.

The first generation of *Ordo* editors was thus held together by strong personal connections.[7] The original *Ordo* editors not only subscribed to the same ideas. The shared liberal credo, as formulated in the introduction to the first volume of the *Ordo* yearbook (Introduction 1948), was also embedded in, and rested upon, a dense network of social ties that had grown out of joint academic and political activities since the mid-1920s and had survived the Nazi years. The nature of these links, the very mixture of academic and other

---

6 v. Dietze and Eucken had known each other since the early 1920s, when both served as lecturers in Berlin. It was Eucken who promoted v. Dietze's appointment in Freiburg after v. Dietze, the last president of the *Verein für Socialpolitik*, the prestigious association of German economists and social scientists, before it dissolved itself in 1935, had fallen into disrepute with the Nazis and had lost his chair at Berlin (BRINTZINGER 1996 pp. 103-109).

7 The list of ties could easily be extended. For instance, Lutz was married to one of Hayek's former research assistants; the first German edition of Hayek's *Road to Serfdom* had an introduction by Röpke and was translated by his wife etc.

ties, facilitates understanding of *Ordo*'s peculiar status as a rare combination of a learned and a political journal.

Over the past 50 years, the composition of *Ordo* editors has been subject to changes. Save for Lenel, the members of the first generation are no longer alive. Eucken died in 1950, Rüstow in 1963, Röpke in 1966, v. Dietze in 1973, Lutz in 1975, Brandt in 1975, Böhm in 1977, Meyer in 1980 and Hayek in 1992. The first infusion of "fresh blood" took place in 1968 when Hans Willgerodt[8] became one of the managing editors. In 1980, Hans-Joachim Mestmäcker was coopted. More far-reaching changes occurred in 1981 and 1990. In 1981, Walter Hamm, Ernst Heuß, Erich Hoppmann and Christian Watrin were made additional editors. In 1990, Helmut Gröner, Wernhard Möschel, Josef Molsberger and Alfred Schüller joined the team. The most recent change in the Editorial Board to date took place in 1998 when Viktor Vanberg and Peter Oberender became editors.

Despite these changes, ties between editors have remained relatively strong. This holds good for ties among the new editors as well as for those between the new and the old editors. A look at the different generations of *Ordo* editors brings fairly close professor–student relationships to the fore. As Figure 5.1 indicates, about two-thirds of the 12 newly coopted editors were students of Böhm, Eucken or of one of their students. Heuß studied in Freiburg during the Second World War and was a doctoral student of Eucken (Heuß 1991); Mestmäcker is perhaps Böhm's best known and most influential student; Gröner, Möschel, Molsberger, Oberender, Schüller and Willgerodt obtained their doctorates under the auspices of various students of Böhm or Eucken. In terms of students, the single most important Eucken student has been Meyer. No fewer than four of his students later became *Ordo* editors. The prominence of a small number of peers also implies that some proportion of the *Ordo* editors temporarily studied or worked together at the same institution. These more academic ties have continued to be supplemented by joint political activities. Like the first generation, most current *Ordo* editors are members of the Mont Pèlerin Society; some of them have also cooperated in the *Kronberger Kreis*, a circle of professors associated with the *Frankfurter Institut – Stiftung Marktwirtschaft und Politik*, a liberal think-tank.

8 He was, incidentally, a nephew of Wilhelm Röpke.

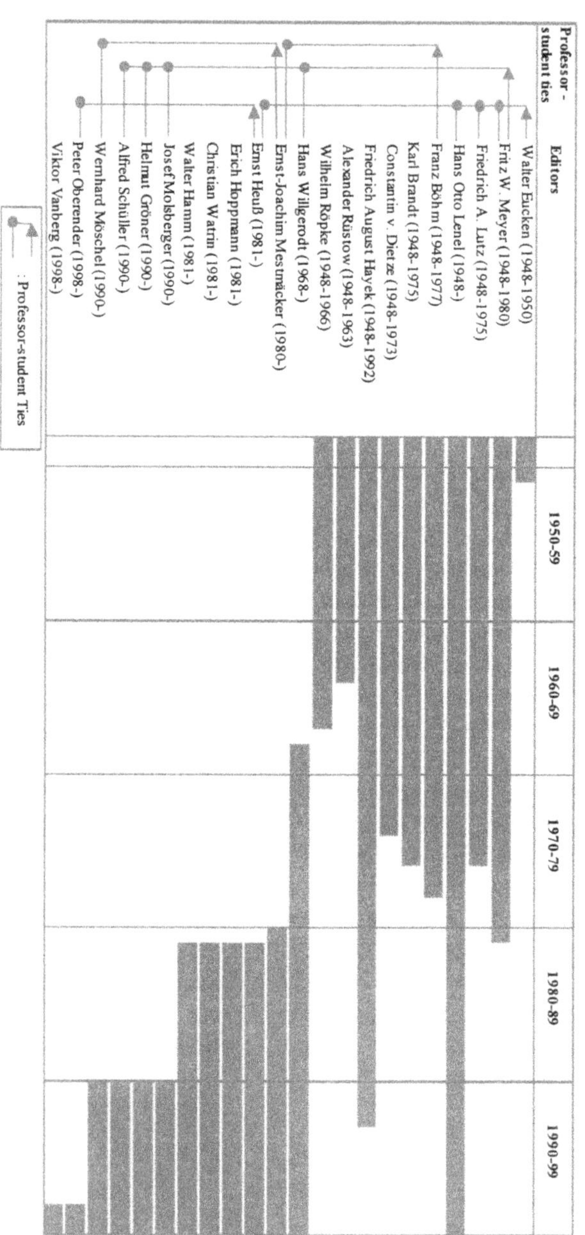

Figure 5.1: Composition of the editorial board of the *Ordo* yearbook

These strong ties among *Ordo* editors are open to different interpretations. From one perspective, the professor–student ties testify to the intellectual appeal of Böhm, Eucken and their students. Seen from this angle, they indicate that ordoliberals continued to attract bright students, produce a number of well-known scholars and succeed in cultivating a tradition. However, the prevailing strength of these ties has another side which suggests some "isolation" of the ordoliberal school. The very fact that the bulk of editors has come from a handful of chairs and universities suggests that the spread and diffusion of Ordoliberalism has been limited. If Ordoliberalism had really conquered German universities, one would have expected editors to come from a more varied range of institutions and to have more diverse backgrounds. This particularly applies to a university system such as that in Germany where the "hierarchy" of universities in terms of reputation is relatively weak. For these reasons, the strong ties among *Ordo* editors do also, in a way, indicate that Ordoliberalism has encountered some problems in expanding and recruiting adherents from outside.[9]

2. Concentration of Ordo contributors

The composition of *Ordo* contributors lends further support to this impression. For an examination of the "pool" of contributors reveals a fairly high concentration together with strong involvement on the part of the editors. To start with, a relatively small number of *Ordo* contributors accounts for a relatively high share in overall articles.[10] Thirty-two authors have published five or more articles each, thereby contributing almost 40 per cent of all articles. Another 37 authors have written three or four articles each and add a further 18 per cent. From a comparative perspective, these figures (which are relatively stable over time) fairly high and clearly exceed the values that are normally found for learned journals (Price 1963/74 pp. 52-57). The high concentration of contributors has gone hand in hand with the strong editorial involvement. With the exception of Brandt, v. Dietze and Rüstow, all *Ordo* editors have published at least four articles in the yearbook.

9 In a way, the strong professor–student-ties among editors validate the popular notion of an ordoliberal school. For in science studies the very notion of a "school of thought", if it is used at all, is normally reserved for relatively "closed" research groups centred around a founding father, built on strong professor–student ties and concentrated at a few places (see STICHWEH 1999).

10 The following calculations refer to the 671 articles published between 1948 and 1998. This figure does not include book reviews or review essays.

Nine of the 22 overall editors have had ten or more articles in the yearbook, two (Willgerodt and Lenel) even more than 20. From a comparative perspective these figures, too, are on the high side.

The strong involvement of leading ordoliberals underscores the importance of the *Ordo* yearbook as the major ordoliberal forum. It confirms *Ordo*'s role as a journal where key ordoliberals publish regularly and where many articles by prominent representatives of the ordoliberal school can be found. At the same time, however, the high concentration of contributors further testifies to the "isolation" of the ordoliberal school. In line with the strong ties among editors, it indicates that the "pool" of ordoliberal authors has remained relatively small and therefore underlines the ordoliberals' problems with opening up and reaching out.

3. Disciplines represented

Ordoliberalism has been an interdisciplinary project. It was the cooperation of economists and lawyers, as exemplified by the famous seminars in the 1930s and "The Order of the Economy" series, which gave rise to the "Freiburg school" (Böhm 1957/60). Moreover, several prominent ordoliberals shared some inclination towards sociology.[11] This is true of Müller-Armack, the late Röpke and, in particular, Rüstow. The latter even held a chair in sociology after 1949 and was the first president of the German Political Science Association (Haselbach 1991 pp. 184-216; Meier-Rust 1993).

The interdisciplinary character of Ordoliberalism is nicely illustrated by the fact that *Ordo* was founded by an economist (Eucken) and a lawyer (Böhm). However, a closer look at the composition of *Ordo* contributors suggests that the *de facto* degree of interdisciplinarity has been limited. While *Ordo* has seen editors and contributors from other disciplines, it has clearly been dominated by economists. To begin with, the number of editors outside of economics has remained small. With Böhm, Mestmäcker and Möschel, there have only been three lawyers among the 22 individuals who have served as editor between 1948 and 1998. Two editors have had some institutional affinities within sociology. Besides Rüstow, there has been Vanberg who started his academic career outside economics and wrote both his doctoral and his post-doctoral thesis in sociology.

---

11 This has led some observers to distinguish a distinct "sociological" strand of Ordoliberalism (BECKER 1965).

The composition of contributors reveals a similar picture. The 32 authors who have published five or more articles in the yearbook include only four non-economists, namely the three lawyers among the editors and Edith Eucken-Erdsiek, Walter Eucken's wife, who had a background in humanities. The number of non-economists (six) among the 37 contributors with three or four articles is similarly small. Four are legal scholars (Volker Emmerich, Thomas Oppermann, Dieter Reuter, Hans-Heinrich Rupp).[12] In addition, we find one sociologist (Michael Zöller) and a single philosopher (Gerard Radnitzky). All this documents that *Ordo* has remained not only an economic but also an economists' journal.

III. Preoccupations and Gaps: Observations on Themes and Issues

The 671 *Ordo* articles that were published between 1948 and 1998 have tackled a wide range of issues. The absolute numbers and relative shares of articles published on various issues provide fresh insights into the intellectual agenda of Ordoliberalism by drawing attention to persistent preoccupations as well as to shifts in emphasis. The following analysis focuses on three issues: the relative weight attached to the issue of economic systems, the coverage of different fields of economic policy and the interest in market economies outside Germany.

1. Economic systems and economic transformation

Ordoliberalism has been concerned with economic systems and the economic order. The title and the subtitle (*Yearbook for the Order of Economy and Society*) of the *Ordo* yearbook clearly express this orientation. It therefore comes as no surprise that, according to our classification, almost a fifth of all *Ordo* articles have focused on the issue of economic systems.

In order to go beyond this aggregate figure, Table 5.1 distinguishes between six sub-periods and five sub-issues. "Comparative Economic Systems" refers to articles that deal with economic systems in general and

12 It further illustrates the salience of professor–student relationships among ordoliberals that two of these four scholars are Mestmäcker students.

contrast and compare the features and the performance of different economic systems or analyse the interrelationship between market and communist economies. In contrast, the categories "Market Economies" and "Communist Economies" include articles that focus exclusively on the functioning and the problems of one particular type of economic system. As for articles on Communist economies, a distinction has been made between those articles that dwell upon communist economies in general ("Communist Economies") and those that focus on the problems of "actually existing" Communism, that is, on particular communist countries ("Communist Countries"). Our final category, "Economic Transformation", comprises all articles devoted to the transition from plan to market or, to use Eucken's terminology, to the transformation from a centrally administrated economy (*Zentralverwaltungswirtschaft*) to an exchange economy (*Verkehrswirtschaft*).

Table 5.1: *Ordo* articles on economic systems, 1948-1998

Sub-Issue	1948 1950	1951 1960	1961 1970	1971 1980	1981 1990	1991 1998	Σ
Comparative Economic Systems	12	8	5	5	9	5	44
Market Economies	-	2	4	10	9	12	37
Communist Economies	-	-	4	3	1	-	8
Communist Countries	-	-	8	3	2	-	13
Economic Transformation	-	-	-	-	4	16	20
Σ	12	10	21	21	25	33	122
Share in Overall Articles (%)	44	11	19	16	19	20	18

The disaggregation shows that interest in economic systems, as documented by the relative number of articles, was strongest in the late 1940s. At that time, the share of articles on economic systems reached its all-time high of than over 40 per cent. These articles generally took an explicitly comparative perspective in that they contrasted different types of economic systems, discussed various theories about economic systems or elaborated upon the "systemic" dimension of various policy issues. The predominance of such a comparative perspective nicely illustrates how contested the case for a market economy was during these years. However, this seems to have changed rather rapidly, as the figures for the 1950s document a dramatically declining interest in economic systems. In this decade, articles on more mundane issues of economic policy gained importance. At the same time, a shift from articles on the comparative performance of market and Communist economies to articles on the functioning of market economies set in which continued in the 1960s and 1970s. This shift in focus can be taken as an indication of the fairly rapid consolidation of the market economy in the Federal Republic of Germany.

Interestingly, the share of articles on economic systems increased once more in the course of the 1960s. Table 5.1 reveals that this surge was largely driven by a new interest in the problems of Communist economies and development in Communist countries. This development can be read in different ways. On the one hand, it shows that Communism had once again become a theoretical and political challenge in the course of the 1960s. From a different perspective, however, it also suggests that some time elapsed before ordoliberals went beyond their typologies and began to analyse Communist economies in an empirical fashion. In the 1970s and 1980s, the share of articles on economic systems fell again. In parallel with the watering-down of reforms and the stagnation of the Soviet bloc, so it seems, the interest in Communist economies dwindled. This decline was only partly compensated for by a higher number of articles on the foundations of market economies. In the late 1980s, then, the breakdown of Communism and the start of the post-Communist economic transformation served as a new stimulus for looking at the elements and the transformation of economic systems.[13]

2. Fields of economic policy

Ordo has published articles on almost all fields of economic policy. Yet coverage has been uneven. As Table 5.2 reveals, some fields of economic policy have been dealt with more extensively than others. The single most important field has been competition policy. The latter has not only drawn the highest number of articles but has also stood out by reason of a fairly steady coverage over time. This finding fits nicely with the prominence that competition policy enjoys within the framework of ordoliberal thinking. The insistence on the need for a strict competition policy is widely regarded as the *differentia specifica* of Ordoliberalism. That is why it comes as no

13 The choice of subperiods underlying Table 1 tends to obscure this stimulus. About half of the 1981-1990 articles on economic systems were published in 1989 and 1990; a different periodization would have thus have made the increasing interest in economic systems after 1989 more visible.

surprise that *Ordo* has featured a vast number of articles on various aspects of competition policy.

Table 5.2: *Ordo* articles on various fields of economic policy, 1948-1998

Field	1948 1950	1951 1960	1961 1970	1971 1980	1981 1990	1991 1998	Σ
Competition Policy	3	12	10	14	8	14	61
Sectoral Regulation	1	5	10	6	16	17	55
Monetary and Exchange Rate Policy	2	7	7	8	7	4	35
Social Policy	1	6	1	7	5	10	30
Trade Policy	-	3	-	3	9	5	20
Fiscal Policy/Public Finance	2	6	1	3	1	3	16
Industrial and Technology Policy	-	1	-	2	5	8	16
Total Number of Articles	27	93	113	134	135	169	671

The fact that the policy field that comes second is regulation of particular economic sectors such as electricity or transport very much points in a similar direction. As sectoral regulation aims at governing monopolies or at exempting sectors from competition, it is closely related to competition policy. One might even characterize most sectoral regulation as a kind of "sector-specific" competition policy. Seen from this perspective, the huge number of articles on sectoral regulation further confirms the primacy of competition policy within ordoliberal thinking. At the same time, the figures reported in Table 5.2 suggest that the "specific" competition policy associated with sectoral regulation has gained importance *vis-à-vis* "general" competition policy over time.

The policy field which scores third in terms of articles is monetary and exchange rate policy. According to our classification, a total of 35 articles on these issues were published between 1948 and 1998. In contrast with most other policy fields, coverage of monetary and exchange rate policy has been relatively even over time. Similar to competition policy, albeit at a lower level of articles, worries about inflation have represented an ordoliberal evergreen. Again, the relatively high weight attached to monetary policy is fully in line with expectations, as the emphasis on stable money has featured prominently among ordoliberals. In their introduction to Ordoliberalism, Peacock and Willgerodt (1989 p. 8) even list "monetary control" as the first element of the ordoliberal policy agenda.

What may be more surprising is the fact that the number of articles on monetary and exchange rate policy has only slightly exceeded the number of articles on social policy. This is somewhat at odds with the view often heard that ordoliberals have focused excessively on competition policy and monetary control, but have neglected other issues (see Peacock/Willgerodt 1989 pp. 8-13). At the same time, figures also show that the interest in social policy has been subject to some fluctuations. While it was relatively strong in the late 1940s and the 1950s, it declined in the 1960s. Between 1961 and 1970, only a single article dealt with social policy. The situation changed in the 1970s. Starting in 1973, coverage of social policy increased once more.

Other policy fields have been dealt with less frequently in *Ordo*. While *Ordo* has published a host of articles on problems of international economic integration, trade policy was not really an issue until the 1970s. Likewise, industrial and technology policy has only been covered in *Ordo* relatively recently. What is most interesting is the case of fiscal policy and public finance. Fiscal policy has clearly been the field most strikingly neglected in the *Ordo* yearbook. Not only that, along with industrial policy, it has attracted the smallest number of articles. Moreover, aggregate figures obscure the fact that half of the 16 articles on public finance were published between 1948 and 1960. The relatively high number and share of articles in the late 1940s and the 1950s contrasts with the extremely meagre record in the following decades. Ever since the 1960s, fiscal policy has become a subordinate issue attracting a mere one to three *Ordo* articles over a whole decade. This far-reaching neglect of fiscal policy and public finance thus emerges as the single most striking gap in ordoliberal policy debates as represented by the articles in the *Ordo* yearbook.

3. Coverage of other market economies

In addition to theoretical articles and articles on the German case, the introduction to the first *Ordo* volume also promised reports on developments in other countries (Introduction 1948 p. ix). As Table 5.3 documents, *Ordo* has in fact featured several articles on other Western market economies. However, the number of such articles has not been overly impressive. With the exception of the United States which ranks first by a wide margin, developments abroad have only rarely been discussed in the *Ordo* yearbook.

A closer look at the articles reveals that the interest in foreign developments has fluctuated. In the beginning, it was fairly weak. The first five volumes of *Ordo* only contained one article which looked abroad (Kronstein 1950). The coverage of other countries increased in the second half of the 1950s. Between 1953 and 1960, more than 10 articles on foreign experiences were published. The new interest stemmed largely from two sources. For one thing, the drafting of the German competition law (*Gesetz gegen Wettwerbsbeschränkungen*), eventually adopted in 1957, triggered interest in foreign experience with competition and competition policy. In the second half of the 1950s, *Ordo* featured articles on competition policy in Switzerland (Mötteli 1955), the United Kingdom (Mestmäcker 1956) and the United States (Heuss 1957, Bernhard 1959, 1960). For another, the approaching end of post-war reconstruction served as a stimulus to take stock and to summarize the different post-war trajectories of economic reform. This led to a group of more broad-brushed articles on economic policy in France (Rueff 1959, 1960), Sweden (Gehnich 1953; Harper 1957) and Switzerland (Mötteli 1958).

Table 5.3: *Ordo* articles on various Western market economies, 1948-1998

Country	1948 1950	1951 1960	1961 1970	1971 1980	1981 1990	1991 1998	Σ
United States	1	5	3	1	-	4	14
France	-	2	1	1	3	-	7
United Kingdom	-	1	-	1	2	-	4
Sweden	-	2	-	-	-	1	3
Switzerland	-	2	-	1	-	-	3
Other Countries	-	-	1	1	3	1	6
Total Number of Articles	27	93	113	134	135	169	671

Interestingly, this curiosity proved short-lived and soon faded away. Between 1961 and 1977, a mere eight articles on developments in other Western countries were published. The situation changed slightly in the late 1970s and early 1980s, when the number of articles on Western countries briefly increased once more. This second "peak" was partly driven by the interest in the shifts in French and British economic policy after the changes in government of 1979 and 1981, respectively. Four of the ten articles that appeared between 1978 and 1983 focused on the effects of these changes (Parkin 1982; Curzon-Price 1982; Rosa 1982; Lozada-Heller 1983). Again, however, the interest did not hold for long. The number of articles on other

Western countries declined after 1983 and has continued to hover at a low level ever since. Between 1984 and 1998, a mere seven articles were published on this subject. Given the ordoliberals' interest in liberal economic reform, this neglect is somewhat surprising. Given the sweeping changes in economic policy since the late 1970s and the recent concentration on "model reformers" (The Netherlands, Chile and so on) and international "benchmarking", one might have expected a broader coverage of foreign developments. To put this "silence" differently, it seems that ordoliberals have continued to derive their policy conclusions from first principles and the good old times of the German "economic miracle" rather than from foreign experience.

IV. Thinking in Orders: Observations on the "Ordoliberal Content" of Ordo Articles

Scrutinizing a journal with a strong presence of economists like *Ordo*, one could expect a profusion of mathematical models and econometric studies. But the yearbook involves very few models and econometrical analyses, a feature which shows a remarkable stability over time: only 13 articles include such items in fifty years. Furthermore, it is worth noticing that the authors using these instruments do not generally belong to the ordoliberal school strictly speaking. Indeed, Ordoliberalism is very little formalized, by contrast with Regulation theory, as Robert Delorme observes in this book (below, chap. 8). This lack of interest in mathematical models is correlated with a preference for typologies, a salient feature which is a sub-product of the ordoliberal approach of "thinking in orders". This leads us to a further question regarding the ordoliberal approach: how is this specificity of "thinking in orders" represented in *Ordo* and how has it evolved over time? Interestingly, in addition to the issues and topics they cover, the articles in the *Ordo* yearbook have also differed in the extent to which they rely on key ordoliberal assumptions and concepts.

1. Looking for the ordoliberal approach in Ordo

In order to tackle this tricky issue we took "thinking in orders" (*Denken in Ordnungen*) as the *differentia specifica* of Ordoliberalism. More specifically, we examined the texts of our 671 articles to see whether they make explicit use of concepts such as Ordo (the desirable order in contrast to the actual), economic order (*Wirtschaftsordnung*) or partial orders (*Teilordnungen*) and pay tribute to ideas such as the coherence and interdependence of orders. If they did so, we classified these articles as "ordoliberal". In order to make allowance for the effects of obituaries, biographical articles and reprints of earlier articles, we counted these articles separately. Figure 5.2 documents the results of our calculations.

According to these figures, the share of "ordoliberal" articles has strongly declined over time. This global trend should not hide the considerable variations over time presented by the share of ordoliberal articles. At the very beginning of the yearbook, the share oscillates between 45 and 88 per cent. From 1956 and 1966, it declines dramatically and falls to 17 per cent. Starting in 1967, the share recovers and peaks at 62 per cent in 1969. However, this renaissance of ordoliberalism proves to be short-lived. Between 1969 and 1988, the share of ordoliberal articles dwindles again and falls to less than 20 per cent. A second revival of ordoliberalism can be observed in 1989 when the share of ordoliberal articles goes up to almost the 1969 level. Once again, however, this increase in "ordoliberal content" is temporary and followed by a drastic reduction.

2. A fading or an open approach?

How should this striking development be interpreted? Part of the trend may simply reflect changes in the policy agenda. This is to say that the "ordoliberal content" has varied in parallel with the interest in economic systems. As Ordoliberalism represents first of all a theory of economic systems, it is quite natural for ordoliberal concepts and ideas to have featured more prominently when there was greater interest in economic systems, but to have been deployed less often when the policy agenda was dominated by more mundane issues of economic policy. The building phase of the "social market economy" in West Germany corresponded simultaneously to the maturation of the ordoliberal approach, which is particularly in touch with the question of building an appropriate economic order.

Figure 5.2: Evolution of the proportions of ordoliberal articles

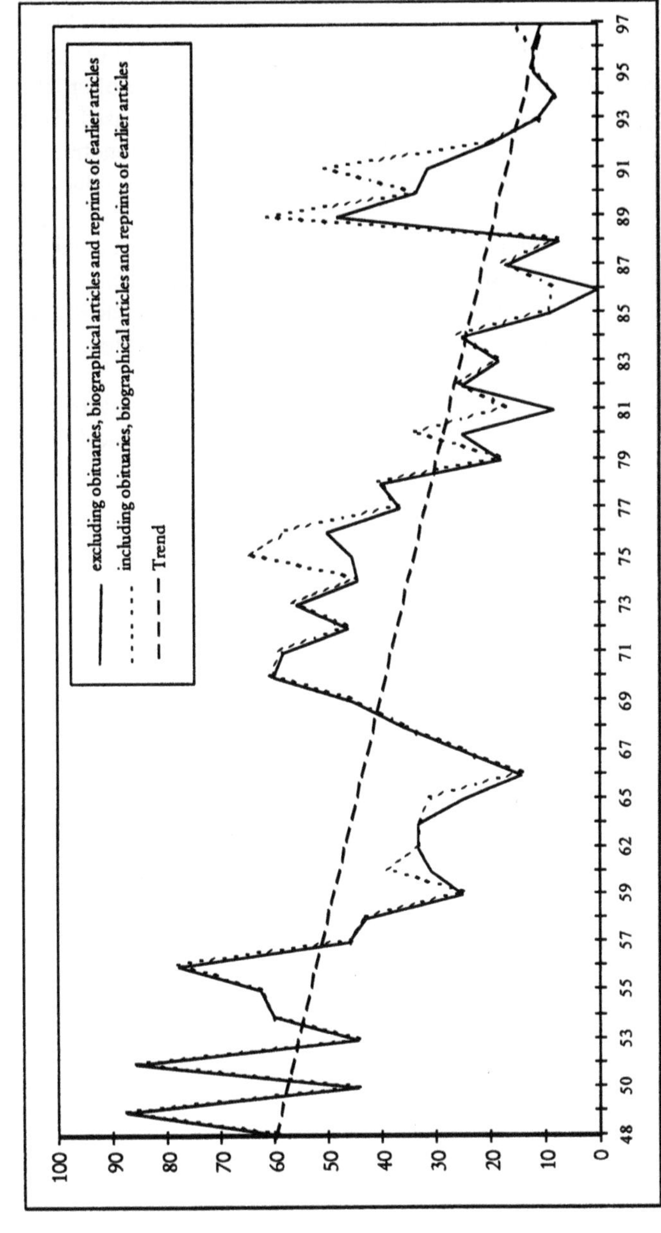

This explains why the "ordoliberal content" was fairly high during the early years of the Federal Republic but fell after a market economy had been established and consolidated. In the same vein, the share of articles increased once again whenever interest in economic systems increased. This applies to the late 1960s and early 1970s, when Communist economies draw more attention, and it also holds true for the transformation-induced renewal of interest after 1989. This "second wind" related to post-socialist transformation, which reminded ordoliberals of the immediate post-war situation, coincided with the 50th anniversary of Eucken's *Foundations of Economics* and the 40th anniversary of the yearbook. It remains questionable whether this peak represents a second youth or a swan song.

The shifting policy agenda appears to be only part of the story. In addition, the declining "ordoliberal content" also suggests that the very specificity of Ordoliberalism has eroded over time. This erosion and the concomitant opening-up to other approaches such as New Institutional Economics, Public Choice or Constitutional Economics can be interpreted differently. From one perspective, it testifies to a dissolution of the original ordoliberal approach and the watering-down of the attempt to rebuild economics on the base of a new compromise between the historical and the neoclassical school (see the contributions by Delorme, Herrmann-Pillath, Weisz in this volume). In such a view, Ordoliberalism seems to have had some difficulties to develop and extend its original approach in order to deal with issues differing from its standard topics, namely economic orders and systems. However, one might also argue the opposite – that the opening-up of Ordoliberalism highlights the openness of Ordoliberalism and documents a successful and much-needed "modernization" and rejuvenation. It goes without saying that these interpretations rest on diverging views of the original ordoliberal approach and that the associated issues cannot be settled within the framework of our bibliometric analysis.

V. Concluding Remarks

This chapter has featured the first-ever bibliometric analysis of the *Ordo* yearbook. It has examined trends in the composition of *Ordo* editors, contributors and articles with a view to investigating some of the "connecting" social and cognitive ties. The various parts of the analysis have converged in their joint focus on the "boundaries" of Ordoliberalism. With regard to the "openness" of Ordoliberalism, the findings point in different directions. On the one hand, Part II indicates that ordoliberals can be described as a relatively "closed" group. As we have shown, the first and the second generation of *Ordo* editors alike were characterized by dense personal, academic and political ties. Moreover, *Ordo* has relied on a limited pool of contributors and has had a rather high concentration of contributors. On the other hand, however, our findings on "ordoliberal content" document an opening-up of Ordoliberalism and indicate that the specificity and distinctiveness of the ordoliberal approach have been eroded over time. This contrast suggests that the ordoliberals have been held together by strong personal ties and shared commitment to a broad economic and political credo rather than by subscription to a narrowly defined approach.

On a more general level, we hope that our chapter has also demonstrated the general potential of bibliometric analyses of the *Ordo* yearbook and of Ordoliberalism. It goes without saying that such analyses are no substitute for other approaches. We also hasten to add that our own investigation represents no more than a first step and that further bibliometric studies can and should be undertaken. Among the most prominent avenues of future research are citation analyses. For instance, it would be interesting to know how often the founding fathers of Ordoliberalism, or other prominent figures, are cited, or to what extent *Ordo* articles serve as stimuli for other *Ordo* articles. Research along these lines would further enlarge our knowledge of Ordoliberalism.

References

BECKER, H. P.: *Die Soziale Frage im Neoliberalismus. Analyse und Kritik*, Heidelberg (Kerle) 1965.

BERNHARD, R. C.: "Schärfere Antitrust-Politik für die Vereinigten Staaten?", *Ordo*, 12 (1960/61), pp. 127-138.

BERNHARD, R. C.: "Unternehmensgröße und Monopol in den Vereinigten Staaten. Nationalökonomische und juristische Kriterien", *Ordo*, 11 (1959), pp. 217-240.

BESNARD, P.: "La Formation de l'équipe de l'Année sociologique", *Revue française de sociologie*, 20 (1979), pp. 7-31. German translation: "Die Bildung des Mitarbeiterstabs der *Année sociologique*", in: W. LEPENIES (Ed.): *Geschichte der Soziologie. Studien zur kognitiven, sozialen und historischen Identität einer Disziplin*, Bd. 2, Frankfurt, M. (Suhrkamp) 1981, pp. 263-302.

BLUM, R.: *Soziale Marktwirtschaft. Wirtschaftspolitik zwischen Neoliberalismus und Ordoliberalismus*. Tübingen (Mohr) 1969.

BLUMENBERG-LAMPE, C.: *Das wirtschaftspolitische Programm der "Freiburger Kreise". Entwurf einer freiheitlich-sozialen Nachkriegsordnung. Nationalökonomie gegen den Nationalsozialismus*, Berlin (Duncker & Humblot) 1973.

BÖHM, F.: "Die Forschungs- und Lehrgemeinschaft zwischen Juristen und Volkswirten an der Universität Freiburg in den dreißiger und vierziger Jahren des 20. Jahrhunderts", in: F. BÖHM: *Reden und Schriften*, Karlsruhe (C.F. Müller), 1960, pp. 158-175. Original: 1957.

BÖHM, F., EUCKEN, W., GROSSMANN-DOERTH, H.: "The Ordo Manifesto of 1936", in: A. PEACOCK, H. WILLGERODT (Eds.): *Germany's Social Market Economy: Origins and Evolution*, London (Macmillan) 1989, pp. 15-26. Original: 1936.

BRINTZINGER, K.-R.: *Die Nationalökonomie an den Universitäten Freiburg, Heidelberg und Tübingen 1918-1945. Eine institutionenhistorische, vergleichende Studie der wirtschaftswissenschaftlichen Fakultäten und Abteilungen südwestdeutscher Universitäten*, Frankfurt am Main (Lang) 1996.

BROCHIER, H.: "Critères de scientificité en économie", in X. GREFFE, J. MAIRESSE, J.L. REIFFERS (Eds.): *Encyclopédie économique*, Paris (Economica) 1990, pp. 25-54.

BURKE, P.: *The French Historical Revolution: The Annales School, 1929-89*, Cambridge (Polity Press) 1990.

EUCKEN, W.: *Grundlagen der Nationalökonomie*, 9th edition, Jena (Springer) 1989. Original: 1939.

EUCKEN, W.: *Grundsätze der Wirtschaftspolitik*, Tübingen (Mohr) 1990. Original: 1952.

GEHNICH, K. G.: "Die Politik der Vollbeschäftigung in Schweden", *Ordo*, 5 (1953), pp. 153-180.

GROSSEKETTLER, H. G.: "On Designing an Economic Order: The Contributions of the Freiburg School", in: A. DONALD, A. WALKER (Eds.): *Perspectives on the History of Economic Thought II: Twentieth-Century Economic Thought*, Aldershot (Elgar) 1989, pp. 38-84.

HAGEMANN, H.: "Learned Journals and the Professionalization of Economics: The German Language Area", *Economic Notes*, 20 (1991), pp. 33-57.

HARPER, F. A.: "Der schwedische 'Mittelweg'", *Ordo*, 9 (1957), pp. 145-156.

HARTWELL, R. M.: *A History of the Mont Pèlerin Society*, Indianapolis (Liberty Fund) 1995.

HASELBACH, D.: *Autoritärer Liberalismus und soziale Marktwirtschaft: Gesellschaft und Politik im Ordoliberalismus*, Baden-Baden (Nomos) 1991.

HAUSMAN, D. M.: "Kuhn, Lakatos and the Character of Economics" in: R. BACKHOUSE (Ed.): *New Directions in Economic Methodology*, London (Routledge) 1994, pp.195-215.

HAYEK, F. A.: "The Transmission of the Ideals of Economic Freedom", in: F. A HAYEK: *Studies in Philosophy, Politics and Economics*, London (Routledge & Kegan Paul) 1951, pp. 195-200. Original: 1951

HAYEK, F. A.: "The Rediscovery of Freedom: Personal Recollections", in: P. G. KLEIN (Ed.): *The Collected Works of F. A. Hayek, Vol. IV: The Fortunes of Liberalism. Essays on Austrian Economics and the Ideal of Freedom*, Chicago/London (University of Chicago Press) 1992, pp. 185-195. Original: "Die Wiederentdeckung der Freiheit - Persönliche Erinnerungen" in: Verein Deutscher Maschinenbau-Anstalten und Institut der deutschen Wirtschaft (Eds.): *Produktivität, Eigenverantwortung und Beschäftigung: Für eine wirtschaftspolitische Vorwärtsstrategie*, Köln (Deutscher Instituts-Verlag) 1983.

HEUß, E.: "Die amerikanische Antitrustpolitik im Lichte der Monopolbekämpfung in Europa", *Ordo*, 9 (1957), pp. 65-98.

HEUß, E.: "Persönliche Erinnerungen an Freiburg während der Kriegszeit", *Ordo*, 42 (1991), pp. 3-10.

INTRODUCTION, 1948: "Die Aufgabe des Jahrbuches", *Ordo*, 1 (1948), pp. vii-xi.

JANSSEN, H.: *Nationalökonomie und Nationalsozialismus. Die deutsche Volkswirtschaftslehre in den dreißiger Jahren*, Marburg (Metropolis) 1998.

KARADY, V.: "Stratégies de réussite et modes de faire-valoir de la sociologie chez les durkheimiens", *Revue française de sociologie*, 20 (1979), pp. 49-82. German translation: "Strategien und Vorgehensweisen der Durkheim-Schule im Bemühen um die Anerkennung der Soziologie", in: W. LEPENIES (Ed.): *Geschichte der Soziologie. Studien zur kognitiven, sozialen und historischen Identität einer Disziplin*, Bd. 2, Frankfurt, M. (Suhrkamp) 1981, pp. 206-262.

KIRCHGÄSSNER, G.: "Wirtschaftspolitik und Politiksystem: Zur Kritik der traditionellen Ordnungstheorie aus der Sicht der Neuen Politischen Ökonomie", in: D. CASSEL, B.-T. RAMB, H. J. THIEME (Eds.): *Ordnungspolitik*, Munich (Vahlen) 1988, pp. 53-75.

KROHN, C.-D.: *Wirtschaftstheorien als politische Interessen. Die akademische Nationalökonomie in Deutschland 1918-1933*, Frankfurt am Main/New York (Campus) 1981.

KRONSTEIN, H.: "Die Politik des Wettbewerbs in den Vereinigten Staaten von Amerika", *Ordo*, 3 (1950), pp. 76-104.

KRÜSSELBERG, H.-G., SCHÜLLER, A.: *Grundbegriffe zur Ordnungsökonomie und Politischen Ökonomie*, Arbeitsberichte zum Systemvergleich, 7, Universität Marburg 1991.

LENEL, H. O.: "Walter Euckens Briefe an Alexander Rüstow", *Ordo* 42 (1991), pp. 11-14.

LENEL, H. O.: "The Life and Work of Franz Böhm", *European Journal of Law and Economics*, 4 (1996), pp. 301-307.

LOZADA-HELLER, R.: "French Socialist Policy and the European Community", *Ordo*, 34 (1983), pp. 141-156.

MEIER-RUST, K.: *Alexander Rüstow: Geschichtsdeutung und liberales Engagement*, Stuttgart (Klett-Cotta) 1993.

MESTMÄCKER, E.-J.: "Der Bericht der englischen Monopolkommission", *Ordo*, 8 (1956), pp. 265-284.

MÖTTELI, C.: "Die Schweiz und das Problem der Wirtschaftsordnung", *Ordo*, 10 (1958), pp. 205-224.

MÖTTELI, C.: "Kartelldebatte auch in der Schweiz. Von der Missbrauchs- zur Verbotsgesetzgebung?", *Ordo*, 7 (1955), pp. 195-214.

NICHOLLS, A. J.: *Freedom with Responsibility. The Social Market Economy in Germany, 1918-1963*, Oxford/New York (Oxford University Press) 1994.

PARKIN, M.: "Mrs Thatcher's Monetary Policy: 1979-1981", *Ordo*, 33 (1982), pp. 61-80.

PEACOCK, A., WILLGERODT, H. (Eds.): *Germany's Social Market Economy: Origins and Evolution*, Basingstoke/London (Macmillan) 1989.

PEACOCK, A., WILLGERODT, H.: "Overall View of the German Liberal Movement", in: PEACOCK, A., WILLGERODT, H. (Eds.): *German Neo-Liberals and the Social Market Economy*, London (Macmillan) 1989, pp. 1-15.

PEACOCK, A., WILLGERODT, H. (Eds.): *German Neo-Liberals and the Social Market Economy*, Basingstoke/London (Macmillan) 1989.

PRICE, DEREK J. DE SOLLA: *Little Science, Big Science. Von der Studierstube zur Großforschung*, Frankfurt, M. (Suhrkamp) 1974. Original: *Little Science, Big Science*, New York/London 1963.

RAPHAEL, L.: "Nationalzentrierte Sozialgeschichte in programmatischer Absicht: Die Zeitschrift 'Geschichte und Gesellschaft. Zeitschrift für Historische Sozialwissenschaft' in den ersten 25 Jahren ihres Bestehens", *Geschichte und Gesellschaft*, 26 (2000), pp. 5-37.

RIETER, H., SCHMOLZ, M.: "The Ideas of German Ordoliberalism 1938-45: Pointing the Way to a New Economic Order", *European Journal of the History of Economic Thought*, 1 (1993), pp. 87-114.

RÖPKE, W.: "Blätter der Erinnerung an Walter Eucken", *Ordo*, 12 (1960/61), pp. 3-19.

ROSA, J.-J.: "Denationalization and Deregulation of Industry: The French Experience 1976-1981", *Ordo*, 33 (1982), pp. 81-94.

RUEFF, J.: "Die französische Wirtschaftsreform. Rückblick und Ausblick", *Ordo*, 12 (1960), pp. 111-126.

RUEFF, J.: "Zur Wirtschaftsreform in Frankreich: Bericht zur Finanzlage", *Ordo*, 11 (1959), pp. 3-68.

STICHWEH, R.: "Zur Soziologie wissenschaftlicher Schulen", in: W. BLEEK, H. J. LIETZMANN (Eds.): *Schulen in der deutschen Politikwissenschaft*, Opladen (Leske and Budrich) 1999, pp. 19-32.

WEINGART, P.: "Wissenschaftlicher Wandel als Institutionalisierungs-strategie", in: P. WEINGART (Ed.): *Wissenschaftssoziologie 2: Determinanten wissenschaftlicher Entwicklung*, Frankfurt am Main (Athenäum Fischer) 1974, pp. 11-35.

Chapter 6

Hayek and Eucken on State and Market Economy

FRANK BÖNKER, HANS-JÜRGEN WAGENER

I. Introduction

Walter Eucken and Friedrich A. Hayek have been two of the most influential proponents of post-war neoliberalism. Despite joint activities and mutual respect,[1] however, they held different views on major economic and political issues and ultimately stand for different brands of liberalism (Dörge 1959; Streit/Wohlgemuth 1999; 2000; Woll 1989). These differences between the two surface in their treatment of the relationship between the state and the market.

Two recent developments have renewed interest in Eucken and Hayek – the post-communist transformation in the East and the on-going renegotiation of the German "social market economy". The demise of Communism has put a gigantic project of political and economic reform on the agenda. Eucken and Hayek alike have featured prominently in the accompanying academic debates, although for different reasons. Eucken has been consulted as a pioneer of comparative economic systems and one of the architects of German post-war economic reform, if not "miracle". By contrast, Hayek has been invoked as one of the most powerful critics of constructivism and a holistic reform design (Voigt 1994; Frei/Nef 1994). Especially in Germany, many scholars have

---

1 Hayek called Eucken "a valuable friend" and elevated him to "probably the most serious thinker in the realm of social philosophy produced by Germany in the last hundred years" (HAYEK 1983/92 p. 190). As professor in Freiburg, Hayek time and again emphasized his respect for, and his overall agreement with, Eucken.

treated Eucken and Hayek as the patron saints, so to speak, of the two basic forms of institutional change, namely pragmatic-constructivist design and organic evolution (Wagener 1992a; 1992b; von Delhaes 1993; Leipold 1997).

The current controversies over the performance and the future prospects of Germany's once famous "social market economy" have further stimulated interest in Eucken and Hayek, as looming discontent with the state of the German economy has stirred the interest in its ideological foundations. In these debates, Hayek is credited for being one of the earliest and most insightful critics of the very idea of a social market economy. Likewise, Eucken is often made the prophet of its authentic, original, version which was, however, later dropped and betrayed. For instance, Jürgen Lange-von Kulessa and Andreas Renner, in a recent contribution (1998), try to drive a wedge between the Ordoliberalism of Eucken and the Freiburg school, on the one side, and Alfred Müller-Armack (who coined the very notion of the "social market economy", worked together with Ludwig Erhard in the Federal Ministry of Economics and founded the so-called Cologne school of neoliberalism), on the other.[2] Conversely, Lange-von Kulessa and Renner stress the affinities between Eucken and Hayek. Departing from the traditional practice, they even make Hayek a member of the Freiburg school (ibid. 80). Other authors have charted the field differently. To cite but one further example, Manfred E. Streit and his collaborators have sought to forge a "new" or "second" Freiburg school whose credo might be summarized as "more Hayek, less Eucken" (Streit 1995).

In this chapter, we aim at outlining, and at putting into perspective, Eucken's and Hayek's thinking about the state and the market economy. In doing so, we concentrate on key differences between the two and seek to set up a kind of imaginary dialogue. We think that such an emphasis on differences is more conducive to identifying the preoccupations, as well as strengths and weaknesses, of the two authors than an ecumenical focus on commonalities. It seems to us that a stylized contrast of the positions adopted by Hayek and Eucken sheds some light on their thinking as well as on the role of the state in the market economy.[3] More specifically, we elaborate upon three major aspects

---

2 This has not passed without comment. Cf. WILLGERODT 1998.

3 It goes without saying that all broad-brush comparisons of Eucken's and Hayek's thinking such as ours run into two major problems. First, which Hayek? As is now well known, Hayek's thinking underwent substantial changes and modifications in the course of his lifetime. The late Hayek took a much more "evolutionary" perspective than the middle one. Secondly, Eucken and Hayek lived in different times. Although only eight years younger, Hayek survived Eucken by forty-two years. This is why Hayek, unlike Eucken, experienced the rise and fall of the

of the relationship between the state and the market economy. In the first part of the chapter, we focus on the role of the state in constituting markets. To put it differently, we ask to what extent Hayek and Eucken view the market economy as a "made" and "constructed", as opposed to a spontaneous, order. Against this background, the second part, then, summarizes and compares Eucken's and Hayek's views on the agenda of economic policy in a market economy. In particular, we look at their criteria for assessing state interventions and their ideas about competition, social and monetary policy. This discussion prompts the question as to what political conditions are needed for the implementation of these policies. This is why the third part of the chapter dwells upon the political preconditions for a functioning market economy and examines how Eucken and Hayek perceive the relationship between the economic and the political sphere. The chapter concludes with a brief summary.

II. Spontaneous Order or Political Artefact? The Role of the State in Constituting a Market Economy

The most fundamental difference between Hayek's and Eucken's thinking about state and market concerns the very character of markets and a market economy. This is to say that Hayek and Eucken hold different views on the role of the state in constituting markets. Whereas Hayek sees markets as spontaneous orders and evolutionary results, Eucken views the competitive order, his synonym for a functioning market economy, as an essentially political creation. Playing with one of Hayek's most famous distinctions (derived, of course, from Ferguson and Menger) (Hayek 1967), one might state that Hayek regards the market economy as the result of human action, whereas Eucken treats it as the result of human design.

For Hayek, the market represents the paradigm case of a spontaneous order. This does not mean that he denies the truism that the functioning of markets presupposes an institutional framework. Quite to the contrary, Hayek explicitly analyses the market, just like any other spontaneous order, as a "two-part

---

Keynesian welfare state and the post-war advances in economics. These differences in background and experience have to be borne in mind when comparing Hayek's and Eucken's economic and political thought.

mechanism" (Kley 1994 p. 49) that combines the observance of certain "rules of conduct" with decentralized, individual adjustments. In "The Constitution of Liberty", Hayek also criticizes the advocates of *laissez faire* for having paved the way for the enemies of the market by simply neglecting the constitutive role of rules and institutions (Hayek 1960/71 pp. 286-287). Compared to Eucken, however, Hayek has tended to downplay the constitutive role of the state.

Hayek's treatment of the market as a spontaneous order reflects his characteristic epistemic scepticism and the associated view of markets. The Humean emphasis on the narrow limits of human understanding not only drove Hayek's opposition to institutional engineering. It also manifested itself in a particular conception of the market which focuses on its ability to handle disparate information. As a consequence, Hayek does not build his case for the market on the assumptions of neoclassical theory and is thus less concerned with deviations from static equilibrium conditions or with monopolies. From the point of view of his theory of the market, it does not make any sense to care about market "inefficiency", "disequilibrium" or "unjust" market outcomes. One corollary of this assumption is that, from a Hayekian perspective, the functioning of markets does not presuppose a particular market structure and a complex institutional framework, but is based on little more than "the rules of property, tort, and contract" (Hayek 1976/82 p. 109).

Yet it is not only the modest institutional requirements which lead Hayek to reduce the role of the state in constituting markets. In addition, the late Hayek increasingly argues that the institutional framework of a market economy itself represents an evolutionary result that is only partly amenable to institutional engineering. As a consequence, the order of rules (*Rechtsordnung*) becomes as much a spontaneous order as the order of actions (*Handelnsordung*). Hayek's evolutionary turn goes hand in hand with a further radicalization of his anti-constructivism. For the late Hayek, the attempt to create the institutional framework for a market economy is prone to become as much a "pretence of knowledge" (Hayek 1975/78) and a "fatal conceit" (Hayek 1988) as the socialist replacement or simulation of markets. While Hayek continues to acknowledge the role of rules in the working of markets, he increasingly questions the view that these rules are political artefacts, let alone the expressions of a comprehensive act of legal-institutional bootstrapping.[4]

In contrast to Hayek, Eucken stresses the "made" character of a market

4 An obvious – and as yet unresolved – problem with Hayek is how to reconcile his verdicts on constructivism with his own suggestions for reform. See e.g. VOIGT 1994.

economy and the constitutive role of the state. It is precisely the emphasis on the state's role in constituting markets that marks the *differentia specifica* of Ordoliberalism and sets it apart from older blends of liberalism and what Alexander Rüstow once called "paleo-liberalism". The setting-up and maintenance of a comprehensive and coherent "economic constitution" by the state represent the very essence of Eucken's distinction between *laissez faire* capitalism and the "competitive order".[5] The elements of this framework, as they are codified in Eucken's famous constitutive and regulatory principles, go beyond the mere guarantee of private property and freedom of contract and include a rigorous competition policy and an elaborate legal–institutional framework (Eucken 1952/90 pp. 254-304). In the words of Grossekettler (1989 p. 63):

"... the Ordoliberals agreed on the rejection of a planned *society*; however, they did not want the economic *constitution* (the 'institutional skeleton') to grow without a plan, influenced by intervention on particular points and by group interests. They wanted it to be *fashioned* according to the model provided by the institutional framework for competition. To use a metaphor: They did not want the *game* to be planned, only the *rules* [original emphasis].[6]"

For Eucken, a functioning market economy neither emerges nor prevails spontaneously, but has to be instituted and protected by the state. It rests on an underlying political decision and demands a comprehensive institutional framework. As his attacks on *laissez faire* capitalism demonstrate, Eucken does not believe in the spontaneous emergence or natural realization of a competitive order (Eucken 1952/90 pp. 26-55, 356-360). For him, guaranteeing private property and freedom of contract does not suffice to prevent otherwise unfettered markets from self-destruction. In particular, Eucken is concerned with the emergence of monopolies and the replacement of "competition through achievement" (*Leistungswettbewerb*) by "competition to prevent competition" (*Behinderungswettbewerb*). Yet he also fears, albeit to a lesser degree, that income inequalities and social hardships may undermine the legitimacy of the market economy. What lurks behind these worries is the ordoliberal concern with private power, a concern that features much less prominently in Hayek (Streit/Wohlgemuth 1999; 2000).

---

5 It seems that Hayek took a similar position in the 1940s. A 1947 address to the Mont Pèlerin Society, e.g., features a distinction between "free economy and competitive order" that comes close to the ordoliberal position (HAYEK 1948).

6 Eucken himself did not distinguish between games and rules, but between processes and forms.

In the case of Eucken, a sceptical attitude towards unfettered markets goes hand in hand with a great optimism regarding the available knowledge about the functioning and the furnishing of markets. Both attitudes flow from a common source, faith in the neoclassical theory of competition and equilibrium. On the one hand, Eucken is sensitive to any deviation from the neoclassical conditions for market efficiency. Given the restrictive character of these conditions, he thus almost necessarily arrives at a critical assessment of many real world markets and at calls for state interventions. On the other hand, Eucken's faith in neoclassical theory leads him to take for granted the availability of a proper model of markets and of clear guidelines for designing economic policy.

In addition to these economic assumptions, Eucken's belief in the constitutive role of the state also reflected his lifelong opposition to historicism. A recurrent motive in Eucken is the critique of any "myth of the inevitability of development" (Eucken 1952/90 p. 200), a preoccupation that must be seen against the narratives of historical necessity that dominated the German intellectual scene in the 1920s and 1930s. The flipside of this critique is a strong decisionist, constructivist and voluntaristic element in Eucken (Riese 1972). From Eucken's point of view, there is such a thing as the intentional introduction of a competitive order; both the creation and the continuous monitoring of the competitive order presuppose a fundamental political decision and reflect a deliberate institutional choice. In contrast to the late Hayek, Eucken opts for pragmatic constructivism writ large and for having organic evolution more as a supplementary device.

It is due to this constructivist perspective that Menger's famous distinction between grown and made orders (*gewachsene* vs. *gesetzte Ordnungen*) features less prominently in Eucken than in Hayek. Whereas Hayek's whole social theory in a way builds on this duality, Eucken mentions it only in passing (Eucken 1939/50 pp. 51-54, 1952/90 pp. 373-374). Instead, his main concern is with the juxtaposition of market and centrally planned economies. It is quite symptomatic that Eucken does not equate the competitive order with a spontaneous order, but treats it as a combination, a kind of mixed system. For him, the chief difference between a planned economy and the competitive order is not that the former is "made" and "constructed", but that the underlying economic constitution is different and that it is not made in accordance with what Eucken regarded as the natural order.[7]

7 By contrast with Hayek, Eucken regards the "competitive order" as the natural order and sees it as "the order which corresponds to the essence of humanity and of

III. Formal or Substantive Requirements? The Agenda of Economic Policy in a Market Economy

It comes as no surprise that these different perspectives on the role of the state in constituting a market economy translate into differing views on the very agenda of economic policy in a market economy. Eucken and Hayek alike seek to constrain the domain of the state and argue for an economic policy guided by principles rather than mere pragmatism. As for state interventions, both effectively propose a kind of two-step procedure which demands an examination of a measure's systemic conformity before the assessment of its expediency. At the same time, however, the particular standards used by Eucken and Hayek differ and they do so in accordance with their different understanding of markets.

Hayek's conception of the state is that of limited government, not of a minimum state (Gray 1989 p. 128). For him, the criteria that state interventions into the market must fulfil are largely formal. State interventions are not confined in substantive terms, but must primarily conform to the rule of law. This is to say that such interventions must be based on general, goal-independent rules rather than on discretionary power and outcome considerations. As Hayek himself has put it (Hayek 1960/71 p. 287), the problem is less the size of the state, but the form of its interventions. As a matter of fact, Hayek's texts offer different and sometimes quite expansive catalogues of legitimate state interventions which range from a universal minimum income to public, subsidized housing and city-planning. Therefore, it is not surprising that Hayek's criterion for assessing state interventions has been criticized as insufficient and that libertarians have attacked Hayek for being a kind of closet social democrat (Hoppe 1994).

In contrast, Eucken arrives at a more substantive criterion for the assessment of state interventions. From his point of view, all state interventions must be in line with the principles of a competitive order. One of the key tenets

---

the object, i. e. the order where measure and equilibrium exist ... It means bringing the multiple facets together into a *meaningful* [*sinnvoll*; original emphasis] whole" (EUCKEN 1952/90 p. 372; our translation). In stark contrast to Hayek, Eucken also argued that "economic policy should aim at implementing the free, natural order intended by God" (EUCKEN 1952/90: 176; our translation). Lange-von Kulessa/Brenner 1998 ignore Eucken's philosophical foundations when they try to set Eucken apart from Müller-Armack and his calls for a "guiding principle" (*Leitbild*) or "an inner orientation of the whole" (*innere Gesamtrichtung*).

of Ordoliberalism is the notion that all measures of economic policy have to be in line with, and to be subordinated to, the maintenance of the coherence of the market economy (Eucken 1952/90 pp. 250-253, 304-308, 380). As a consequence, Eucken's criterion for assessing state interventions is both narrower and broader than Hayek's. On the one hand, it is narrower in that it is not purely formal, but instead refers to a substantively defined economic order. On the other, it is broader in that it allows for types of state intervention which transcend general rules *à la* Hayek.

These differences between Hayek and Eucken are most visible in the case of competition policy. In line with his general credo, Hayek supports general restrictions on competition-preventing practices to be enforced by private sues (Hayek 1979/82 pp. 65-93). In contrast, Eucken advocates much more far-reaching, outcome-oriented state interventions in that he calls for an independent antitrust authority, patterned upon an independent central bank and in charge of breaking up monopolies and of regulating those that remain (Eucken 1952/90 pp. 264-270, 291-299). Eucken's belief that these remaining monopolies can and should be made to act "as if" neoclassical perfect competition was operating, illustrates his commitment to neoclassical market theory and his more substantive notion of a market economy. At the same time, Eucken's fantasies of an anti-trust authority setting prices at marginal costs once more reveal his epistemic optimism and come close to market socialism Lange-style.

With regard to social policy, the differences between Hayek and Eucken are less striking. Here, their specific proposals are not that different. The theoretical argument, however, is much more fully developed in Hayek than in Eucken. As many observers have noted, Hayek, at least the young Hayek (The Road to Serfdom, 1944) and the middle Hayek (The Constitution of Liberty, 1960/71), left some scope for social policy.[8] In particular, he accepted anti-poverty measures and mandatory social insurance. According to him, it is crucial to keep these measures outside of the market and not have them interfere with the price-setting function, above all of the labour market. Eucken's attitude towards social policy was rather reserved, too, and did not really go beyond classical liberal positions (Volkert 1991). Compared to Müller-Armack and others, as well as to the factual expansion of the welfare state, Eucken's social policy agenda, as sketched in an unfinished and preliminary part of *Grundsätze der*

8 As Albert O. HIRSCHMAN (1991 p. 111) once observed with regard to Hayek's *The Road to Serfdom*, "today's neoconservatives would be shocked on rereading it, for Hayek goes surprisingly far in endorsing what was later to be called the Welfare State".

Wirtschaftspolitik (1952/90 pp. 312-324), remained relatively modest. Instead, he argued times and again that the best social policy is the introduction or maintenance of the competitive order. What remains, however, is a major conceptual difference between Hayek and Eucken. This difference stems from the fact that Eucken, even though he did not see a need for an expansive social policy, basically adopted an outcome-oriented perspective on social policy. This strongly contrasts with Hayek's rebuttal of any notion of social justice and his rejection of outcome-oriented state interventions (Hayek 1976/82). "Social", in the eyes of Hayek, is a code-word for a universal hierarchy of values – that would be in conflict with individual freedom – and for the task of society as a subject – which does not exist – to implement it.

A third field of economic policy which highlights the differences between Eucken and Hayek is monetary policy. Both converge in their fierce opposition to monetary discretion and Keynesian-style managed money. Yet this shared preference for "automatic", de-politicized money is expressed in different ways. In order to bind monetary policy, the late Eucken subscribed to a proposal made by Hayek in the 1940s (Hayek 1943) and favoured a commodity-based kind of currency board as a substitute for the gold standard (Eucken 1952/90 pp. 255-264; Wagener 1996). Yet Eucken was not only concerned with discretionary behaviour of the monetary authorities. In addition, he feared that price stability might be endangered by the autonomous creation of money through commercial banks and argued for a 100 per cent reserve requirement in order to control bank credit supply. This proposal again documents Eucken's limited belief in the functioning of markets. Needless to say, this distrust in markets marks an important difference from Hayek who, in the 1970s, went so far to advocate the privatization of money through free banking and the denationalization of money (Hayek 1976).

At the end of the day, both authors were convinced that any deviation from a pure market system would start a vicious circle ending in perfect planning. In Hayek the conviction can be traced back to Mises (1922) who spoke of all government interventions into the economy as infringement of the state called economic policy. *The Road to Serfdom* is exactly this vicious circle which implies that any half-way position of welfare state or social market economy is inherently unstable (see Chaloupek 1998). The mechanism generating the instability is to be found in the recurrent elections where any "underprivileged" group will successfully press for equal treatment. Precisely the same idea of the instability of intermediate systems can be found in Eucken: "Has central administration of the economic process been introduced in *one* point, it has the tendency to expand" (Eucken 1952/1990 p. 154; original emphasis, our translation). There are two mechanisms propelling this instability. The first, which was already Mises's idea, is due to the complex interdependence of the economy: an intervention at one point will have besides its intended effects those which were unintentional as well. To compensate for these, the next intervention becomes necessary. The second is a lust for power which can be satisfied by extending the scope of power. At least in the case of Hayek, the stubbornness with which this impossibility theorem is upheld against all evidence gives rise to the conjecture that it is meant to serve as a memento for policy-making: *principiis obsta*!

IV. Cognitive Delusions or Elitist Illusions? The State and the Political Preconditions for a Functioning Market Economy

The differing views on the role of the state are also associated with a disagreement over the "political preconditions" (Eucken 1932/97 p. 19) for a functioning market economy. As a matter of fact, both Hayek and Eucken emphasize the close relationship between the economic and the political system or the "interdependence of orders", as Eucken called it (1952/90 pp. 332-334). Both warned against the danger that economic planning poses in potentially undermining freedom and democracy. This is the main message of Hayek's famous book *The Road to Serfdom* (Hayek 1944) which is approvingly cited by Eucken (1952/90 p. 334 n. 1). Both Hayek and Eucken have also voiced concerns about the negative effects of democracy on the market economy and have argued strongly for constraining democracy. At the same time, however, their emphases differ. Whereas Hayek's focus is on the dangers of majoritarianism, Eucken is concerned with the weakening of the state by interest groups. To put it differently: whereas Hayek primarily seeks to rule out "bad" state interventions, Eucken aims at making "good" ones possible. These differences reflect different political predilections, as well as differing views on the role of the state in a market economy.

Hayek basically takes a classical liberal position in that his main concern is with the tension between constitutionalism and democracy.[9] While he

9 For this theme, see the contributions in Elster/Slagstad 1988.

acknowledges a number of virtues of democracy such as peaceful change of government, popular education, protection against arbitrary rule, he fears that unconstrained democracy might result in an infringement of individual freedom and the spontaneous order of the market (Hayek 1960/71 chap. 7). For Hayek, the primary danger stems from popular majorities misled by false ideas, much more than from a majority coalition of the poorer strata of society or from rent-seeking activities by interest groups.[10] In line with classical liberalism, Hayek's constitutional proposals, including his famous proposal for a bicameral legislature, basically aim at "dethroning politics" by restricting popular sovereignty and the domain of majority rule (Bellamy 1994). Hayek's "negative constitutionalism" (Holmes 1995 pp. 135, 238) documents his firm commitment to liberalism and the impact of Anglo-Saxon thinking. His political-constitutional program is closely intertwined with his views on the role of the state, as it clearly documents his conviction of a largely negative role of state interventions.

Hayek's political concerns also reflect his deeply felt belief in the power of ideas. For him, economic policy was governed by ideas rather than by interests.[11] This is why he so often refers to Keynes's famous remarks on the impact of ideas on policy-making at the very end of *The General Theory* (Keynes 1936/73 pp. 383-384), and it also explains Hayek's almost Gramscian obsession with intellectuals and their influence on society (Hayek 1949/97). At the same time, this emphasis on ideas rather than interests, his belief in "the Keynesian cliché" (Barry 1984), sets Hayek apart from most of positive Public Choice theory (Peacock 1992 pp. 59-60).

Compared to Hayek, Eucken is much more concerned with the power of interest groups and what he describes as the "neo-feudal erosion of state authority" (Eucken 1952/90 p. 334). His main fear, closely related to his concern with private power, is that interest groups, in their struggle for privileges, will undermine the competitive order. Similar to that of other ordoliberals, Eucken's position is hence characterized by a strong sentiment against pluralism and organized interests. This "anti-pluralist liberalism" (Megay 1970) goes hand in hand with the call for a "strong" state capable of

10 It should be noted, though, that the late Hayek is increasingly concerned with organized interest groups (see e.g. HAYEK 1979 pp. 143-145)

11 When Schumpeter, in his review of *The Road to Serfdom*, aptly observed that Hayek "hardly ever attributes to opponents anything beyond intellectual error" (SCHUMPETER 1946 p. 269), Hayek responded by stressing that this was not "politeness to a fault" (as Schumpeter had suggested), but in fact his "profound conviction" (HAYEK 1949/97 p. 227 n. 4).

imposing and guarding the market economy. This call features prominently in Eucken's famous 1932 article on "Staatliche Strukturwandlungen und die Krisis des Kapitalismus" (Eucken 1932/97) and is basically repeated in Eucken's post-war book on economic policy (Eucken 1952/90 pp. 325-338). When Eucken speaks of the mutual interdependence of the economic and the political order, he refers first and foremost to the complementarity between the competitive order and a strong state and not to the congruence between the market economy and democracy. For Eucken, a strong state represents as much a precondition for the creation and maintenance of a competitive order as a limited state agenda is a precondition for state strength.

Eucken's call for a strong state has been interpreted quite differently. Borchardt (1981 p. 45) and Kirchgässner (1988), to name but two authors, have been quick to identify Eucken's strong state with benevolent dictatorship of welfare economics and to point at Eucken's elitist and authoritarian leanings. Fischer (1993 p. 176) accused Eucken of "(neo)absolutist thinking" (Fischer 1993 p. 176; our translation). Haselbach (1991) adopts Hermann Heller's term "authoritarian liberalism", coined in an argument with Carl Schmitt (Heller 1933/71), to characterize the position taken by Eucken and other ordoliberals. In contrast, other authors, most notably Streit and Wohlgemuth (1999; 2000) have suggested a more sympathetic, constitutionalist reading of Eucken which elevates him to an advocate of democratic self-constraint and limited government.

It goes without saying that the different assessments partly flow from differences in the authors' own political preferences. In addition, there is a striking imbalance in Eucken's treatment of economic and political issues. While Eucken calls for comprehensive limits on the scope of state activity and deals extensively with the economic constitution, he is conspicuously silent on the design of political institutions (Dörge 1955 p. 92; Friedrich 1955 p. 518; Vanberg 1988 p. 24). Unlike Hayek, he never came up with detailed institutional suggestions, let alone a model constitution. This silence renders it difficult to distil Eucken's exact position. However, a historical approach favours the "authoritarian" interpretation. Eucken's lament for the capturing of the state by interest groups, modern as it may sound, clearly echoes prominent themes of anti-democratic thinking in the late Weimar Republic. In particular, a parallel reading suggests the deep influence of Carl Schmitt on Eucken (Haselbach 1991). Eucken's writings also document strong elitist inclinations. The latter come out nicely in a passage of the Ordo Manifesto, one of the founding documents of Ordoliberalism published by Franz Böhm, Eucken and Hans Grossmann-Doerth in 1936 (Böhm *et al.* 1936/89 p. 15):

"Men of science, by virtue of their profession and position being independent of economic interests, are the only objective, independent advisers capable of providing true insight into the intricate interrelationships of economic activity, and therefore also providing the basis upon which economic judgements can be made."

Remarks such as these remind one of Plato and Keynes and smack of what Roy Harrod (1951 pp. 192-193) once famously dubbed "the presuppositions of Harvey Road":

"Keynes tended till the end to think of the really important decisions being reached by a small group of intelligent people, like the group that fashioned the Bretton Woods plan."

Harrod, then, immediately addresses the apparent dilemma,

"how to reconcile the functioning of a planning and interfering democracy with the requirement that in the last resort the best considered judgement should prevail. It may be that the presuppositions of Harvey Road were so much of a second nature to Keynes that he did not give this dilemma the full consideration which it deserves."

Similar comments could be made about Eucken, too. Like Keynes, and only eight years his younger, he grew up in the house of a well-known university professor, the nobel-prize winner Rudolf Eucken – Harvey Road in Jena was called *Wildstraße*. Favoured by this background, Eucken shared with Keynes a deeply felt belief in the intelligence of experts and a detached attitude towards mass democracy. Eucken's and Keynes's confidence in expert judgement set both apart from Hayek. Although Hayek had an academic background, too, and strongly believed in the power of ideas, his epistemic scepticism led him to question the intelligence of Eucken's men of science and Keynes's small group of intelligent people.

V. Concluding Comparative Assessment

So far, we have confined ourselves to contrasting the different positions Hayek and Eucken have taken, but have largely refrained from discussing their relative merits and from awarding marks. This is not, in fact, the place to engage in a comprehensive discussion of the two approaches. We do think, however, that, by contrasting Hayek and Eucken, our analysis has indirectly illuminated the strengths and weaknesses of both authors.

Hayek and Eucken are both "extremists" who sharpen and stretch arguments, thereby pushing them to their very limit. It is precisely this radicalism which makes for their intellectual appeal. The other side of the coin, however, is a certain overshooting and one-sidedness. Contrasting Hayek and Eucken, in a way, unearths the respective exaggerations. To cite but a few examples, Hayek's theory of markets highlights Eucken's constructivism, just as Eucken's analysis of the legal–institutional preconditions for functioning markets undermines Hayek's faith in spontaneous orders. Hayek's plea for organic evolution corrects Eucken's emphasis on pragmatic-constructivist design and vice versa. Hayek's formal criteria for assessing state interventions point at dubious substantive and outcome-oriented elements in Eucken, whereas Eucken's analysis indirectly reveals the vagueness of Hayek's standards. Thus, there is no point in playing off one against the other, but still a lot to learn from both.

References

BARRY, N.: "Ideas versus Interests: The Classical Liberal Dilemma", in: N. BARRY *et al.*: *Hayek's "Serfdom" Revisited. Essays by Economists, Philosophers and Political Scientists on "The Road to Serfdom" after 40 Years*, London (Institute of Economic Affairs), 1984, pp. 45-64.

BELLAMY, R.: "'Dethroning Politics': Liberalism, Constitutionalism, and Democracy in the Thought of F. A. Hayek", *British Journal of Political Science*, 24/4 (1994), pp. 419-441.

BÖHM, F, EUCKEN, W., GROSSMANN-DOERTH, H.: "The Ordo Manifesto of 1936", in: A. PEACOCK, H. WILLGERODT (Eds.): *Germany's Social Market Economy: Origins and Evolution*, London (Macmillan) 1989, pp. 15-26. First edition: 1936.

BORCHARDT, K.: "Die Konzeption der Sozialen Marktwirtschaft in heutiger Sicht", in: O. ISSING (Ed.): *Zukunftsprobleme der Sozialen Marktwirtschaft*, Berlin (Duncker & Humblot) 1981, pp. 33-53.

CHALOUPEK, G.: "Die Dauerhaftigkeit der `Zwischenform': Hayeks Verdikt gegen den Wohlfahrtsstaat nach fünfzig Jahren", in: F. BALTZAREK, F. BUTSCHEK, G. TICHY (Eds.): *Von der Theorie zur Wirtschaftspolitik - ein österreichischer Weg. Festschrift zum 65. Geburtstag von Erich W. Streissler*, Stuttgart (Lucius & Lucius) 1998, pp. 85-108.

DELHAES, K. v.: "Aktive Ordnungspolitik in der Transformation: Konstruktivismus oder Voraussetzung freiheitlicher Entwicklung?", *Ordo*, 44 (1993), pp. 307-318.

DÖRGE, F.-W.: "Menschenbild und Institution in der Idee des Wirtschaftsliberalismus bei A. Smith, L. v. Mises, W. Eucken und F. A. von Hayek", *Hamburger Jahrbuch für Wirtschafts- und Gesellschaftspolitik*, 4 (1959), pp. 82-99.

ELSTER, J., SLAGSTAD R. (Eds.): *Constitutionalism and Democracy*, Cambridge (Cambridge University Press) 1988.

EUCKEN, W.: "Staatliche Strukturwandlungen und die Krisis des Kapitalismus", *Ordo* 48 (1997), pp. 5-24. First published: 1932.

EUCKEN, W.: *Die Grundlagen der Nationalökonomie*, Berlin etc. (Springer) 1989. First edition: 1939.

EUCKEN, W.: *Grundsätze der Wirtschaftspolitik*, Tübingen (Mohr) 1990. First edition: 1952

FISCHER, T.: *Staat, Recht und Verfassung im Denken von Walter Eucken. Zu den staats- und rechtstheoretischen Grundlagen einer wirtschaftsordungspolitischen Konzeption*, Frankfurt am Main etc. (Lang) 1993.

FREI, C., NEF, R. (Eds.): *Contending with Hayek: On Liberalism, Spontaneous Order and the Post-Communist Societies in Transition*, Bern etc. (Lang) 1994.

FRIEDRICH, C. J.: "The Political Thought of Neo-Liberalism", *American Political Science Review*, 49/4 (1955), pp. 509-525.

GRAY, J.,: "Hayek on the Market Economy and the Limits of State Action", in: D. HELM (Ed.): *The Economic Borders of the State*, Oxford/New York (Oxford University Press) 1989, pp. 127-143.

GROSSEKETTLER, H. G: "On Designing an Economic Order: The Contributions of the Freiburg School", in: D. A. WALKER (Ed.): *Perspectives on the History of Economic Thought II: Twentieth-Century Economic Thought*, Aldershot (Elgar) 1989, pp. 38-84.

HARROD, R. F.: *The Life of John Maynard Keynes*, New York (Harcourt, Brace & Co.) 1951.

HASELBACH, D.: *Autoritärer Liberalismus und soziale Marktwirtschaft: Gesellschaft und Politik im Ordoliberalismus*, Baden-Baden (Nomos) 1991.

HAYEK, F. A.: "A Commodity Reserve Currency", *Economic Journal*, 53 (1943), pp. 176-184.

HAYEK, F. A.: *The Road to Serfdom*, London (Routledge and Kegan Paul) 1944.

HAYEK, F. A.: "'Free Enterprise' and Competitive Order", in: F. A. HAYEK: *Individualism and Economic Order*, Chicago (University of Chicago Press) 1948, pp. 107-118.

HAYEK, F. A.: "The Intellectuals and Socialism", in: B. CALDWELL (Ed.): *The Collected Works of F.A. Hayek, Vol. X: Socialism and War. Essays, Documents, Reviews*, Chicago/London (University of Chicago Press) 1997, pp. 221-237. First edition: 1949.

HAYEK, F. A.: *Die Verfassung der Freiheit*, Tübingen (Mohr) 1971. First edition: 1960.

HAYEK, F. A.: "The Results of Human Action but not of Human Design", in: F. A. HAYEK: *Studies in Philosophy, Politics and Economics*, Chicago (University of Chicago Press) 1967, pp. 96-105.

HAYEK, F. A.: *Law, Legislation and Liberty. Vol. 1: Rules and Order*, London (Routledge) 1982. First edition: 1973.

HAYEK, F. A.: "The Pretence of Knowledge", in: F. A. HAYEK: *New Studies in Philosophy, Politics, Economics and the History of Ideas*, Chicago (University of Chicago Press) 1975/78, pp. 23-34.

HAYEK, F. A.: *The Denationalization of Money. An Analysis of the Theory and Practice of Concurrent Currencies*, London (Institute of Economic Affairs) 1976.

HAYEK, F. A.: *Law, Legislation and Liberty Vol. 2: The Mirage of Social Justice*, London (Routledge) 1982. First edition: 1976.

HAYEK, F. A.: *Law, Legislation and Liberty Vol. 3: The Political Order of a Free People*, London (Routledge) 1979/82. First ed.: 1979.

HAYEK, F. A.: "The Rediscovery of Freedom: Personal Recollections", in: Peter G. KLEIN (Ed.): *The Collected Works of F. A. Hayek, Vol. IV: The Fortunes of Liberalism. Essays on Austrian Economics and the Ideal of Freedom*, Chicago-London (University of Chicago Press) 1992, pp. 185-195. First edition: 1983.

HAYEK, F. A.: *The Fatal Conceit: The Errors of Socialism*, London (Routledge) 1988.

HELLER, H.: "Autoritärer Liberalismus", in: H. HELLER: *Gesammelte Schriften*, vol. 2: *Recht, Staat und Macht*, Leiden (A. W. Sijthoff) 1971, pp. 643-653. First ed.: 1933.

HIRSCHMAN, A. O.: *The Rhetoric of Reaction. Perversity, Futility, Jeopardy*, Cambridge MA/London (Belknap) 1991.

HOLMES, S.: *Passions and Constraints. On the Theory of Liberal Democracy*, Chicago/London (University of Chicago Press) 1995.

HOPPE, H.-H.: "*F. A. Hayek on Government and Social Evolution: A Critique*", in: C. FREI, R. NEF (Eds.): *Contending with Hayek: On Liberalism, Spontaneous Order and the Post-Communist Societies in Transition*, Bern etc. (Lang) 1994, pp. 127-159.

KEYNES, J. M.: *The General Theory of Employment, Interest and Money. Collected Writings, Vol. 7*, London/Basingstoke (Macmillan) 1973. First edition: 1936.

KIRCHGÄSSNER, G.: "Wirtschaftspolitik und Politiksystem: Zur Kritik der traditionellen Ordnungstheorie aus der Sicht der Neuen Politischen Ökonomie", in: D. CASSEL, B.-T. RAMB, H. J. THIEME (Eds.): *Ordnungspolitik*, Munich (Vahlen) 1988, pp. 53-75.

KLEY, R.: *Hayek's Social and Political Thought*, Oxford (Clarendon Press) 1994.

LANGE-VON KULESSA, J., RENNER A.: "Die Soziale Marktwirtschaft Alfred Müller-Armacks und der Ordoliberalismus der Freiburger Schule: Zur Unvereinbarkeit zweier Staatsauffassungen", *Ordo*, 49 (1998), pp. 79-104.

LEIPOLD, H.: "Der Zusammenhang zwischen gewachsener und gesetzter Ordnung: Einige Lehren aus den postsozialistischen Reformerfahrungen", in: D. CASSEL (Ed.): *Institutionelle Probleme der Systemtransformation*, Berlin (Duncker & Humblot) 1997, pp. 43-68.

MEGAY, E. N.: "Anti-Pluralist Liberalism: The German Neoliberals", *Political Science Quarterly*, 85/3 (1970), pp. 422-442.

MISES, L. v.: *Die Gemeinwirtschaft*, Jena (Gustav Fischer) 1922.

PEACOCK, A.: *Public Choice Analysis in Historical Perspective*, Cambridge (Cambridge University Press) 1992.

RIESE, H.: "Ordnungsidee und Ordnungspolitik: Kritik einer wirtschaftspolitischen Konzeption", *Kyklos*, 25/1 (1972), pp. 24-48.

SCHUMPETER, J. A.: Review of Hayek 1944, *Journal of Political Economy*, 54/3 (1946), pp. 269-270.

STREIT, M. E.: *Freiburger Beiträge zur Ordnungsökonomik*, Tübingen (Mohr) 1995.

STREIT, M. E, WOHLGEMUTH, M.: "Walter Eucken and Friedrich A. von Hayek: Initiatoren der Ordnungsökonomik", *Diskussionbeitrag 11-99*, Jena (Max-Planck-Institut zur Erforschung von Wirtschaftssystemen) 1999.

STREIT, M. E, WOHLGEMUTH, M.: "The Market Economy and the State: Hayekian and Ordoliberal Conceptions", in: P. KOSLOWSKI (Ed.): *Economic Ethics and the Theory of Capitalism in the German Tradition of Economics*, Berlin etc. (Springer) forthcoming 2000.

VANBERG, V.: "'Ordnungstheorie' as Constitutional Economics: The German Conception of a 'Social Market Economy'", *Ordo*, 39 (1988), pp. 17-31.

VOIGT, S.: "Der Weg zur Freiheit. Mögliche Implikationen Hayekscher Hypothesen für die Transformation der Wirtschaftssysteme Mittel- und Osteuropas", in: J. HÖLSCHER, A. JACOBSEN, H. TOMANN, H. WEISFELD (Eds.): *Bedingungen ökonomischer Entwicklung in Zentralosteuropa, Bd. 2: Wirtschaftliche Entwicklung und institutioneller Wandel*, Marburg (Metropolis) 1994, pp. 63-105.

VOLKERT, J.: "Sozialpolitik und Wettbewerbsordnung: Die Bedeutung der wirtschafts- und sozialpolitischen Konzeption Walter Euckens für ein geordnetes sozialpolitisches System der Gegenwart", *Ordo*, 42 (1991), pp. 91-114.

WAGENER, H.-J. (1992a): "Wieweit ist Systemtransformation planbar?", in: H. ALBECK (Ed.): *Wirtschaftsordnung und Geldverfassung. Symposion zum 65. Geburtstag für Norbert Kloten*, Göttingen (Vandenhoeck & Ruprecht) 1992, pp. 99-115.

WAGENER, H.-J. (1992b): "Pragmatic and Organic Change of Socio-Economic Institutions", in: B. DALLAGO, H. BREZINSKI, W. ANDREFF (Eds.): *Convergence and System Change: The Convergence Hypothesis in the Light of Transition in Eastern Europe*, Dartmouth (Aldershot etc.) 1992, pp. 17-37.

WAGENER, H.-J.: "Stabiles Geld oder neutrales Geld: Eine ordnungspolitische Diskussion", in: P. BOFINGER, K.-H. KETTERER (Eds.): *Neuere Entwicklungen in der Geldtheorie und Geldpolitik. Festschrift für Norbert Kloten*, Tübingen (Mohr) 1996, pp. 71-90.

WILLGERODT, H.: "Die Liberalen und ihr Staat: Gesellschaftspolitik zwischen Laissez-faire und Diktatur", *Ordo*, 49 (1998), pp. 43-78.

WOLL, A.: "Freiheit durch Ordnung: Die gesellschaftspolitische Leitidee im Denken von Walter Eucken und Friedrich A. von Hayek", *Ordo*, 40 (1989), pp. 87-97.

Chapter 7

The Present Relevance of Ordnungstheorie for the Politics and the Economics of the Social Order

MICHAEL WOHLGEMUTH*

I. Introduction: Germany's "Social Market Economy" Fifty Years On

Are *Ordnungstheorie* and *Ordnungspolitik* together with their German trademark – the *Soziale Marktwirtschaft* – still relevant today? The year 1998 saw a number of solemn celebrations of the 50th birthday of the *Soziale Marktwirtschaft*. Politicians from all major parties will come to praise

* I wish to thank Thoralf Erb, Thráinn Eggertsson, Werner Mussler, Pavel Pelikan, Axel D. Schulz and Stefan Voigt for valuable criticism and suggestions including some diverging views. I also want to thank Thomas Welsch for compiling the data and producing the figures.

Ludwig Erhard and the enactment of *Ordnungspolitik* which is today regarded as a major cause of the so-called German *Wirtschaftswunder* of the 1950s. Some may even remember that social scientists like Walter Eucken, Franz Böhm or Alfred Müller-Armack have made their contributions. Thus, for an outside observer the answer to the initial question seems evident: yes, the basic tenets of *Ordnungstheorie* and *Ordnungspolitik* "made in Germany" must be highly effective today; they seem to have become a sturdy German standard. And indeed, there is no longer any serious political opposition to the basic idea of a social market economy – especially in Germany. The only serious opponent, however, is the political and social reality – especially in Germany. Several indicators can support this account of what has become of ordoliberalism in its country of origin:

- The German national accounts show an almost uninterrupted rise in the government expenditure share of GDP at the expense of the market economy share. Between 1950 and 1962 less than 35% of national income was in one way or another directed by agencies of the state and of social security, but since 1975 it has remained at about 50% (see Graph 7.1 in the appendix). The increase is mainly due to a sharp increase in the "social budget" which in 1960 accounted for 21.7% of GDP and in 1995 amounted to more than 34%. These figures could be summarized to the effect that the "social market economy" is becoming synonymous with what might be called a "social state economy".
- This development, however, is not reflected in the opinion of the "man on the street". According to a recent Allensbach poll, 54% of Germans today have the impression that the "market" element is clearly dominating the "social" element in the German "social market economy". Only 10% seem to realize that the "market share" of the system has actually been on the retreat for quite some time (*Wirtschaftswoche* 4/1997).[1]
- Of course, the common impression that the social situation in Germany has deteriorated cannot be countered by pointing to the increase in the social budget. Quite the contrary: the feeling of a receding "social element" can be corroborated with dramatic increases of unemployment. During the 1960s and early 1970s well below 2% of the labour force was unemployed, but proportion has now reached alarming heights at about 13% (Graph 7.2).

1 A similar result is shown in a very recent poll, in which only 23% out of 1,165 Germans questioned agreed that in Germany today market competition and social equilibration are sucessfully combined. A majority (58.8%) feels that the "social" element is neglected. In the new *Länder*, this proportion even reaches 70.5% (see FIO 1998).

- These diverse symptoms seem to allow for only one conclusive diagnosis: the social market economy *as a whole* is in a state of sickness and disorder. This verdict can be further illustrated by another part of the Allensbach poll which shows that Germany's economic order has lost the approval of many Germans during recent years: in 1994 53% of Germans expressed a "favourable opinion" of their economic system; in 1996 this figure dropped to only 40%. The most dramatic decline of declared support can be observed in the *Neue Länder*. Here, these questions were already asked in 1990. And here, the proportion of favourable opinions of the new system has collapsed from the peak of 77% to only 24% in 1996, while a relative majority of 41% of East Germans now declares they "have no good opinion" of the German economic system (see Graph 7.3).

Thus, the present state of the social market economy in Germany does not look healthy. Not only has the market element of the "social market economy" been in almost constantly decline – the feeling of social stability which is intimately related to the threat of lasting unemployment also shows signs of decay. Neither the market nor the "social" component are actually dominant. Both of the components, so ordoliberals would argue, have more and more been superseded by state interventions in the name of the so-called *Sozial* (social issues) which have, however, been quite ineffective in reaching their very goals.

- It cannot be argued that these developments are parts of an overall international trend. The "German disease" is home-made and not contagious. This can be illustrated by pointing to another telling indicator which shows Germany to be falling behind a growing number of countries in terms of economic freedom. In the newest annual report on "Economic Freedom of the World" (Gwartney/Lawson 1997), Germany is now ranked 25th (tied with Argentina, Bolivia and Chile, see Graph 7.4). This "world league of economic freedom" tries to capture "the extent to which individuals are free to choose for themselves and engage in voluntary transactions with others, and have their rightly acquired property protected from invasions by others" (ibid. p. 2). It consists of 17 components mainly based on objective indicators like inflation, government consumption, tax rates and so on. Unfortunately, we have no data from the early years of German *Ordnungspolitik*. Gwartney and Lawson's figures indicate, however, that from 1975 until 1990 Germany remained among the 15 most market-oriented and open societies in the world. Its falling to 25th in 1995 is mainly due to other countries' successful liberalization policies. At the same time, Germany remained – especially in the areas conducive to its high rate of government involvement – stuck in political inertia that is discussed under labels such headings as *Reformstau* ("reform jam": e.g. Mummert/ Wohlgemuth 1998) or *Politik-verflechtungsfalle* ("joint decision trap": Scharpf 1998).[2]

Before we go on to discuss the intellectual and political relevance of *Ordnungstheorie*, we have to make clear what we mean by this term. To begin with, there is no sole paradigm and school of *Ordnungstheorie*. Rather, there is a general and wide research programme based on various German traditions which are each associated with different conceptions of the market economy and its politics. For example, there are good reasons to distinguish between a Freiburg and a Cologne mode of (ordo)liberal thought (see also Sally 1996 p. 248 ff.; Vanberg 1988 pp. 20 ff.). Even if both "schools" had (and still have) largely unofficial and overlapping memberships, the former can be said to be more sceptical towards attempts to combine individual freedom on markets with social balance through government intervention, as highlighted in the politically effective term of a "social market economy" (Müller-Armack 1956). We will come back to this major difference. At any rate, it would be wrong to equate any ordoliberal conception with the *practice* of the "social market economy". To make such differences as clear as possible (while trying to avoid exaggeration), we will concentrate our presentation of *Ordnungstheorie* on Walter Eucken and Franz Böhm, founding fathers of the "Freiburg school of Law and Economics" (Streit 1994).

II. Historical and Intellectual Roots of Ordnungstheorie

As we will argue later, *Ordnungstheorie* failed to become an internationally successful (or even known) research programme. Some reasons for this might go back to the peculiar origins of German *Ordnungsdenken* and the fact that *Ordnungstheorie* was inspired by a

2 In a similar vein, critical observers have come to the conclusion that the social market economy has experienced a "decay of style" (Streit 1998 p. 52). The original concept, they argue, has become "a phaseout model" in its country of origin while, at least to some extent, it has become an "export model" from the perspective of some of the core principles of the EU's economic constitution and of some eastern European countries in transformation (see Streit 1997).

political and intellectual atmosphere which to an important degree reflected singular experiences of German history.

Politically, two experiences were dominant: (a) the highly cartelized and regulated German economic order together with a "weak" state that was captured by vested interests during the Weimar republic, and (b) the destruction of civil liberties and the rule of law during the Nazi regime. The traumatic experience of both interventionism and collectivism created the ordoliberals' programmatic interest in active economic policy. The task of reorganizing the institutional structure and relationship of state and society as a whole was a responsibility deeply felt by German ordoliberals who had already, during the Second World War, secretly discussed the principles of a new institutional order to be established immediately after the breakdown of Nazi-Germany. Their programme reflects the understanding that a mutual interpenetration of state and society had been the recurrent cause of German "roads to serfdom". Thus, when the Supreme Court of the German empire permitted cartel agreements as legal expressions of the freedom of contract (in 1897), ordoliberals later concluded that it incidentally helped create "powerful interest groups and a new dependence of workers on their employers, of consumers on monopolists and of retailers on combines and cartels" (Watrin 1998 p. 18). This experience might also help to explain the ordoliberals' scepticism towards invisible-hand explanations of institutional change such as were most notably taken up by Hayek in his theory of the spontaneous order. We shall return to these distinctions.

Intellectually, the task (as the ordoliberals saw it) was to overcome the antagonism between two methodological tendencies that dominated German social sciences: (a) the Historical school, in so far as it represented a strict avoidance of or even hostility towards theoretical reasoning and abstractions; and (b) the marginalist school, in so far as it represented purely abstract modelling, without taking historical experiences or institutional preconditions of the economic system into account.

Within the Freiburg school, especially Eucken (e.g. 1940/92 pp. 34 ff.) struggled to overcome this "great antinomy" that found its expression in the acrimonious *Methodenstreit* between Gustav Schmoller and Carl Menger. It induced him to develop a methodology of "isolating abstraction" which starts with the observation of historical cases and ends with classifications of theoretical "ideal-types".[3] Some of those ideal-types (such as Eucken's "complete competition", Böhm's "private law society" or Müller-Armack's "social market economy") not only describe purely theoretical conceptions, but also normative positions seeking approval and political realization. And indeed, it is a characteristic trait of traditional *Ordnungstheorie* that it is often impossible to see any clear-cut distinction between its positive-theoretical and its normative-political components. The fact that this was openly admitted and even demanded by some ordoliberals has been the cause of great deal of suspicion within the scientific community right up to the present day. We will have to come back to this discussion.[4]

III. The Central Messages of Ordnungstheorie and Ordnungspolitik

The central messages can be summarized as follows: (1) it is the central task of legal design backed by a "strong" *and* constitutionally limited state to create and cultivate an institutional framework that makes possible an order of decentralized coordination among free individuals. (2) The main problem to be addressed by *Ordnungspolitik* is the problem of power in its two dimensions: (a) the economic power of cartels and monopolies as well as (b)

3 Eucken's method of "isolating abstraction" (1940/92 p. 107) refers to the phenomenologist philosophy of Husserl who developed the principle of "phenomenological reduction". Instead of "generalizing abstraction", which seeks to identify common traits of different phenomena, Eucken argued that one should identify recurrent elementary forms in economic life. This is how he went about isolating, e.g., the centrally directed and the exchange economy as conceptual answers to the following question: "Does one central authority direct everyday life, or do countless single individuals make their own decisions?" (ibid. p. 81).

4 Compared to the interdisciplinary programme of the Freiburg school, Müller-Armack's intellectual programme was even more ambitious. Whereas the former relied mainly on a division of labour between economists and jurists combining their efforts in the solution of particular problems of economic policy, Müller-Armack takes an encompassing sociological and philosophical view of the society as a whole. Basically, he tried to reconcile and even unify almost all different forms of *Weltanschauung*, ideologies and religious denominations into one formula or *Sozialidee* according to his concept of social peace-making, or *Soziale Irenik* (1950/81), as he called it.

the political power of governments and special interest groups. (3) The main solution to the problem of the economic order is the creation and strict political defence of competition as a major source of economic progress and as a device to curb positions of economic power (*Wettbewerbsordnung*). (4) The main solution to the problem of political power is the creation of legal commitments based on the idea of the rule of law (*Rechtsstaatlichkeit*). These basic ideas will now be looked at individually.

The theoretical and political programme of the Freiburg school can be described in the words of Böhm (1957/60, p.2) as follows:

"The question that preoccupied us all, was ... the question of private power in a free society. This leads by necessity to the further question of what an order of a free economy is made of. From there one arrives at the question, what kinds and possibilities of an economic order are at all feasible, what role is played by power in each, in fact the power of government as well as the power of private persons and groups, and what obstructions of order arise if a distribution of power emerges within state and society which differs from that which conforms to the respective economic system."

Note that these questions referring to "power" and "order" as the key-notions of ordoliberalism differ from typical economic concepts as well as typical legal concepts. For the Freiburg school, they define the fields of common theoretical interest and political concern which centre around the basic concept of the competitive order (*Wettbewerbsordnung*). The defining concern of ordoliberalism was to establish "order" as a set of legal rules for a society of essentially self-reliant decision makers whose actions are controlled and co-ordinated by market competition. It is a key message of the Freiburg tradition that private (market) power not only reduces the freedom of the many in favour of domination by the few in the economic system, but that it also penetrates and impairs the political system. We now turn to the first, and, according to ordoliberalism, original, aspect of the problem of social power.

1. Ordoliberals and the market: The ideal-type, *Wettbewerbsordnung*

This question of "how an order of a free economy is constituted" considered above was understood by Böhm as a problem of an adequate legal order. Free market exchange requires a system of rules which define a protected domain of private action (property rights) and allow cooperation among equals (private contracts) where they seek it and settlement of conflicts (arbitration) where they need it. A society that builds on these principles is justly called a private law society (*Privatrechtsgesellschaft*: Böhm 1966/89). However, the basic provisions of private law still provide no complete answer to the question of private power. This was a central lesson drawn from German history: private attempts to close markets (for example, through the formation of cartels) were considered legitimate uses of the freedom of contract, and boycotts or collective discrimination applied against outsiders received support from the courts. As a consequence the freedom to compete of third parties was reduced, while economic power was now based on vested rights within the body of private law. Therefore, ordoliberals insisted on supplementing the private law society with an institutional guarantee of open markets in order to ensure that market competition could display its central function as "the most genial instrument of emasculating power" (Böhm 1961 p. 22).

This is the most important and original contribution of the Freiburg school to a political economy of the market: competition is not only (and not even primarily) regarded as a means to achieve "economic" goals like growth and efficiency. It is mainly advocated as a procedure to curb the power of economic agents and organizations. Economic power, in turn, is regarded as evil not only because it cripples the price mechanism and its allocative potential, but also, and primarily, because it allows infringements on the liberty of others which is regarded as the fundamental precondition of moral behaviour. In accordance with this broader view of competition, ordoliberal economists and – above all – legal scholars developed a broader and more articulate assessment of the preconditions of competition than the neoclassical formalists of their time. Where the latter employed rules of decision logic and equilibrium conditions to deduce abstract results, ordoliberals emphasized legal conditions and rules that create and structure real market processes. We will come back to these characteristic differences.

The institutional framework of the competitive order is described by the famous "principles" laid down by Eucken (1952/90 pp. 254 ff.). As "fundamental principle" the creation of a workable price system is postulated. As "constituent principles", conducive to a well-functioning price system, he enumerates the stability of the monetary system, open markets, private property, freedom to contract, liability for one's commitments and actions, and steadiness of economic policy. The necessity for political intervention that might still arise should follow certain "regulating principles", such as monopoly control, income policies and correction for technological external effects (ibid. pp. 291 ff.). All these principles, of course, have no addressee other than government because "a private law society cannot function without authority ... it requires a support, which it cannot produce from within its own resources, in order to function at all" (Böhm 1966/89, p. 51). By conferring such authority upon government, a second source of power necessarily emerges: political power.

2. Ordoliberals and the state: The ideal-type, *Rechtsstaat*

As we have seen, ordoliberals assigned to competition a role which went beyond its economic importance. It deserved protection by the law mainly because of its capacity to curb lasting power relations in society. In addition, active competition policy was considered a genuine task of the state. This implied, of course, that the state had to refrain from activities which could restrain competition. Considered from this point of view, it seems rather a matter of course for ordoliberals to advocate a strong but limited government. It must be strong in order to hold out against monopolies and pressure groups, thus safeguarding the "economic constitution" (*Wirtschaftsverfassung*). At the same time, it must be limited to pursue only this genuine task and use only liberty- and market- compatible means. Thus, it became the double task of the political constitution (1) to grant independence to those who are entrusted and legitimized to make laws and to govern, thus allowing for political neutrality or "strength" against economic power groups; and (2) to provide a sophisticated combination of checks and balances, thus preventing and limiting an arbitrary use of political coercive power.

The belief that a strong and limited state is no contradiction in terms, that political authority and constitutional limitations actually complement each other, became a central tenet of ordoliberalism. That an unlimited state is always in danger of being weakened and corrupted by economic power groups was a central conclusion ordoliberals drew from German history, especially from the period of "experimental economic policies". Eucken (1932 p. 307) observed that the "expansion of government activities ... not at all meant a strengthening, but to the contrary, a weakening of the state". The corruption of the political order and unreliability of economic institutions combined with politicians' subsequent dependence upon economic power groups is a prominent example of the general idea of an "interdependence" of the legal, political and economic order (e.g. Eucken 1952/90 pp. 332 ff.).

Together, these analyses of the collusion of public and private power in cartel-like, corporativistic arrangements, or the "capturing" of the state by vested interests (Böhm 1950 p. xxxvi) can still today serve as rather apt descriptions of the weaknesses of modern welfare states.

It should be evident that this conception of the *Rechtsstaat* is a far cry from the image of a "benevolent dictator" that still looms large in welfare economics. Eucken (1952/90 p. 338) made this unmistakably clear: "It is wrong to see the existing state as an all-knowing, all-powerful guardian of all economic activity. But it is also incorrect to accept the existing state which is corrupted by interest groups as irreversibly given and consequently to despair of mastering the problem of building a proper political-economic order." Eucken, and no less Böhm, evidently opposed unlimited government; they clearly exposed the institutional framework of what later became known as the "rent-seeking society" (see Tollison 1982); and they proposed institutional precautions to prevent the "wild refeudalization of society" (Böhm 1958/80 p. 258) which they witnessed in German history (see also Eucken 1932; Böhm 1933). At best, therefore, they can be "criticized" for not having produced an "economic theory of politics" which was only later to become a theory of public choice. We return to this line of argument.

3. The Cologne twist: *Soziale Marktwirtschaft*

Besides the Freiburg school, a somewhat different tradition of *Ordnungstheorie* which might be labelled the "Cologne school" can be discerned. Reference must be made to one scholar as the leading and founding figure, Alfred Müller-Armack. Like the ordoliberals, he put much emphasis on the political, legal and cultural conditions of the economic order which he tried to integrate into a concept of "totality of the economic order" (*Wirtschaftliche Gesamtordnung*). This was meant more or less explicitly to represent a "liberal analogue to the idea of political economy developed in the Marxist tradition" (Koslowski 1998 p. 78). The guiding idea for Müller-Armack (1956 p. 390) was "to combine the principle of freedom on the market with that of social balance" ("das Prinzip der Freiheit auf dem Markte mit dem des sozialen Ausgleichs zu verbinden"). It was developed as "a dialectical concept, in which social goals are just as important as economic goals" (1965/98 p. 258). To make sure that "social equilibration" was not pursued in a way that would violate the "principle of the market", members of the "Cologne school" laboured with various definitions of "conformity" principles which were to operate as guidelines for selecting appropriate instruments from the politicians' tool-kit and for framing the judgements of German courts. To be sure, as we will see, neither of these addressees was willing to subscribe to any such conformity rule and none of these rules made their way into German public law or even the German *Grundgesetz*.[5] Thus, the idea to harmonize market freedom and social balance could not successfully restrain actual German politics. Rather, some of Müller-Armack's (e.g. 1948/81 pp. 100 ff.) declared objectives of social and economic policy (such as the codetermination of workers, progressive taxation, social transfers, minimum wages, subsidies for small and medium-sized firms, subsidies for housing, or business-cycle stabilization) were liable to be taken up by interest groups and vote-seeking politicians – without any concern for the abstract principle of "free markets". Not surprisingly, the resulting policies revealed innate tensions between the self-coordinating character of markets within the "private law society" and attempts to (re)direct this spontaneous order with the use of politically centralizable knowledge and of politically controllable means of public law.[6] We will come back to these problems.

5 As it was, the Supreme Court made it clear at a very early date that the German constitution must be regarded as "neutral" regarding a specific economic order, and that a social market economy was just one of many conceptions in conformity with constitutional basic rights and fundamental principles (see also NÖRR 1998).

6 There is at least a difference in emphasis between the idea of "social irenics" of Müller-Armack and the account of the "social question" given by ordoliberals: for Eucken social justice was not a countervailing political target *vis-à-vis* a competitive order - it was rather its most natural consequence. Hence, he argues (1952/90 p. 317) that "social justice should be produced ... mainly through submitting the creation of incomes to the strict rules of competition, risk, and liability". Another difference lies in the realization of legal problems created by the tension between a liberal *Rechtsstaat* or protective state on the one hand and a politicized welfare state (*Sozialstaat*) on the other, as highlighted e.g. in the analysis of BÖHM (1953/60).

IV. The Intellectual Relevance of Ordnungstheorie

1. *Ordnungstheorie* today: Old traditions and new combinations in German academia

As a research programme, *Ordnungstheorie* can be characterized as the attempt to create empirically sound knowledge about the preconditions and working properties of economic orders that can be translated into political principles, the adherence to which is expected to support an order of a free people. Methodologically, this implies somehow overcoming what Eucken at an early date (1940/92 pp. 34 ff.) described as the "great antinomy" of the social sciences which there then was between "individual-historical" approaches stressing real institutional circumstances on the one hand and "general-theoretical" approaches that relied on simplifying abstractions from reality on the other. At least within the economics profession, this old *Methodenstreit* has long since been settled, producing a clear predominance for abstract modelling. Neoclassical orthodoxy is now the international standard. With it, however, the meaning of social institutions as preconditions of market systems, the overall view of social orders and their structural interdependence, as well as the claim to be politically relevant and generally comprehensible has often vanished (see e.g. Albert 1998; Blaug 1998; Boettke 1997).

As a consequence, *Ordnungstheorie* can be argued to have become as academically uninfluential as never before. It has not held out against an ever-growing mainstream which is neoclassical, formal "tight prior equilibrium" economics. This mainstream consists of various forms of "social mathematics" (Blaug 1998 p. 3) which basically rely on the rational choice paradigm applied to equilibrium settings, be it in general equilibrium theories, game theories or rational expectations macroeconomics. *Ordnungstheorie*, as it were, has had strictly no influence on these developments which have since the 1950s became international standards of economics and hence the major content of economic journals and textbooks.

On the other hand, however, there have been alternative currents within and outside the neoclassical mainstream which have taken up some issues that have long since been at the heart of the *Ordnungstheorie* enterprise. Especially prominent are the new currents which originated largely within the neoclassical mainstream and which can be summarized under the label of

"New Institutional Economics".[7] Most widely defined, this research programme embraces such different fields as economic theories of public (political) choice, property rights analysis, contractarian constitutional choice theories or transaction-cost theories of the firm, of the market and of historical institutional change. Although these theories mostly still rely on rational maximizing choices of individuals and on equilibrium settings, other elements of mainstream economics have been abandoned or changed.[8] The set of restrictions or objects of (individual or collective) choice now actually takes into account various kinds of social institutions. Furthermore, the concept of rationality is now often modified by the use of alternative assumptions such as "bounded rationality" (e.g. Simon 1978).

Not so much from within, but beside the mainstream a rather small brook of alternative theories has kept flowing which consists of various currents of (Neo-)Austrian Economics. Several of these currents are most prominently represented by Friedrich Hayek and have attracted not a vast number of academics but a respectable group of heterogeneous scholars interested in increasing the heritage of classical political economy. Following the "classical" or "Austrian" path, however, meant more rigorously abandoning basic assumptions of the neoclassical orthodoxy in order to arrive at useful assessments of the meaning of competitive processes, of the use of knowledge in society or of the institutional preconditions of prosperous societies (see below).

With the passing away of the first generation of ordoliberals both neoinstitutional and Austrian economics have become increasingly influential in shaping the course of *Ordnungstheorie*. The intellectual transfers, however, have been distressingly one-sided. While many proponents of *Ordnungstheorie* have been eager to learn and integrate the ideas of Hayek, Coase, Buchanan or North (to name just a few) and to combine these with German lessons in *Ordnungstheorie*, I know of no considerable export of German *Ordnungstheorie* across the Atlantic. Indeed,

7 For an instructive overview of New Institutional Economics, see EGGERTSSON (1990). The programmatic closeness of modern institutional economics and *Ordnungstheorie* has (to my knowledge) only been recognized by German scholars such as FELDMANN (1995), TIETZEL (1991) or SCHMIDTCHEN (1984).

8 Some pioneers of the New Institutional Economics have increasingly given the impression that what they were doing and what they wanted done in the future was much more building a "new construct" rather than simply an "extension of the neoclassical model" (see FURUBOTN 1994).

as a research programme, *Ordnungstheorie* not only failed to become internationally successful; it even failed to become internationally known – if one takes the approval and knowledge of American scholars (and their dominant organizations) as major indicators of international success. I leave open the question as to whether the fact that "institutional economics made in Germany" has not been successfully exported is due to deficiencies in the product or its academic marketing or whether it is due to entry barriers in the relevant American "market place of ideas".[9]

In its "home market", however, *Ordnungstheorie* has more-or-less retained its position not as a dominant area of research and teaching, but as one that is concomitant and generally accepted. Here, it continues to attract a core of scientists (mostly economists) devoted to various refinements and amendments within the research area of *Ordnungstheorie*. Within this core, (at least) two broad tendencies can be observed:

(1) traditional ordoliberal theories of competition and of social orders have been further developed under the influence of Austrian economics and especially of Hayekian concepts like "competition as a discovery procedure" and the "spontaneous order", but also preliminary approaches to "evolutionary economics" (see, for instance, the contributions of Erich Hoppmann, Manfred E. Streit, Jochen Röpke, Ulrich Fehl, Ulrich Witt).

(2) theories of economic and political systems have been supplemented and modified with modern property rights analysis and other branches of the New Institutional Economics, such as Constitutional Economics or Public Choice (see e.g. contributions of Alfred Schüller, Helmut Leipold, Hans-Jürgen Wagener, Viktor Vanberg, Manfred Tietzel, Roland Vaubel).

As a firmly established interdisciplinary group and politically dominant "think tank" however, ordoliberalism has even in Germany virtually ceased to exist. In the law faculties, ordoliberalism seems to have almost vanished (if it ever was influential); it only remains relevant in the work of a handful of scholars (like Christian Kirchner, Erich Mestmäcker, Wernhard Möschel, Knut W. Nörr). Along the lines sketched above, ordoliberal ideas are, however, still held in tender remembrance and/or used in various new research programmes within the economics profession in various universities, organizations and academic journals (see appendix).

9 For a critical examination of the political concept of a "market place of ideas", see WEISSBERG (1996); for an account of the dismal American market for economic research(ers), see BLAUG (1998).

I shall now briefly discuss some of the basic challenges to ordoliberal *Ordnungstheorie*. I do this by way of contrasting it with (1) Hayekian concepts and (2) economic theories of politics. The integration of these theories in new blends of *Ordnungsökonomik* has in some instances lead to rather mild changes in or supplements to the ordoliberal heritage, while in others it took the form of a (friendly) "take-over" of the old ordo-enterprise.

2. Hayekian and ordoliberal *Ordnungstheorie*[10]

Hayek, Eucken and Böhm developed conceptions of the market economy and of the state that have much more in common than is indicated by the number of explicit references to each others' work. Although they started from different angles and with different intellectual backgrounds, they very often met on common ground and arrived at quite similar policy conclusions. Admittedly, however, there is also a number of substantial differences. It is the main point of Streit/Wohlgemuth (1998) that most of these differences can be attributed to the ordoliberals' predominant occupation with the problem of private (economic) power and Hayek's preoccupation with the problem of private (subjective) knowledge. Three examples may serve to illustrate the point:

(1) Hayek's view of competition as a "discovery procedure" forced him to abandon conventional equilibrium theory, whereas the ordo-conception of competition as an "emasculating instrument" might have led some ordoliberals to stick to ideal-types that were not unlike, for that matter, perfect competition which is characterized by the absence of social power (all "actors" are price-takers), but which is also unable to account for any kind of entrepreneurial action. Such theoretical differences also had consequences for their respective views on competition policy. Thus, Eucken (1952/90 p. 295) argued that it was the aim of a governmental monopoly agency to act "as if there was complete competition", which means that "the price has to be fixed such that supply and demand reach their equilibrium and at the same time equals marginal costs" (ibid. p. 295). This was exactly the view that Hayek (1940/94) challenged in another context, when he attacked Oskar Lange's proposed market socialism as totally beside the point as a result of "an excessive preoccupation with problems of the pure theory of

10 A more detailed discussion of these matters can be found in STREIT/WOHLGEMUTH (1999).

stationary equilibrium" (ibid. p. 240). It was Hoppmann (e.g. 1967) who later developed the ordoliberal view of competition and competition policies along Hayekian lines. Further amendments were made, for example by Streit/ Wegner (1992), using transaction costs concepts and taking seriously the epistemological consequences of Hayek's (1968/78) idea of "competition as a discovery procedure".

(2) Hayek's concept of institutions as "an adaptation to our ignorance" (1976 p. 39) led most naturally to his evolutionary concept of the selection and expansion of social institutions and nurtured his scepticism regarding conscious institutional changes. By contrast, the ordo-view of institutions as instruments to reduce private power produced the task of "fashioning the legal instruments for an economic constitution", which itself has to be "understood as a general political decision as to how the economic life of the nation is to be structured" (Böhm *et al.* 1936/89 p. 24). Both conceptions of institutional change face methodological and empirical problems of their own which cannot be discussed here. Some differences can be illustrated with reference to a distinction between a "French" and a "British" view of the social order. For Hayek (1970/78 p. 3) the idea "that man 'created' his civilization and its institutions" and that therefore "he must also be able to alter them at will so as to satisfy his desires or wishes", commonly held in the French tradition of rational liberalism, was the expression of a most dangerous abuse of reason. In the British tradition of evolutionary liberalism, he took the opposite view that modern civilization with its fine-structured layers of customs, conventions and legal principles could only have grown as the unplanned result of human interaction and not of human design. These orders which have not been created at will can only be destroyed by attempts to redesign them on the basis of holistic blueprints of an ideal society. Of course, this does not imply that Hayek would have opposed the ordoliberal attempt to provide guiding principles of an economic constitution for Germany after the war. However, a Hayekian evolutionist must frown at parts of the underlying ordoliberal *Weltanschauung* and the extent of its faith in human intelligence and knowledge to design a new order for state and society.

(3) Hayek's concept of "catallaxy" leaves no room meaningfully to pose "the social question" in so far as it aims at distributive justice and his concept of "nomocracy" precludes political attempts to answer this question by employing any discriminatory means. Ordoliberals, in turn, cannot simply discard the social question once they define economic power also in material terms which require judgement to be passed on market results and the adoption of a somewhat instrumentalist approach towards the economic

constitution. To be sure, in terms of political pragmatism, Hayek and many ordoliberals are quite close, both favouring a collective protection against severe deprivation in the form of a guaranteed minimum income which should, as far as possible, be secured "outside of and supplementary to the market" (e.g. Hayek 1976 p. 87).[11] The difference in substance is at the theoretical stage. Hayek did not follow the tempting ordoliberal slogan that a well-functioning competitive market order would in itself be the pre-eminent "social" device or even a major precondition for the solution of the "social question" (Eucken 1952/90 p. 314). And he would certainly not follow Eucken in arguing that "the realization of social justice ... depends on the realization of the general principle of a competitive economy" (ibid. p. 315) if this implied a concept of material justice. In this case Hayek (1978 p. 140) would argue to the contrary that the realization of social justice required "a kind of order of society altogether different from that spontaneous order which will form itself if individuals are restrained only by general rules of just conduct".

Summing up, one can say that whereas for Hayek the problem of the division of knowledge permeates all his theory of the spontaneous order, this problem remained largely neglected by the early ordoliberals. Even if Hayek could subscribe to most of the liberal principles defining a *Wettbewerbsordnung*, his conception of "catallaxy" represents a fundamentally different research programme.

3. Ordoliberal political economy and positive economics of politics

In terms of pre-scientific interests (basic value judgements) and normative conclusions, the ordoliberal concept of an economic constitution and Hayek's "constitution of liberty" (1960) do not differ dramatically from modern constitutional economics as represented especially by Buchanan.[12]

---

11 Hayek's rejection of the German term and concept of "social market economy" as "most confusing and harmful" (HAYEK 1965/67 p. 83) or as "a real danger" and "camouflage for aspirations that certainly have nothing to do with the common interest" (HAYEK 1957/67 pp. 238 ff.) may be a critique of German politics rather than a critique of ordoliberal concepts. Nevertheless, even if only in a more modest form, the critique also applies to the latter.

12 For a programmatic delimitation of the domain of constitutional economics, see BUCHANAN (1990). The relation between ordoliberalism and constitutional economics is discussed by VANBERG (1988) and LEIPOLD (1990). The relation between Hayek and constitutional economics is analysed e.g. by VANBERG

However, in terms of the "economics" employed, the approaches differ considerably. Obviously, neither Böhm nor Hayek would claim that "rational choice" or stylized maximizing behaviour as used within mainstream economics (including most of constitutional economics) belong to the "hard core" of their research programmes, whereas these elements are assumed to be indispensable for modern theories of public choice or constitutional economics (e.g. Buchanan 1990 pp. 12 ff). In short, ordoliberals and Hayek are political economists in a traditional (say, Smithian) sense; they command no "economics of politics" or economic theory of the state if economics is defined by the above hard core-elements (see Tietzel 1991).

As far as the research programme of the ordoliberals and their principles of public policy is concerned, the lack of a positive theory of politics may pose a serious problem of theoretical incompleteness.[13] Since ordoliberals strictly rely on the state, and hence on elected politicians, to enact the comprehensive institutional reforms, the incentive structures and competence allocations[14] of politicians, voters and interest groups become empirically important and merit the most serious analysis. That said, I hasten to add that Eucken and Böhm were not ignorant euphoricists who disregarded the workings of real political processes. As we have seen, their historical observations and political demands were rich in considerations of vested interests, of their institutionally permitted influence on the political process and of the damaging effects this must have on the workability of the economic order.

In this tradition, ordoliberals and Hayekians (even if they tend to disagree on the conceptual role of social justice within a constitution of liberty) provide similar insights into the political economy of the welfare state as well as into its disruptive effects on the economic and the political constitution. For both schools of thought, the welfare state is "captured" by economic interest groups and forced to displace the autonomy of the individual and the

---

(1989) or BUCHANAN (1989) and – in an interesting attempt to build a synthesis – by LESCHKE (1993).

13 See also KIRCHGÄSSNER (1988 pp. 55 ff.). In fact, EUCKEN (1940/92 pp. 213 ff.) states that the emergence and changes (but not the effects) of "the social and legal organization" and hence politics themselves ought to be treated as "data" for the analysis. It is here that he proposed that "the theoretical explanation has to break off".

14 See PELIKAN (1995) for a discussion of a neglected field within theories of politics (and economics): the allocation of scarce competence which determines the quality of decisions.

rule of private law with corporativism and the rule of public law. In ordoliberal terms, this leads to the destruction of the competitive order as a means to emasculate economic power; in Hayekian terms it (also) means the abandonment of equality before the rules of just conduct as a precondition of a spontaneous order of free actions. Thus, one can argue that Eucken and Böhm to some degree also anticipated the "rent-seeking" insight that was later to emerge within the Public Choice movement (e.g. Krueger 1974; Olson 1982).

Nevertheless, ordoliberals and (more-or-less explicitly and consistently) Hayekians continue to regard politicians as addressees of programmatic demands rather than as objects for a behavioural analysis. This is inevitable: how would the early ordoliberals have reacted to the dilemma posed by their well-intentioned principles while knowing both that rent-seeking interest groups have constantly thwarted their coming into existence and that vote-seeking politicians would be ill advised to enact them? Their problem corresponds to that of the "endogenous public choice theorist" (Witt 1992) who faces the "seemingly paradoxical situation ... whenever public choice theorists, or political economists in general, who go normative, leave it open how, as a result of their normative plea for change and reform, forces other than those that cause the criticised development can be activated" (ibid., p. 118).

This is still no firm basis for despair, however. The knowledge and arguments of political economists can become politically relevant by slowly creeping through communication channels, through networks of communication circles and thus finally having an impact upon the perceptions of boundedly rational individuals – perceptions which form public opinion and (sometimes) influence government positions. Special circumstances can help to shorten these channels and master the decisive circles. The early history of *Ordnungspolitik* is proof of the fact that general, market-oriented ideas and conceptions do not necessarily have to fail. While the gradual growth of sclerosis within the German *Wirtschaftsordnung* is relatively easy to grasp with the help of modern explanations of the "Leviathan" and "rent-seeking" type, it is somewhat more demanding to produce an equally convincing "public choice explanation" for this early stage of German politics. Let us now sketch the political career of *Ordnungstheorie*.

V. The Political Relevance of Ordnungspolitik

1. 1948: Historical "moments"

The initial success of the concept of the social market economy was a surprising event. Neither the political nor the intellectual mood of the time particularly favoured reliance on markets and competition as major elements of an economic order under construction (Streit/Kasper 1992). Indeed, it can be argued that the significant political impact of *Ordnungspolitik* from 1948 until, say, the late 1960s, was the result of politically favourable circumstances rather than a general acceptance of the theoretical underpinnings of *Ordnungstheorie*. These favourable conditions included:

(a) The "historical momentum" created by the need for an institutional reconstruction of the economic and political constitution. At the same time, because of the destruction or initial weakness of pressure groups and the encompassing character of the trade unions, neither the established vested interests nor log-rolling among the established political factions could obstruct a political "big bang" transformation along ordoliberal signposts.[15]

(b) The political task was facilitated by the work of ordoliberals in their capacity as politically committed scholars who retained their personal integrity during the Nazi period and supplied policy conclusions that could be put to work by government officials and which could even be translated into the language of private and public law.

(c) Ludwig Erhard acted as a courageous political entrepreneur who knew how to use both the historical moment and the theoretical advice in order to push through a comparatively radical market-oriented policy. Besides integrity and political power he had abundant charisma which helped him successfully to apply moral pressure in order to create popular confidence in the welfare-enhancing properties of a market system (*"Wohlstand für alle"*).

The honeymoon of *Ordnungspolitik* lasted quite some time but gradually its attraction dwindled. Many instances in German history were cited as examples of the decline of ordoliberal discipline in actual politics which led to the situation depicted in my rather gloomy assessment of the social market economy fifty years on (see Introduction). One such instance (which clearly coincides with the drastic increase of the government expenditure share: see

---

15 The weakness or encompassing character of organized vested interests is the central argument that OLSON (1982) uses for the relative success of German or Japanese economies after the Second World War.

Figure 1) was the attempt to introduce Keynesian-like macroeconomic steering together with corporatist decision procedures in the 1970s.

2. The 1970s: Changes of style

As early as 1971, Hans-Otto Lenel, a second-generation German ordoliberal, was led to ask: "Does Germany still have a social market economy?" It was the time when Karl Schiller, then Minister of Economic Affairs aimed at what he called an "enlightened" economic policy with the task of steering macroeconomic variables by way of targets which were intended to direct the coordinated actions of politicians, unions and employers. In the context of traditional *Ordnungspolitik*, this was more than just a slight change in style in the wake of the vogue for Keynesianism. The so-called *Globalsteuerung* (which Watrin 1998 p. 23 translates as the notorious French term *planification*) was introduced in 1967 and, from the very outset, for many ordoliberals it implied abandoning substantial principles of *Ordnungspolitik* – more so for adherents of the "Freiburg imperative" and less so for permissive liberals such as Müller-Armack or Röpke (see Nörr 1998 p. 231). Thus Hoppmann (1973) argued in typically Hayekian vein that the constructivistic idea of a planned economic organization was about to be installed instead of the spontaneous order of market co-ordination which alone could guarantee individual freedom and the decentral use of knowledge in society. For the lawyer Hans Rupp (1971), the new forms of bargaining between representatives of state and society did not only lead to agreements at the expense of unrepresented third parties (like consumers or the unemployed). He also made the point that the corporatistic decision-making processes endangered the very basis of the *Rechtsstaat* which requires political power to be exercised within bodies that are responsible to voters and accessible to the control of public law.[16]

---

16 On the other hand, during there is find ample evidence from Karl Schiller's "reign" as "Super-Minister" of at least continued rhetorical support for the social market economy. Thus, one can also argue with NÖRR (1998 p. 213) that it "is controversial whether the direction that Schiller took still can be ascribed to the concept of Social Market Economy or not. Schiller himself would have given an affirmative answer. The macroeconomic approach of Keynesian provenance was not meant to destroy but to complement the microeconomics of the market." NÖRR (ibid. p. 230) offers another variant reading of the rationale of the development of German politics. According to his understanding, this change came about both because within the European Union and also at home "political and academic circles began to flirt with the idea of planning in many directions".

These ordoliberal criticisms made little impression on politicians or public opinion. In particular, the idea of making policy more "effective" and "fair" by bringing "all parties concerned" to "round tables" has never lost its attraction for the German media and for public opinion. *Solidaritätspakte* or *Beschäftigungspakte* are perhaps even more fashionable today than ever before – notwithstanding the dangers of concerted exploitation of unrepresented third parties and of collective irresponsibility towards voters or courts. It is now widely acknowledged by economists and some politicians that, after a period of comparative success, *Globalsteuerung* failed to attain its self-imposed economic goals. Many would argue that it even has some responsibility for leading Germany into the country's first serious recession during the mid-seventies (e.g. Streit 1981; Watrin 1998). Some would even assert that the ensuing business cycles were largely a product of economic policy itself, aggravated by political election cycles or intervention cycles.[17] One of the first public organizations to reach such conclusions was an offspring of the demand-management era, the Council of Economic Experts (*Sachverständigenrat*, founded in 1963). In the mid-70s it turned against its political patrons and, having built up considerable reputation for independent analysis and advice, from 1976 onwards it can be regarded as an unswerving advocate of supply-side economic policies. However, the independence of such agencies cuts both ways: politicians more often than not acted in defiance of the experts' proposals. As a consequence, sectoral protectionism and redistributionist interventions persisted and often even increased independently of the "good" advice of the *Sachverständigenrat* and the *Bundesbank*, and in some cases ministerial permissions were used to veto recommendations of the *Kartellamt*.

3. Towards a social state economy

With Keynesian *Globalsteuerung* reaching the end of its intellectual and political career, "social justice" entered the political stage with unprecedented vigour. Political actors on the supply side and interest groups on the demand side of the political market for special favours began increasingly to employ the epithet "social" to describe and justify their

In this situation, Nörr argues, moving in the direction of more redistributionist and interventionist policies "could serve as a kind of antidote against planification; the field was to be occupied before the rival will be prepared to enter".

17 Not surprisingly, theories of the political business cycle originated at that time (e.g. NORDHAUS 1975; FREY 1978).

endeavours. Since no particular mention is made in the German *Grundgesetz* of the functional properties of a market economy or of *Wettbewerbsordnung*, whereas it is particularly stressed that Germany should adhere to the principle of a "social state" (*Sozialstaatsprinzip*), redistribution in all directions and without any regard for economic side-effects encountered no effective legal constraints. In contrast to the liberal provisions of the constitution that act as a protection for individual rights against state interference (*Grundrechte als Abwehrrechte*), this principle assigns to politicians an "unqualified authorisation" (Benda, cited in Streit 1998 p. 55) to interfere with private property rights - mainly by way of correcting market results via redistributional measures. As can be observed in many other welfare states, such policies were not particularly effective in helping the poorest. Rather, they were aimed at supporting the loudest (that is, those who command the most powerful "voice" because of their organizational power, which can be used to obstruct government policies or to influence voters). The politico-economic logic of the virtually irreversible growth of government expenditure and state intervention, the ineffectiveness of even well-meaning social policies to help the poor and unemployed and the damaging effects of state interference with competitive processes on the economy's potential for growth and structural change have been exhaustively analysed by economists (and not only those with an ordoliberal background). This is not the place to repeat these well-known arguments.[18]

In brief conclusion, then, we can observe a roughly inverse correlation between the appearance of *Ordnungspolitik* as a stock phrase in formal speeches and the decline in its real political clout. With the state now controlling over 50% of national income, Germany has become a welfare state with dimensions none of the first-generation ordoliberals could possibly have imagined or would ever have approved. Moreover, if the institutional details of Germany's economic constitution are examined, it is clear that ordoliberal principles have been violated to a considerable degree: the present state of social insurance systems, massive and ever-growing transfers directed at all sorts of economic activities and social circumstances, labour regulations and market regulations in many fields have become largely irreconcilable with ordoliberal principles. Here and in other areas, the fundamental ordoliberal demands of stability, neutrality and subsidiarity of

18 For a more detailed analysis of the effects and a public choice explanation of "Sozialpolitik" in Germany, see e.g. VAUBEL (1991). The damaging influence of vested interests on the political and economic order is discussed in STREIT (1987).

economic policy have been obviously (but never explicitly) violated by the acts of those whom Hayek would have called the "socialists of all parties".

4. *Ordnungspolitik* in the process of German unification[19]

The most important German historical event, unification, is not actually proof of an *Ordnungspolitik* victory or of a consummation of the "social market economy" as the founding fathers of *Ordnungstheorie* used the term. It is easier to argue for the opposite. At least in retrospect, one is inclined to lament the fact that, with German unification, a "historical moment" in the Euckenian sense has remained unexploited. The almost total transfer of the whole raft of taxes, regulations and social security systems has proved too much for the new *Länder* to bear. While no other Central or Eastern European country in transformation was as fortunate as the former GDR in having a complete system of institutions ready for use and a rich elder brother willing to pay for the costs of economic infancy and upbringing, on the other hand none of these countries was forced to comply with a comparable degree of institutional sclerosis ready to suffocate the emergent competitive market structures from the very outset. And indeed, even now, the new *Länder* are falling behind many other societies in transformation in terms of unemployment and growth rates. It can be argued that this would not have happened had West Germany not, in recent decades, added more and more interventionist policies to its initially more market-oriented economic constitution - policies which were then put into operation in the former GDR together with the institutional framework necessary for a social market economy.

To be sure, East Germans obviously had no wish to become the playing-field for further political experimentation and they were mostly looking forward to securing their share in the "social" component of what they anticipated the social market economy would be. It can additionally be argued that the geo-strategical dimension of the moment of "historical opportunity" did perhaps indeed amount to a comparatively short "moment", or time-span. This was the time to act quickly and reliably, not to call into question the established economic order of West Germany and offer East Germans a new and unprecedented institutional "package". However, in

19 For a detailed examination of these matters, see e.g. the early ordoliberal warnings of STREIT (1991/95) and WILLGERODT (1990), or the retrospective assessments by SCHÜLLER/WEBER (1998) or MUMMERT/WOHLGEMUTH (1998).

order to provide the flexibility and openness the political and economic order badly needed to meet the massive demands of structural change, the use of more "opening clauses" (*Öffnungsklauseln*) would have been welcome, not only in the area of collective wage bargaining but also in those of administrative law and political regulation.[20] Finally, there is always a time to initiate the necessary reforms not only to create *Aufschwung Ost* but also to strengthen *Standort Deutschland* now that a relatively small distinction has to be made between East-German and West-German needs. At least in some areas (such as collective wage determination) it may well be that West Germany will in the near future have to face problems similar to those that East Germany is currently struggling to overcome. Consequently; the "West" may have to learn a lesson from the "East" about such qualities as endurance and the ability to adapt to new circumstances (see, e.g. Truhn 1998).

5. The europeanization of *Ordnungspolitik*: A success story?

Ordoliberalism has not been obliterated in every sphere. It could be argued that at least in the fields of competition policy and monetary policy, Germany has remained comparatively loyal to ordoliberal principles up to the present day. It is no coincidence that it was precisely in these fields that responsibility for *Ordnungspolitik* was delegated to politically independent agencies, the *Bundeskartellamt* and the *Bundesbank*[21]. Their policies have proved much more immune to the assaults of rent-seekers and spendthrifts than all those under the auspices of German ministers (including the *Bundeswirtschaftsministerium* which - rhetorically - considers itself a committed stronghold of *Ordnungspolitik*). Indeed, the *Bundeskartellamt* and the *Bundesbank* were soon admired and envied in many parts of the world.

Both agencies are now being either totally or in part replaced by European substitutes. For ordoliberal optimists (who trust in the European

20 Such proposals are discussed under the label of "experimentary clauses" in the realm of law (*Experimentierklauseln*). Arguably, they would allow for decentral solutions to local problems and for recurrent occasions to reassess the properties and side-effects of state regulations. For a discussion of such clauses, see e.g. HILL/KLAGES (1996); EHRICKE (1998).

21 It must be noted that in the early days of *Ordnungstheorie* there was some degree of scepticism about the workability of an independent central bank which some ordoliberals regarded as a second-best solution compared to different forms of very strict reserve requirements (BERNHOLZ 1989).

continuation of both German rules *and* their ordoliberal spirit), this could mean a final triumph. For ordoliberal pessimists who observe or expect institutional and cultural differences between German and foreign anti-trust and anti-inflationist policies, ordoliberalism has now lost its last standard-bearers. This is not the place to make a substantial contribution to these shaky speculations. Positive net results may be anticipated, however, because some deviation from the high standards of ordoliberal ideals and (possibly) even from German practice can be more than compensated for by an internationalized acceptance of even somewhat compromised principles of *Ordnungspolitik.* In consequence, the advantages of European economic integration (the common market) would more than compensate for the disadvantages of European political cartelization (in common policies such as agriculture).

VI. Ordnungsökonomik in Perspective

What might *Ordnungstheorie* in its modern form – let us call it *Ordnungsökonomik* – have to offer in the future? So far I have given a rather mixed picture of the present state of the academic and political reception of ordoliberal issues and demands. Since I started with a provocatively gloomy account of the present state of the social market economy after fifty years, let me end with an attempt to find some promising signs for the future of *Ordnungsökonomik.*

1. Theoretical rigour vs. practical relevance: Looking for promising signs

The present state of economics can be characterized using the idea of a trade-off between relevance/importance on the one side and rigour/precision on the other (Mayer 1993 p. 7). Traditional *Ordnungstheorie* and – to a lesser degree – modern *Ordnungsökonomik* certainly do not tip the scales on the rigour/precision side. Consequently, they are either little esteemed or simply ignored within the dominant mainstream of economics. Modern economics consists mainly of formal representations of rational maximization and "tight prior equilibrium". To be sure, different links in a chain of arguments can be dealt with and sometimes brought to an original conclusion using these tools. This method, however, does not easily lend itself to the consideration of many different links within one framework of analysis. Rather, it compels its users to follow the "principle of the strongest link" (ibid. p. 57) – that is, to single out one theoretical puzzle where one expects to be able to apply the most vigorous model and work hard to strengthen and polish this one link. Problems with this procedure arise when it is more-or-less implicitly claimed that the whole argument is rigorous because one link is. All too often, the additional links, especially those in the direction of policy implications, remain extremely weak and vague. The whole chain of arguments then ends in practical irrelevance and political inapplicability. Following this strategy, economists seem to have created a dismal science not unlike that crushingly described by Mark Blaug (1998 p. 3): "modern economics is sick; economics has increasingly become an intellectual game played for its own sake and not for its practical consequences; economists have gradually converted the subject into a sort of social mathematics in which analytical rigour as understood in math departments is everything and empirical relevance (as understood in physics departments) is nothing; if a topic cannot be tackled by formal modelling, it is simply consigned to the intellectual underworld; to pick up a copy these days of *American Economic Review* or the *Economic Journal* not to mention *Econometrica* or the *Review of Economic Studies*, is to wonder whether one has landed on a strange planet in which tedium is the deliberate objective of professional publication."

Ordnungsökonomik, by contrast, gives a more wholesome impression. It has always been much more comprehensive and empirically relevant, following what might be called a "principle of the conclusive chain". It aims at combining various links (such as historical experience, comparative analyses of social systems, possibilities and incentives for acting within different types of institutions and using different property rights) with conclusive links which may take the form of political advice or at least political warning-signs. Very different from econometric models or mathematical puzzles, these chains of arguments can also be used within various, and even broad, arenas of public discourse and, finally, their conclusions can sometimes even be usefully translated into the language of the law.[22]

---

22 Note that in some works of modern "Law and Economics" we often find that very singular elements of legal systems are isolated and translated into the language of economics. In the process of translation into rational choice strategies and variants of Nash-equilibria, however, much legal content, empirical and political relevance is lost. Sometimes (but certainly not systematically), the reader of "Law and Economics"-affiliated papers has the impression that the authors are trying to solve puzzles which are mathematical

To summarize, both *Ordnungstheorie* in its traditional form and *Ordnungsökonomik* often appear weak in theoretical terms if the strength of the chain of arguments is solely judged by a comparison of the "strongest link". If the quality of alternative theories is judged by the strength of the entire chain of arguments and their conclusiveness in terms of empirical relevance and political applicability, *Ordnungsökonomik* possesses a certain virtue when compared to the work of economists who currently dominate the most prestigious journals and academic chairs. Even today, economists bored or irritated by the extent to which economists are engaged in the idle pursuit of formal rigour and logical precision without regard for empirical relevance or political significance might find that many interesting topics have already been dealt with in the great works of German *Ordnungstheorie* and are now increasingly treated in the international New Institutional Economics movement.

But is such a "paradigm change" likely to occur? Perhaps not. The self-enforcing logic of the academic system with its built-in path-dependencies seems irreversible. However, I trust that there will continue to be sufficient demand for politically and empirically relevant research to nourish small groups of scholars devoted to the enterprise of *Ordnungsökonomik.* At the same time, the academic system is not totally closed and capable of only nourishing and reproducing itself. And indeed, there are some indications that outside demand (public and private sponsors of research, for example) is not totally inelastic. The demand for scholastic "social mathematics" seems to be declining – to about the same degree that " students vote with their feet and go elsewhere to learn about the practical problems of the economic order" (Blaug 1998 p. 19).

2. Political sclerosis vs. new political agility: Looking for promising signs

As far as the future of *Ordnungspolitik* is concerned, there are grounds for both an optimistic and pessimistic assessments. The current German situation might lend more support to the latter. But here, again, we find the necessary impulses for change stemming from outside market forces. Germany is not a closed system in which a "social state economy" could

---

rather than practical, thus reaching conclusions which are not actually related to human problem-solving with the help of legal institutions.

nourish itself. And the demand for market institutions which come closer to the ordoliberal vision is not totally inelastic either.

Furthermore, outside Germany there is a growing supply of competitive alternatives. Many parts of the globalizing world around us can be argued to have become, as it were, *de facto* more "ordoliberal" than Germany. Besides the macroeconomic goals, the institutional aspects especially of what today is called a "Washington consensus" (Williamson 1990 chap. 2), and which inspired political reforms in many different countries, are basically identical with what some decades ago was called the "Freiburg imperative". Thus, the wave of reforms which ("encouraged" in part by the strict conditionality of the IMF and the World Bank) have since the 1980s seized Australia, Chile, Colombia, Mexico, New Zealand, Poland, Portugal and Spain[23], but also Great Britain and the Czech Republic[24] can be viewed as an impressive success of *de facto Ordnungspolitik.* The observation that – after a time of painful and costly transition – most of the reforms have been successful in producing economic growth can also serve as a corroboration of the very basic hypotheses of *Ordnungsökonomik.* The fact that the German *Standort* is competing with a growing number of increasingly market-oriented economies might induce a political return to the basic virtues of *Ordnungspolitik.* Here, not unlike typical market processes, competition can serve as a "discovery procedure" that produces both the possibilities of learning ordoliberal lessons and the incentives to do so (see Wohlgemuth 1998).

To sum up: with a little effort, promising signs indicative of a persistent or even a growing relevance of both *Ordnungsökonomik* and *Ordnungspolitik* can be found. Certainly, however, this expectation relies fundamentally on a process of pathological learning.

23 These countries are dealt with in the collection of politico-economic case studies by WILLIAMSON (1994).

24 The latter cases are also instructive because they can be regarded as differentiating reactions to the German model of a social market economy. In her rhetoric at least, Lady Thatcher has claimed to be informed by Hayekian liberalism - especially in its rejection of combining notions of "social justice" with the institutions of a spontaneous market order. Vaclav Klaus even explicitly claimed that he sought out a "market economy without attributes", hence denying policies of redistribution an importance comparable to the creation and preservation of private market institutions.

Appendix 1

Graph 7.1: German government expenditure as % of GDP 1950 – 1997

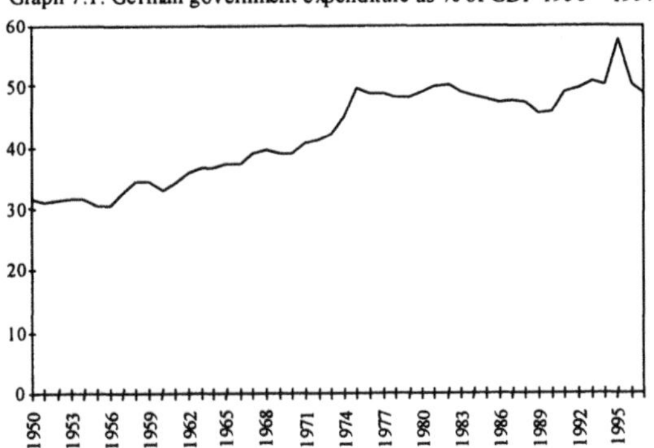

Source: Statistisches Bundesamt (1998). From 1960 inclusive of Saarland and Berlin (West); from 1991 inclusive of the new *Bundesländern* and Berlin (East); provisional details from 1995.

Graph 7.2: Unemployment rates in Germany 1950 - 1997

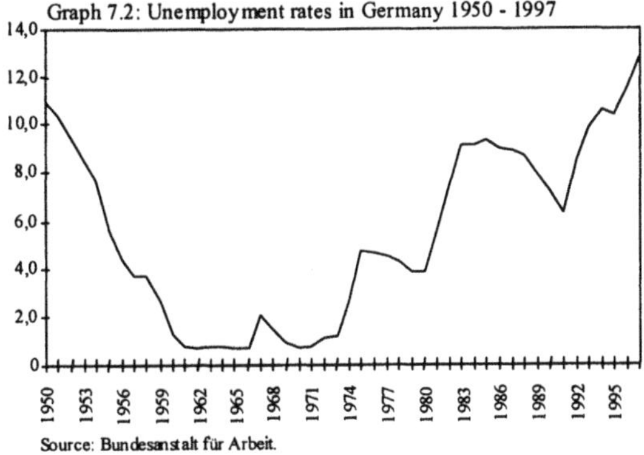

Source: Bundesanstalt für Arbeit.

Graph 7.3: Share of those interviewed expressing a "good opinion" of the social market economy

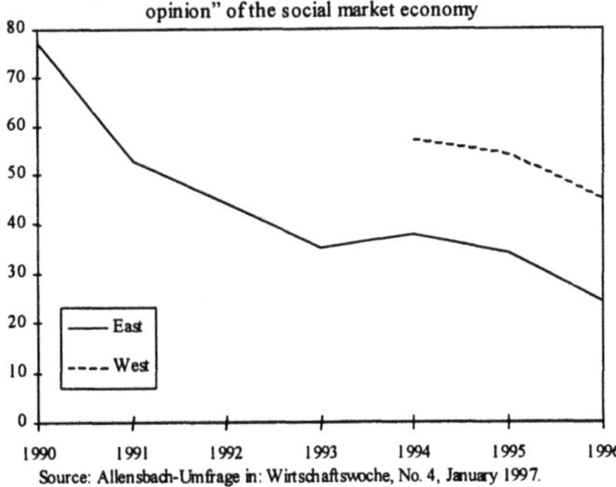

Source: Allensbach-Umfrage in: Wirtschaftswoche, No. 4, January 1997.

Graph 7.4: "Economic Freedom of the World" Ranking

Source: J. D. Gwartney / R. A. Lawson: *Economic Freedom of the World-1997 Annual Report,* Vancouver B.C. (Fraser Institute) 1997, p. 27.

Appendix 2:
A sketch of German centres working in the field of *Ordnungsökonomik*

This list does not claim to be comprehensive and is by no means to be regarded as a "ranking". It is not based on systematic criteria of selection but solely on my subjective choice according to my limited knowledge of the academic scene in Germany.[25]

• At the University of **Freiburg**, the chair formerly held by Hayek has at least remained in the hands of scholars related to *Ordnungstheorie.* It was subsequently held by Erich Hoppmann, then Manfred E. Streit and, finally, Viktor Vanberg. Along this line Hayekian issues such as knowledge and the market process, spontaneous orders or legal principles of a "Constitution of Liberty" have been continuously taught and elaborated. The "Freiburg school of Law and Economics" as a "Forschungs- und Lehrgemeinschaft von Juristen und Ökonomen" (Böhm 1957/60) has, however, ceased to exist. This loss can only be compensated for in part by the activities of the Freiburg Walter-Eucken-Institut which continues to publish, and to organize lectures and conferences.

• At the University of **Marburg**, the Institute for the Comparison of Economic Systems ("Forschungsstelle zum Vergleich wirtschaftlicher Lenkungssysteme") in particular carries on the basic research programme of *Ordnungstheorie.* The Institute was founded in Freiburg (in 1954) by K. Paul Hensel, who was an active member of the Freiburg school. In 1957 Hensel and his Institute went to Marburg, where the Institute has since attracted many leading ordoliberal scholars, such as Alfred Schüller, who has headed the institute since 1975. The research programme concentrates on the analysis of economic property rights and social institutions (laws, customs, conventions) as fundamental characteristics of economic orders or systems which, in turn, influence the possibilities and incentives for individual behaviour and the emergence of social coordination in a comprehensive order of actions. Besides refinements in property rights economics, the Institute engages in various empirical studies on comparative economic

25 For another account of institutional economics in Germany see PASCHA (1994). For a list of "German Economists working on the New Institutional Economics"see the homepage of the Saarbrücken Center for the Study of the New Institutional Economics (www.uni-sb.de/rewi/fb2/eichberger/institut). I must apologize for not having systematically traced *Ordnungstheorie* in neighbouring countries.

systems. The transformation of formerly socialist systems may have reduced the pool of conflicting economic systems, but it has certainly confirmed the importance of studies on comparative economic systems.

• At the recently founded European University Viadrina in **Frankfurt an der Oder**, the Frankfurt Institute for Transformation Studies (FIT) has responded to the impulses to, and challenges for, institutional economic research created by the transformation of socialist countries. Under the direction of Hans-Jürgen Wagener (who also holds a chair of Ordnungspolitik) various social scientists work on projects in the fields of "cognitive changes as a condition for and a consequence of transformation ... setting the course for a functioning market economy ... structural change and communication across borders [and] transformation of law".[26] As far as the "theory of systemic change and transformation" (Wagener 1993) is concerned, many different approaches – traditional *Ordnungstheorie* together with institutional economics avenues as well as Austrian paths – are combined.

• At the private University of **Witten/Herdecke** another centre of institutional economics, comparative economic systems and also moral philosophy has been established which is characterized by a variety of methods and ideas unheard of in other German economics departments. Its major proponents, who clearly work "off the beaten track" of orthodox economics, are Carsten Herrmann-Pillath and Birger Priddat.

• At the University of **Saarbrücken**, New Institutional Economics has become an established discipline. This is mainly due to Rudolf Richter, who was among the first to make known some of the new developments in this field (e.g. Richter 1994) and (as editor of the *Journal of Institutional and Theoretical Economics*) academically attractive in Germany. Together with Jürgen Eichberger and Eirik Furubotn, Richter is director of the Centre for the Study of the New Institutional Economics in Saarbrücken which has been hosting international seminars on subjects related to new institutionalism since 1983. A Center for the Study of Law and Economics was also founded in 1993, under the directorship of Dieter Schmidtchen, which uses variants of rational choice theory in order to study the effects and changes of legal constraints. The *foci* of their research are the role of law in an international context, constitutional and institutional questions concerning the European

26 See the respective pages on http://fit.euv-frankfurt.de

Union, as well as, for example, problems relating to the legal protection of intellectual property.

• At various other German universities prominent scholars (most of them current or former holders of chairs in departments commonly called "theory of economic politics") were, or still are, teaching and working on aspects related to *Ordnungstheorie*. The places that most readily come to my mind are those of **Cologne** (Christian Watrin, Gernot Gutmann, Hans Willgerodt), **Münster** (Heinz Grossekettler, Holger Bonus), **Duisburg** (Dieter Cassel, Manfred Tietzel), **Bayreuth** (Peter Oberender; Helmut Gröner), **Würzburg** (Norbert Bertold), and **Bochum** (Karl-Hans Hartwig, Wolfgang Kerber, Gerhard Wegner). In addition, the internationally renowned scholar Herbert Giersch (long time president of the *Institut für Weltwirtschaft* in **Kiel**) can be argued to be very close to ordoliberal as well as Hayekian *Ordnungstheorie.*

• It may be interesting to note that in at least two organizations backed by the Catholic church, the University of **Eichstädt** and the **Hannover** Institute of Philosophical Research, there are scholars devoted to combinations of the ordoliberal programme with new elements of social philosophy and ethical economics.

Within the former, Karl Homann is the prominent figure. His idea of *Ordnungsethik* (Homann 1994; Homann/Kirchner 1995) tries to reconceptualize the normative task of *Ordnungstheorie.*[27] This is done with the aid of contractarian constitutional economics (and hence criteria of consent that basically follow James M. Buchanan, e.g. 1975) together with a comparative analysis of the effects of institutional settings on individual incentives to act, thus developing an interesting blend of *Diskursethik* and (as they call it) *Anreizethik*. In similar ways, Ingo Pies (e.g. 1993) proceeds to develop what he calls "normative institutional economics" (*Normative Institutionenökonomik*).

The Koslowski programme has an equally ambitious agenda for recombining different strands of thought. Compared to the above-mentioned

27 The vicinity to the ordoliberal task is expressed in the intention of a "Wiederaufnahme der normativen Fragen nach der Ordnung der Wirtschaft und der Gesellschaft" (HOMANN/KIRCHNER 1995 p. 193). At the same time, authors like Homann or Pies miss no opportunity to stress that they break with what they describe the "dualistic paradigm" ("dualistische Paradigma der Ordnungsformen") which it is argued characterizes many ordoliberal studies and policy implications and supposedly find its expression in simplistic notions like democracy vs. liberty or capitalism vs. socialism.

authors, however, Koslowski is much less reluctant to accept ordoliberal ideas at face value, as when he states (1996 p. 4): "With respect to this interest in a synthesis of social philosophy, political economy, economic ethics and theory of the economic order the author considers himself to be close to the Freiburg school of political economy, especially to the works of Walter Eucken and Friedrich August von Hayek, whose work, in a wider sense, belongs also to this school." One consequence of accepting these traditions is that for Koslowski (ibid.) the "theory of the economic order (*Ordnungstheorie*) of the social market economy has to be sought in a synthesis of economic liberalism and the older tradition of natural right". For precisely this "neo-Aristotelian synthesis" (Koslowski 1988) is heavily criticized by Homann and his followers. They argue that such a combination would obstruct their attempts to deduce individual liberty from certain procedures of collective action (e.g. Homann/Kirchner 1995 pp. 193 ff.).

Remaining within the field of social philosophy, mention must be made of two other scholars with affinities to ordoliberal ideas, although more preoccupied with epistemological questions of the social sciences and of the social order: Hans Albert (e.g. 1986; 1998) and Gérard Radnitzky (e.g. 1991). Important differences in their respective work notwithstanding,[28] both are similarily close to Popperian concepts of critical rationalism.

• As far as academic journals are concerned, the yearbook, *Ordo*, which has been the major ordoliberal journal since its foundation by Böhm and Eucken in 1948, is still published today and continues to attract writers of all ages (mostly economists) who are mainly concerned with normative issues of economic policies but also with positive theories, for example, of market processes or institutional change. Taking greater account of the standards of theoretical excellence such as now characterize the international economic mainstream is the *Journal of Institutional and Theoretical Economics*, formerly *Zeitschrift für die gesamten Staatswissenschaften*. It is no longer the truly interdisciplinary journal its old name (*Staatswissenschaften*) implied. Its editors (Rudolf Richter and Ekkehard Schlicht) mostly insist on the use of economic formalism while encouraging applications of rational choice models in fields other than economic markets. Hence it is here that we find interesting contributions to Law and Economics, Public Choice, Social Choice and the like. Authors such as Böhm or Hayek who relied on the power of verbal argumentation, however, would today find it hard to be

28 Politically, Radnitzky takes an undisguised classical liberal stance whereas Albert is rather critical of ideological contents in academic work while subscribing to Popper's idea of the open society.

published in *JITE* today. Next, there is the *Jahrbuch für Neue Politische Ökonomie* which is based on yearly conferences of economists adhering to various subdisciplines of New Institutional Economics to which scholars close to ordoliberal traditions continue to be invited. Other German outlets for *Ordnungstökonomik* include journals of public finance or economic policy such as *List Forum*, *Zeitschrift für Wirtschaftspolitik* or *Wirtschaftsdienst* which are basically "free of the style and method limitations of the leading mainstream economics journals" (Pascha 1994 p. 11).

• Last but not least, let me mention my employer. At the Max-Planck-Institute for Research into Economic Systems in Jena (founded in September 1993), one of two research units thus far established,[29] is devoted to questions that are basically identical with the themes of traditional *Ordnungstheorie*: (1) How do institutions operate and how exactly does institutional change take place? (2) How do market processes operate? (3) What options do politicians have to influence the economic development?"[30] A summary introduction to the answers to these questions - answers which most members of the unit share in principle – can be found in the encyclopaedia article, *Ordnungsökonomik* (Streit 1996a).[31] This neologism has been coined in order to signal that it is programmatically intended to be a blend of traditional *Ordnungstheorie*, various disciplines of the New Institutional Economics and Austrian classical political economy. One of the various applications of such a blend has lately been in the field of institutional, interjurisdictional competition. Within this field, several members of the unit worked on institutional competition "as a defence of liberty" (Streit 1996b),

29 A third unit is yet to be established. Head of the unit on Institutional Economics and founding director of the Institute is Manfred E. Streit. He comes from the University of Freiburg, where he followed Erich Hoppmann on the chair that was formerly held by Hayek. This genealogy also illustrates a continuity of scientific interests. The director of the unit on Evolutionary Economics is Ulrich Witt who also came from Freiburg. While he can not be called a confessing ordoliberal, he did contribute to the theories of the market process, of institutional change, spontaneous orders and other areas within the issue-space of Austrian and Institutional Economics, albeit with a more pronounced evolutionary (non-neoclassical) emphasis.

30 The quotations are taken from the Institute's documentation "Programme - Persons – Projects" which is available on request. Please feel free to contact me.

31 A longer version of this paper can be ordered as Discussionpaper 04-95. In English, basic aspects of *Ordnungsökonomik* will be published in a forthcoming textbook (KASPER/STREIT 1998).

a "discovery procedure" (Wohlgemuth 1998) or a strategy for European integration (Mussler 1996), while others showed the manifold limitations of this peculiar kind of competitive process (Kiwit/Voigt 1998) or detected forms of institutional competition in history, for example, of the Holy Roman Empire (Volckart, ed. 1998). Other research programmes include studies on the "limits to institutional reforms" (Eggertsson 1998), the spontaneous development of private rules for cross-border transactions (Streit/Mangels 1996), the relation of internal and external institutions and its influence on institutional change (Kiwit/Voigt 1995), the role of informal institutions in shaping transformation processes (Mummert 1995), the character and evolution of the economic constitution of the European Union (Mussler 1998), a new area of research named "positive constitutional economics" (Voigt 1997), the role of institutional competition in history (Volckart 1997), as well as the renewed debate on market socialism and analysis of political competition along Austrian guiding ideas (Wohlgemuth 1995; 1997).

References

ALBERT, H.: *Freiheit und Ordnung. Zwei Abhandlungen zum Problem einer offenen Gesellschaft*, Tübingen (J.C.B. Mohr [Paul Siebeck]) 1986.

ALBERT, H.: *Bemerkungen zur Wertproblematik. Von der Bewertung des Sozialprodukts zur Analyse der sozialen Ordnung*, Jena (Max-Planck-Institute for Research into Economic Systems) 1998 (=Lectiones Jenenses, Vol. 15).

BERNHOLZ, P.: "Ordo-Liberals and the Control of Money", in: A. PEACOCK, H. WILLGERODT (Eds.): *German Neoliberals and the Social Market Economy*, London (Macmillan) 1989, pp. 191-215.

BLAUG, M.: *The Disease of Formalism in Economics, or Bad Games that Economists Play*, Jena (Max-Planck-Institut for Research into Economic Systems) 1998 (= Lectiones Jenenses, Vol. 16).

BÖHM, F.: "Die Idee des ORDO im Denken Walter Euckens - Dem Freunde und Mitherausgeber zum Gedächtnis", *Ordo*, 3 (1950), pp. xv-lxiv.

BÖHM, F.: "Der Rechtsstaat und der soziale Wohlfahrtsstaat", in: F. BÖHM: *Reden und Schriften*, Karlsruhe (Müller) 1960, pp. 82-156. First published: 1953.

BÖHM, F.: "Die Forschungs- und Lehrgemeinschaft zwischen Juristen und Vokswirten an der Universität Freiburg in den dreißiger und vierziger Jahren des 20. Jahrhunderts", in: F. BÖHM: *Reden und Schriften*, Karlsruhe (Müller) 1960, pp. 158-175. First published: 1957.

BÖHM, F.: "Wettbewerbsfreiheit und Kartellfreiheit", in: F. BÖHM: *Freiheit und Ordnung in der Marktwirtschaft* (edited by Ernst-Joachim Mestmäcker), Baden-Baden (Nomos) 1980, pp. 233-262. First published: 1958.

BÖHM, F.: "Demokratie und ökonomische Macht", in: INSTITUT FÜR AUSLÄNDISCHES UND INTERNATIONALES WIRTSCHAFTSRECHT (Ed.), *Kartelle und Monopole im modernen Recht*, Karlsruhe (Müller) 1961, pp. 1-24.

BÖHM, F.: "Rule of Law in a Market Economy", in: A. PEACOCK, H.WILLGERODT (Eds.), *Germany's Social Market Economy: Origins and Evolution*, London (Macmillan) 1989, pp. 46-67. First published: 1966.

BÖHM, F., W. EUCKEN, H. GROSSMANN-DOERTH: "The Ordo Manifesto of 1936", in: A. PEACOCK, H. WILLGERODT (Eds.): *Germany's Social Market Economy: Origins and Evolution*, London (Macmillan) 1989, pp. 15-26. First published: 1936.

BOETTKE, P. J.: "Where did Economics go Wrong? Modern Economics as a Flight from Reality", *Critical Review*, 11 (1997), pp. 11-64.

BUCHANAN, J. M.: *The Limits of Liberty - Between Anarchy and Leviathan*, Chicago (University of Chicago Press) 1975.

BUCHANAN, J. M.: "The Domain of Constitutional", *Constitutional Political Economy*, 1 (1990), pp. 1-18.

DOWNS, A.: *An Economic Theory of Democracy*, New York (Harper & Row) 1957.

EGGERTSSON, T.: *Economic Behaviour and Institutions*, Cambridge (Cambridge University Press) 1990.

EGGERTSSON, T.: "Limits to Institutional Reforms", *Scandinavian Journal of Economics*, forthcoming 1998.

EHRICKE, U.: "Experimentierklauseln in der Gesetzgebung und Entlastung der Justiz - eine Kritik im Schnittpunkt von Rechtssoziologie und Jurisprudenz aus der Perspektive des Marktes", in: U. MUMMERT, M. WOHLGEMUTH (Eds.), *"Aufschwung Ost" im Reformstau West*, Baden-Baden (Nomos) 1998, pp. 230-265.

EUCKEN, W.: "Staatliche Strukturwandlungen und die Krisis des Kapitalismus", *Weltwirtschaftliches Archiv*, 84 (1932), pp. 297-331.

EUCKEN, W.: *The Foundations of Economics - History and Theory in the Analysis of Economic Reality*, Berlin etc. (Springer Verlag) 1992. First published: 1940.

EUCKEN, W.: "What Kind of Economic and Social System?", in: A. PEACOCK, H. WILLGERODT (Eds.): *Germany's Social Market Economy: Origins and Evolution*, London (Macmillan) 1989, pp. 27-45. First published: 1948.

EUCKEN, W.: *Grundsätze der Wirtschaftspolitik*, Tübingen (J.C.B. Mohr [Paul Siebeck]) 1990. 6th revised edition; first published: 1952.

FELDMANN, H.: *Eine institutionalistische Revolution? Zur dogmenhistorischen Bedeutung der modernen Institutionenökonomik*, Berlin (Duncker & Humblot) 1995.

FiO (Forschungsinstitut für Ordnungspolitik Köln): "Die Soziale Markt-wirtschaft in der Konsenskrise?", *FiO-Brief*, 1/98 (1998).

FREY, B. S.: "Politico-Economic Models and Cycles", *Journal of Public Economics*, 9 (1978), pp. 203-220.

FURUBOTN, E. G.: *Future Development of the New Institutional Economics: Extension of the Neoclassical Model or New Construct*, Jena (Max-Planck-Institute for Research into Economic Systems) 1994 (= Lectiones Jenenses, Vol.15).

GWARTNEY, J. D., LAWSON R. A.: *Economic Freedom of the World. 1997 Annual Report*, Vancouver, B.C. (Fraser Institute) (copublished by Liberales Institut, Bonn) 1997.

HAYEK, F. A.: "Socialist Calculation III: The Competitive 'Solution'", in I. M. KIRZNER (Ed.): *Classics in Austrian Economics - A Sampling in the History of a Tradition, Vol. III: The Age of Mises and Hayek*, London (William Pickering) 1994, pp. 235-257. First published: 1940.

HAYEK, F. A.: *The Counterrevolution of Science - Studies on the Abuse of Reason*, Indianapolis (The Free Press) 1979. First published: 1952.

HAYEK, F. A.: "What is 'Social' - What Does It Mean?", in: F. A. HAYEK: *Studies in Philosophy, Politics and Economics*, Chicago (University of Chicago Press) 1967, pp. 237-247. First published: 1957.

HAYEK, F. A.: "Kinds of Rationalism", in: F. A. HAYEK: *Studies in Philosophy, Politics and Economics*, Chicago (University of Chicago Press) 1967, pp. 82-95. First published: 1965.

HAYEK, F. A.: *The Constitution of Liberty*, Chicago (University of Chicago Press) 1960.

HAYEK, F. A.: " Competition as a Discovery Procedure ", in: F. A. HAYEK: *New Studies in Philosophy, Politics, Economics and the History of Ideas*, London (Routledge) 1978, pp. 179-190. First published: 1968.

HAYEK, F. A.: " The Errors of Constructivism ", in: F. A. HAYEK: *New Studies in Philosophy, Politics, Economics and the History of Ideas*, London (Routledge) 1978, pp. 3-22. First published: 1970.

HAYEK, F. A.: *Law, Legislation and Liberty, Vol. II: The Mirage of Social Justice*, Chicago (The University of Chicago Press) 1976.

HAYEK, F. A.: " Liberalism ", in: F. A. HAYEK: *New Studies in Philosophy, Politics, Economics and the History of Ideas*, London (Routledge) 1978, pp. 119-151.

HILL H., KLAGES H. (Eds.): *Jenseits der Experimentierklausel*, Stuttgart etc. (Raabe) 1996.

HOMANN, K.: " Ethik und Ökonomik ", in: K. HOMANN (Ed.), *Wirtschafts-ethische Perspektiven I: Theorie, Ordnungsfragen, Internationale Institutionen*, Berlin (Duncker & Humblot) 1994, pp. 9-30 (= Schriften des Vereins für Socialpolitik, Vol. 228/I).

HOMANN, K., C. KIRCHNER: " Ordnungsethik ", *Jahrbuch für Neue Politische Ökonomie*, 14 (1995), 189-211.

HOPPMANN, E.: " Wettbewerb als Norm der Wettbewerbspolitik ", *Ordo*, 18 (1967), pp. 77-94.

HOPPMANN, E.: " Soziale Marktwirtschaft oder konstruktivistischer Interventionismus? ", in: E. TUCHTFELDT (Ed.), *Soziale Marktwirtschaft im Wandel*, Freiburg im Breisgau (Rombach) 1973, pp. 27-68.

KASPER, W., STREIT M. E.: *Institutional Economics - Social Order and Public Policy*, Cheltenham (Elgar) 1998.

KIRCHGÄSSNER, G: "Wirtschaftspolitik und Politiksystem: Zur Kritik der traditionellen Ordnungstheorie aus der Sicht der Neuen Politischen Ökonomie", in: D. CASSEL, B. RAMB, H. J. THIEME (Eds.), *Ordnungspolitik*, Munich (Vahlen) 1988, pp. 53-75.

KIWIT, D., VOIGT S.: " Überlegungen zum institutionellen Wandel unter Berücksichtigung des Verhältnisses interner und externer Institutionen ", *Ordo*, 46 (1995), pp. 117-148.

KIWIT, D., VOIGT S.: " Grenzen des institutionellen Wettbewerbs ", *Jahrbuch für Neue Politische Ökonomie*, Vol. 17 (1998).

KLEIN, D. B.: *On the Character and Social Contribution of Economists*, unpublished manuscript, mimeo 1998.

KOSLOWSKI, P.: *Prinzipien der Ethischen Ökonomie*, Tübingen (J.C.B. Mohr [Paul Siebeck]) 1988.

KOSLOWSKI, P.: *Ethics of Capitalism and Critique of Sociobiology. Two Essays with a Comment by James M. Buchanan*, Berlin etc. (Springer) 1996.

KOSLOWSKI, P.: "The Social Market Economy: Social Equilibration of Capitalism and Consideration of the Totality of the Economic Order. Notes on Alfred Müller-Armack", in P. KOSLOWSKI: *The Social Market Economy. Theory and Ethics of the Economic Order*, Berlin etc. (Springer) 1998, pp. 73-95.

KRUEGER, A. O.: "The Political Economy of the Rent-Seeking Society", *American Economic Review*, 64 (1974), pp. 291-203.

LEIPOLD, H.: "Neoliberal Ordnungstheorie and Constitutional Economics - A Comparison between Eucken and Buchanan", *Constitutional Political Economy*, 1 (1990), pp. 47-65.

LENEL, H.-O.: "Does Germany Still Have a Social Market Economy?", in: A. PEACOCK, H. WILLGERODT (Eds.), *German Neo-Liberals and the Social Market Economy*, London (Macmillan) 1989, pp. 16-39. First published: 1971.

LESCHKE, M.: *Ökonomische Verfassungstheorie und Demokratie*, Berlin (Duncker & Humblot) 1993.

MAYER, T.: *Truth Versus Precision in Economics*, New York (Cambridge University Press) 1993.

MÜLLER-ARMACK, A.: "Vorschläge zur Verwirklichung der Sozialen Marktwirtschaft", in: A. MÜLLER-ARMACK.: *Genealogie der sozialen Marktwirtschaft. Frühschriften und weiterführende Konzepte*, Bern/Stuttgart (Paul Haupt) 1981, pp. 90-109. 2nd edition; first published: 1948.

MÜLLER-ARMACK, A.: "Soziale Irenik", in: A. MÜLLER-ARMACK.: *Religion und Wirtschaft. Geistesgeschichtliche Hintergründe unserer europäischen Lebensform*, Bern/Stuttgart (Paul Haupt) 1981. 3rd edition; first published: 1950.

MÜLLER-ARMACK, A.: "Soziale Marktwirtschaft", in: *Handwörterbuch der Sozialwissenschaften*, Bd. 9., Stuttgart etc. 1956, pp. 390-392.

MÜLLER-ARMACK, A.: "The Principles of the Social Market Economy", in: P. KOSLOWSKI (Ed.): *The Social Market Economy. Theory and Ethics of the Economic Order*, Berlin etc. (Springer Verlag) 1998, pp. 255-274. First published: 1965.

MUMMERT, U.: *Informelle Institutionen in ökonomischen Transformationsprozessen*, Baden-Baden (Nomos) 1995.

MUMMERT, U., WOHLGEMUTH, M.: "Ordnungsökonomische Aspekte der Transformation und wirtschaftlichen Entwicklung Ostdeutschlands", in: U. MUMMERT, M. WOHLGEMUTH (Eds.): *"Aufschwung Ost" im Reformstau West*, Baden-Baden (Nomos) 1998, pp. 15-39.

MUSSLER, W.: "Integration durch Wettbewerb oder durch Intervention? Zur Zukunft zweier Integrationsstrategien in der Europäischen Union", in: G. PROSI UND C. WATRIN (Eds.): *Gesundung der Staatsfinanzen - Wege aus der blockierten Gesellschaft*, Köln (Hanns-Martin-Schleyer-Stiftung) 1996, pp. 318-326.

MUSSLER, W.: *Die Wirtschaftsverfassung der Europäischen Gemeinschaft im Wandel - Von Rom nach Maastricht*, Baden-Baden (Nomos) 1998.

NORDHAUS, W.D.: "The Political Business Cycle", *Review of Economic Studies*, 42 (1975), pp. 1969-1990.

NÖRR, K. W.: "Symbiosis with Reserve: Social Market Economy and Legal Order in Germany ", in: P. KOSLOWSKI (Ed.): *The Social Market Economy. Theory and Ethics of the Economic Order*, Berlin et al. (Springer Verlag) 1998, pp. 220-251.

OLSON, M.: *The Rise and Decline of Nations. Economic Growth, Stagflation, and Social Rigidities*, New Haven (Yale University Press) 1982.

PASCHA, W.: "Institutional and Evolutionary Economics in Germany", *Diskussionsbeiträge des Fachbereichs Wirtschaftswissenschaft der Gerhard-Mercator-Universität - Gesamthochschule Duisburg*, No. 205 (1994).

PELIKAN, P.: "Competitions of Socio-Economic Institutions: In Search of Winners", in: L. GERKEN (Ed.), *Competition among Institutions*, London (Macmillan) 1995, pp. 177-205.

PIES, I.: *Normative Institutionenökonomik. Zur Rationalisierung des politischen Liberalismus*, Tübingen (J.C.B. Mohr [Paul Siebeck]) 1993.

RADNITZKY, G.: "Marktwirtschaft: frei oder sozial?", in: G. RADNITZKY, H. BOUILLON (Eds.), *Ordnungstheorie und Ordnungspolitik*, Berlin etc.. (Springer Verlag) 1991, pp. 47-75.

RICHTER, R.: *Institutionen ökonomisch analysiert. Zur jüngeren Entwicklung auf dem Gebiet der Wirtschaftstheorie*, Tübingen (J.C.B. Mohr [Paul Siebeck]) 1994.

RUPP, H. H.: "Konzertierte Aktion und freiheitlich-rechtsstaatliche Demokratie", in: E. HOPPMANN (Ed.), *Konzertierte Aktion - Kritische Beiträge zu einem Experiment*, Frankfurt am Main. (Athenäum) 1971, pp. 1-18.

SALLY, R.: "Ordoliberalism and the Social Market: Classical Political Economy from Germany", *New Political Economy*, 1 (1996), pp. 233-257.

SCHARPF, F. W.: "The Joint Decision Trap", in: COHEN, FUNG (Eds.), *Constitution, Democracy and State Power*, London (Elgar) 1998 (forthcoming).

SIMON, H. A.: "Rationality as a Process and as a Product of Thought", *American Economic Review*, 68 (1978), pp. 1-15.

SCHMIDTCHEN, D.: "German 'Ordnungspolitik' as Institutional Choice", *Zeitschrift für die gesamte Staatswissenschaft (Journal of Institutional and Theoretical Economics)*, 140 (1984), pp. 55-70.

STREIT, M. E.: "Demand Management and Catallaxy - Reflections on a Poor Policy Record", *Ordo*, 32 (1981), pp. 17-34.

STREIT, M. E.: "Economic Order and Public Policy - Market, Constitution, and the Welfare State", in: R. PETHING, U. SCHLIEPER (Eds.): *Efficiency, Institutions, and Economic Policy*; Berlin etc. (Springer Verlag) 1987, pp. 1-21.

STREIT, M. E.: "Ordnungspolitische Defizite der deutschen Vereinigung", in: M. STREIT: *Freiburger Beiträge zur Ordnungsökonomik*, Tübingen (J.C.B. Mohr [Paul Siebeck]) 1995, pp. 342-356. First published: 1991.

STREIT, M. E.: "The Freiburg School of Law and Economics", in: P. J. BOETTKE (Ed.), *The Elgar Companion to Austrian Economics*, Aldershot (Edward Elgar) 1994.

STREIT, M. E. (1996a): "Ordnungsökonomik", in: *Gabler Volkswirtschaftslexikon*, Vol. 2, Wiesbaden (Gabler) 1996, pp. 814-843. Extended version in: "Ordnungsökonomik. Versuch einer Standortbestimmung, Max-Planck-Institut zur Erforschung von Wirtschaftssytemen", Diskussionsbeitrag 04-95.

STREIT, M. E.(1996b): "Competition among Systems as a Defense of Liberty", in: H. BOUILLON (Ed.), *Libertarians and Liberalism - Essays in Honour of Gerard Radnitzky*, Aldershot (Avebury) 1996, pp. 236-252.

STREIT, M. E.: "Die Soziale Marktwirtschaft - Auslauf- oder Exportmodell?", in: KONRAD ADENAUER STIFTUNG (Ed.), *Historisch-Politische Mitteilungen. Archiv für Christlich-Demokratische Politik*, 4 (1997), pp. 239-259.

STREIT, M. E.: "Has the Market Economy Still a Chance? - On the Lack of a Disciplining Challenge", in: P. KOSLOWSKI (Ed.), *The Social Market Economy. Theory and Ethics of the Economic Order*, Berlin etc. (Springer) 1998, pp. 50-63.

STREIT, M. E., W. KASPER: *The Institutional Foundations of Freedom and Prosperity - Lessons from the Freiburg School*, Sydney (Center for Independent Studies) 1992.

STREIT, M. E., A. MANGELS: "Privatautonomes Recht und grenzüberschreitende Transaktionen", *Ordo*, 45 (1996), pp. 105-135.

STREIT, M. E., G. WEGNER: "Information, Transactions and Catallaxy: Reflections on some Key Concepts of Evolutionary Market Theory", in: U. WITT (Ed.): *Explaning Process and Change - Approaches to Evolutionary Economics*, Ann Arbor (University of Michigan Press) 1992, pp. 125-149.

STREIT, M. E., M. WOHLGEMUTH: "The Market Economy and the State. Hayekian and Ordoliberal Conceptions", in: P. KOSLOWSKI (Ed.), *Economic Ethics and the Theory of Capitalism in the German Tradition of Economics: Historism, Ordo-Liberalism, Critical Theory, Solidarism*, Berlin etc. (Springer) forthcoming 2000.

TIETZEL, M.: "Der Neue Institutionalismus auf dem Hintergrund der alten Ordnungsdebatte", *Jahrbuch für Neue Politische Ökonomie*, 10 (1991), pp. 3-37.

TOLLISON, R. D.: "Rent Seeking: A Survey", *Kyklos*, 35 (1982), pp. 575-602.

TRUHN, J. P.: "Korreferat zu Roland von Lilienfeld-Toal und Frank Schneegaß", in: U. MUMMERT, M. WOHLGEMUTH (Eds.): *"Aufschwung Ost" im Reformstau West?*, Baden-Baden (Nomos) 1998.

VANBERG, V.: "Spontaneous Market Order and Social Rules: A Critical Examination of F. A. Hayek's Theory of Cultural Evolution", *Economics and Philosophy*, 2 (1986), pp. 75-100.

VANBERG, V.: "'Ordnungstheorie' as Constitutional Economics - The German Conception of a 'Social Market Economy'", *ORDO*, 39 (1988), pp. 17-31.

VANBERG, V.: "Hayek as Constitutional Political Economist", *Wirtschaftspolitische Blätter*, 36 (1989), pp. 170-182.

VAUBEL, R.: "Der Mißbrauch der Sozialpolitik in Deutschland: Historischer Überblick und Politisch-ökonomische Erklärung", in: G. RADNITZKY, H. BOUILLON (Eds.): *Ordnungstheorie und Ordnungspolitik*, Berlin etc. (Springer Verlag) 1991, pp. 173-201.

VOLCKART, O.: *Stabilität und Wandel der Ständegesellschaft aus intitutionenökonomischer Perspektive*, Max-Planck-Institut zur Erforschung von Wirtschaftssystemen, Diskussionsbeitrag 07-97 (1997).

VOLCKART, O. (Ed.): *Frühneuzeitliche Obrigkeiten im Wettbewerb: Institutioneller und wirtschaftlicher Wandel zwischen dem 16. und dem 18. Jahrhundert*, Baden-Baden (Nomos) 1998.

VOIGT, S.: "Positive Constitutional Economics - A Survey", *Public Choice*, 90 (1997), pp. 11-53.

WAGENER, H.-J.: "Some Theory of Systemic Change and Transformation", in his (Ed.), *On the Theory and Policy of Systemic Change*, Heidelberg (Physica) 1993, pp. 1-20.

WATRIN, C.: "The Social Market Economy: The Main Ideas and Their Influence on Economic Policy", in: P. KOSLOWSKI (Ed.), *The Social Market Economy. Theory and ethics of the Economic Order*, Berlin etc. (Springer Verlag) 1998, pp. 13-28.

WEISSBERG, R.: "The Real Marketplace of Ideas", *Critical Review*, 10 (1996), pp. 107-121.

WILLGERODT, H.: "Probleme der deutsch-deutschen Wirtschafts- und Währungsunion", *Zeitschrift für Wirtschaftspolitik*, 39 (1990), pp. 311-323.

WILLIAMSON, J.: *Latin American Adjustment: How Much Has Happened?*, Washington (Institute for International Economics) 1990.

WITT, U.: "The Endogenous Public Choice Theorist", *Public Choice*, 73 (1992), pp. 117-129.

WOHLGEMUTH, M.: "Economic and Political Competition in Neoclassical and Evolutionary Perspective", *Constitutional Political Economy*, 6 (1995), pp. 71-96.

WOHLGEMUTH, M.: "Has John Roemer Resurrected Market Socialism?", *The Independent Review*, 2 (1997), pp. 201-224.

WOHLGEMUTH, M.: "Institutioneller Wettbewerb als Entdeckungsverfahren. Zur Rolle von Abwanderung und Widerspruch im Europäischen Binnenmarkt", in: T. KÖNIG, E. RIEGER, H. SCHMITT (Eds.), *Europa der Bürger? Voraussetzungen, Alternativen, Konsequenzen*, Frankfurt am Main (Campus) 1998, pp. 64-88 (= Mannheimer Jahrbuch für Europäische Sozialforschung, Vol. 3).

Part C

The Methodological Foundations Compared: The Issue of Complexity

Chapter 8

Ordnungstheorie and Theory of Regulation Compared from the Standpoint of Complexity

ROBERT DELORME

I. Introduction

Wirtschaftsordnungstheorie, or simply *Ordnungstheorie*, and the French Regulation theory are two theoretical constructions in which the connection between the historical character of the economy and the necessarily theoretical dimension of economics has been and is most investigated.[1] How to articulate these two aspects has been a permanent subject of debate since the nineteenth century. This question was already addressed in the *Methodenstreit* and by Thorstein Veblen (Veblen 1919). However, both *Ordnungstheorie* and the theory of Regulation are uniquely articulated schemes in their attempts at providing theoretical alternatives to what they perceive to be an unsatisfactory state of scientific economics. This basic dissatisfaction arises in short from the lack of integration of history and theory. The aim of this contribution is to identify a common salient feature related to this issue and developed independently in these theories. It is this feature and the argumentation deriving from it which contribute mostly to the design of these theories as alternative research programmes to what they perceive as an economics mainstream. Although they do not explicitly name

---

1 In using the original German word for the theory of Economic Order and the English expression for the *théorie de la Régulation*, I am following what appears to be majority English-language practice in the literature on this subject.

and conceive it as "complexity", its substance is what a noticeable part of present-day complexity theory deals with. It even reaches such a high degree of complexity that it compels a change in the method of theorizing. Such a kind of complexity may be called *essential* complexity.

The aim pursued here is not to assess the general merits and disadvantages of these theories. Nor is it to compare them systematically. It is to concentrate on one aspect whose detection and identification in these theories does not seem to have been exploited, to the author's knowledge. This aspect is essential complexity. In the analysis of these theories taken individually and in isolation from one another, I experienced personally how the observer's attention tends to be attracted by the specific interest he or she has in mind, whereas comparison between the theories leads one to think about criteria and sort out those which are most important, and it can hardly avoid the irreducibility issue present in both of them. Essential complexity stems from it and from the kind of solutions the proponents of both theories strive for, although they never identify explicitly this notion as essential complexity.

I do not discuss whether these theories represent genuine scientific paradigms or research programmes. They are general theoretical settings or research programmes in a generic sense and are thought of by their founders as alternatives to what they consider as the mainstream in economic theory.

My presentation of *Ordnungstheorie* relies entirely on Walter Eucken's book, *The Foundations of Economics*, which was first published in 1939 and whose edition in English, which I am using, came out in 1950. *The Foundations of Economics* is the undisputed core exposition of the economic theoretical framework of what is known as the ordoliberal doctrine. Ordoliberalism is itself diversified (Wohlgemuth, below, chap.7). It contains a normative part oriented towards economic policy-making that I am leaving aside here. And the theoretical part pertains to law and economics. The *Foundations* address the economic side of it. The so-called French "theory of Regulation" emerged in the 1970s independently, and even in ignorance, of *Ordnungstheorie*, among a group of economists based in Paris.

These theoretical bodies have naturally no monopoly in treating institutions. And complexity was felt to be an important problem for economic theory much earlier, notably by Veblen, Marshall and Keynes. However, it is the consequences drawn from complexity and the method elaborated to take them into account which single out these two theories, as we attempt to demonstrate in the following sections. In the next section I stress the morphological analysis of the economy developed by these theories. It appears as the kind of solution devised in both of them in order to integrate essential complexity, a notion presented in section III. This integration is, however, not entirely satisfactory. I suggest in conclusion that improving it may depend on the elaboration of a theory of essential complexity in its own right.

II. Two Theories of Morphological Formations and Processes

1. *Ordnungstheorie* in Eucken's Foundations of Economics

The *Foundations of Economics* is the most complete statement about *Ordnungstheorie* understood from its economic side. From our standpoint, the *Foundations* are divided into two parts. First a problem is identified for which available theories are considered unsatisfactory. It is the question of taking into account in an integrated way two opposite notions, the individuality attached to history and the generality necessary for theory, what Eucken calls the "Great Antinomy". It is the source of complexity, since they are irreducible to one another. The second part develops Eucken's answer and solution to this issue.

Eucken proposes to extend the analytical apparatus of economics and to include in it a morphological analysis of economic phenomena. In his repeated plea for basing study on actual, everyday economic life, he contends that this demands an understanding of the different forms in which economy activity takes place, "and therefore that a morphological analysis must precede a theoretical analysis" (p. 11),[2] since "the morphological study of economic history reveals a limited number of pure forms out of which all actual economic systems past and present are made up. To work out these pure forms, and at the same time provide a basis for theoretical analysis which will explain the course of economic process, are our two tasks" (p. 10). Eucken's ambitious goal is to construct "a morphological and theoretical

2 All references are to W. Eucken, *The Foundations of Economics*, 1950.

system which is able to comprehend *all* (original emphasis) economic life [...] and which is able to catch, as in a net, the changing shape of economic reality" (pp. 10-11). At this stage it is necessary to set out briefly the strategy followed by Eucken since it determines the substance of *Ordnungstheorie*.

a) The Great Antinomy

A cornerstone of the *Foundations* is what Eucken describes as the "Great Antinomy", introduced with capital letters. This notion runs throughout the book and commands Eucken's striving to overcome the difficulty it presents to scientific enquiry in social sciences, especially in political economy.

This is the central theme of the *Foundations*. Economic problems have a dual aspect, which has led to a dual approach to them, one historical, the other theoretical. This is not new. It culminated in the *Methodenstreit*. A separation established between individual-historical economics developed in Schmoller's pure empiricism and a general-theoretical economics illustrated by Menger's dualism of theoretical and historical economics.

In this debate "both parties were wrong, nor was the truth somewhere in the middle between the two. Neither Menger's dualism, of which Schmoller perceived the danger, nor Schmoller's pure empiricism, the failure of which Menger foresaw, does justice to economic reality. A new start is necessary" (pp. 324).

Yet the subject-matter of economic reality has a dual aspect according to Eucken. "The economist has to see economic events as part of a particular individual-historical situation if he is to do justice to the real world. He must see them also as presenting general-theoretical problems if the relationships of the real world are not to escape him" (p. 41). Then Eucken asks: "How can he [the economist] combine these two views? If he does only the one or only the other, he is out of touch with the real world" (p. 41). Here lies a deep tension, what Eucken calls the Great Antinomy. Eliminating or overcoming it can hardly be achieved by getting historians and theorists to work together (p. 43). This antinomy is larger in recent decades than in the past. The structure "of our social economy [is] becoming more and more complex, thus making theoretical analysis more and more clearly indispensable" (p. 44).

The uniformity of chemical reactions or of the movement of bodies or growth of plants "makes it possible to formulate theoretical questions and generally valid physical, chemical, or biological laws. No such uniformity exists in the economic world, which exhibits an immense variety of forms and historical processes" (p. 42). How, then to integrate the individual-historical nature of economic life with the general-theoretical study without which "there can be no scientific experience in this field, just as there cannot be without individual-historical study" (p. 42)?

According to Eucken, "economic reality compels the economist to formulate his first main problem as a historical one, but it also forces him in quite another direction" (p. 37). He must understand the interrelations of which every activity is a part. And this cannot be achieved "*simply by looking directly at contemporary economic reality* (original emphasis) or by "the simple direct contemplation of the facts of economic history" (p. 38) even when the economist has experience of economic reality (p. 39). The usual historical method fails. Quoting Lotze, the author of *Logik*, published in 1874, Eucken states that theorizing enables man "to transform what is given to us as *happening* together into what is *connected* together" (p. 40, original emphasis).

Thus economic life presents the economist with a "complex phenomenon": "There is only one way out of this situation. We must try to break down and analyse the complex phenomenon into its different components" (p. 40). Understanding reality and its relationships requires to put the problem in a general form accessible to theoretical investigation if one wishes to achieve "scientific experience" in place of "everyday" experience" and treat the problem as a general-theoretical issue.

Due to this fundamental duality, economic life appears as a complex phenomenon and cannot be reduced either to its individual-historical character or to a general theoretical problem. The historical method or direct observation provides a description of a phenomenon but without enabling to understand it since it lacks the theoretical concepts and does not contain the tools for establishing relationships. The theoretical method may enable to formulate abstract relationships but at the risk of loosing contact with the real world of historical variety and contextualization of facts and events: "We would no longer see anything of the variety of actual historical phenomena and of individual facts" (p. 42).

The Great Antinomy presents irreducibly two sides and, according to Eucken, has not been treated satisfactorily by economists, either because only one side is treated or because, when the antinomy is taken into account, it is theoretically flawed. In this last case, Eucken criticizes the theories of the stages of economic development (from List to Sombart) and of the styles of economic development (Spiethoff). In his view, these notions characterize moments of reality but cannot be applied to other contexts or to past phenomena.

The remainder of the *Foundations* is devoted to Eucken's answer to this problem of integrating the two sides of the antinomy.

b) The answer to the Great Antinomy

Eucken takes a radical position: "Because established doctrines fail before the Great Antinomy, we must make a completely new approach to the subject-matter itself. Simply to continue on existing lines, either "historical" or "theoretical", is impossible. From now on we shall disregard for the time being all existing economic doctrines and hold quite radically to this point" (p. 101).

However, this does not mean rejecting theorizing. The solution he proposes is to start by looking at "everyday economic life and asking questions about it" (p. 101), notably how the economic process, of which each individual economic fact is a part, hangs together as a whole (p. 63). The *Foundations* proposes nothing less than a reconstruction of economics based on a redefinition of its object and its method. The object of economics is stated by going back to the classical economists: how does a multitude of autonomous decisions and actions hang together in a more-or-less unified whole? The method is confronted by the problem well described by the subtitle of the *Foundations*: "History and Theory in the Analysis of Economic Reality". It is the integration of history and theory which leads to the issue of the Great Antinomy. Eucken proposes to solve it primarily by a morphological analysis. Morphological analysis is conceived of as preparing for the next step of theoretical analysis.

Form and process appear as two central notions in this argumentation. The differing structure of the problem of the forms in which economic life unfolds and of the course of the daily process within these forms, determines the character of economics (p. 298). Identifying the basic forms through the observation of everyday life of say, the household, the farm, the firm in a given place and at a given time is the only way to remain grounded in what Eucken calls "economic reality". The immense variety of everyday economic life can be rendered tractable thanks to abstraction. But here it is a specific kind of abstraction. Eucken develops a method of "isolating abstraction" which starts with the observation of a historical fact or phenomenon whose individual features are extracted, and ends with building ideal-types out of these individual features. Isolating abstraction is the abstraction of "specially significant characteristics" (p. 326), of the "distinguishing or significant characteristics" (p. 107), of "significant salient characteristics" (p. 332).

Isolating abstraction is contrasted with the "generalizing" abstraction "which seeks to fasten on to what is common to *many* (original emphasis) phenomena, and with which the constructors of "stages" and "styles" of development work". (p. 107) Eucken insists repeatedly on this distinction. He seeks to identify recurrent elementary forms of economic life from which ideal types are built. It can thus be said that the phenomenological unity of a fact is maintained in isolating abstraction whereas generalizing abstraction involves a "withdrawal from actual economic phenomena" since it leads to identifying common traits of different phenomena. Generalizing abstraction "must take second place" in the definition of economic systems (p. 299). The influence of Husserl's principle of phenomenological reduction is manifest in this reduction to the salient features of an individual phenomenon.

Everyday economic life "everywhere and at all times" shows that men "act on the basis of economic plans for overcoming their shortages of goods". Economic plans and the data which influence planning are "the point of entry into the real economic world". This is where the study "of either the structure of any actual economic system" or "of everyday economic processes" has to begin, "and this beginning decides the rest of the path" (p. 303).

The notion of plan occupies a central place in Eucken's construction. He tackles it through the unique criterion of individual freedom and is thus led to differentiate between the two ideal-types of a centrally directed economy, controlled by a single authority, and the exchange economy composed of many independent agents, each with her/his own plans which have therefore to be coordinated. This coordination will itself depend on the forms of the market and of the monetary system.

By starting from the degree of freedom of planning in the exchange economy, one discovers a multitude of cases. But it results from the different ways in which the elements are combined. Indeed, the number of pure formative elements or basic forms is limited. A morphological scheme is analogous to the individual letters of an alphabet out of which a huge variety of words can be formed.

Different economic systems can be identified through applying this morphological apparatus. Indeed, an economic system or order "comprises the totality of forms through which the everyday economic process at any particular time or place, past or present, is actually controlled" (p. 227).

The understanding of the different economic systems is a first moment of the scientific understanding of economic reality, to paraphrase Eucken. We can interpret it as a kind of bottom-up move from the observation of facts of everyday economic life to abstracting pure forms and their particular combination in an economic order. At this "top level" the economic order (or system) is an ideal-type. It remains to understand the actual course of economic events in a kind of top-down move back to the interrelationships of everyday economic activity, by applying theoretical analysis. Here Eucken seems to rely on the tools provided mainly by microeconomic and business cycle theory. He advises application of the "relevant theory" (p. 237) but also seems to pay little attention as to how we discover it or make sure that it is the relevant theory. Then comes the question of power. Eucken sees economic power as a question "of the greatest importance" (p. 263). Power is the opposite of freedom and may be characterized by applying morphology and theory to situations grasped with a historical perspective. On the first account, the more the form of the market approaches that of a monopoly or a monopsony, the larger is the power of the economic unit. With respect to theory, the larger the elasticity of demand, the less powerful is the position of the supplier. The less elastic supply is, the smaller is the power of the supplier. Beyond these particular instances complete competition is the form of market in which power is minimized, which explains the central place given to competition in Eucken's writings on economic policy.

Finally, Eucken's book offers an instance of an approximate systemic theory before systems theory developed after the 1950s. "Thinking in orders" (or systems) is the motto of *Ordnungstheorie*. From the start Eucken quotes approvingly a statement saying that economic life is an organic whole. He claims that for this reason the attempt to formulate independent theories of production, exchange, distribution and consumption must be abandoned (p. 320) and that problems must be unified and complete if justice is to be done to economic reality (p. 28). The modern theory of complex systems contains the central proposition that a complex system can be modelled only by a complex system (Le Moigne 1990). An almost similar statement was made by Eucken: " ... because economic events make up an interdependent whole, economics itself must form an interconnected body of knowledge. It does this by developing a morphological scheme and a systematic theory. The structure and interrelationships of events, and the way they fit together, has to be matched by the interrelationships in the system of our scientific knowledge. Otherwise scientific knowledge is incomplete. To be systematic it must be organised as an interrelated unity" (p. 304).

This sets out clearly what Eucken's ambition was. Did he succeed in explaining "genuinely" and "free of all bias and subjectivism" the "interrelationships of everyday economic life"? (p. 33). Another distinctive feature is the interdependence of orders (or systems). It applies first to the relationships between partial orders (today, we would say subsystems) within an economic order. It pertains also to the connections with other orders, notably the legal system. And we can extend it to a defence of interdisciplinarity when Eucken emphasizes the points of contact between economics and other sciences, such as history, business administration and law.

2. The theory of Regulation

Regulation in the theory of Regulation denotes the process of mutual adjustment of production and demand at a global level. It emerges from local economic adjustments occurring within a given configuration of institutional forms.[3]

What appeared *ex post* as a rather consistent programme arising from the findings of economists sharing similar basic dissatisfactions with the then-prevailing approaches in economics originated in a pragmatic and rather disparate fashion among economists mainly based in Paris in the early seventies and became progressively perceived as the Regulation school. It started practically simultaneously with the study of long-run and structural change in the French and the US economies based on a macroeconomic, sectoral and historical, empirical approach. Michel Aglietta analysed the growth of the US economy in the long run (Aglietta 1976). A simultaneous enquiry into the secular evolution of output and prices in France was conducted at CEPREMAP (Centre d'études prospectives d'économie mathématique appliquées à la planification, Paris) by several researchers. It

3 This presentation of the theory of regulation is inspired by Delorme (2000). See also the contributions by Boyer and Vidal (above, chaps. 1, 2).

gave rise to several publications, the first of which was Robert Boyer and Jacques Mistral (1978).

The approaches then dominant in France were Keynesian, neoclassical and Marxist. Keynesian-based modelling, with its emphasis on aggregate supply and demand in the short run, was felt too limited for dealing with mid-term and long-run changes, notably those concerning production. Neoclassical theory, with its emphasis on rationally substantive agents, on coordination obtained exclusively through markets and on equilibrium, was considered too narrow and static. Last but not least, those who engaged in the Regulation programme shared a basic interest in the way Marx introduced an analysis of the long-run dynamics of capitalism with an emphasis on social relations and on the process of accumulation. But they rejected the somewhat mechanical and deterministic interpretations of Marxism and the idea of a predefined end state to the evolution of capitalist economies. These latter criticisms are a key to understanding what makes Regulation distinctive and also what makes it belong to the broader contemporary thrust towards an open-ended evolutionary-institutional-socioeconomic stance. It was these premises, and the convergence of my own findings on the long-run growth of public spending in France with those of early Regulationists, that made me join the Regulation perspective.

It is worth adding that Regulation has diffused and raised interest among a growing number of scholars. A recent survey of the theory of Regulation contains 54 chapters written by 46 authors (Boyer/Saillard 1995).

Indeed, two main insights arise. First, macroeconomic theorizing in general is often criticized for not taking institutions properly into account. What makes the theory of Regulation truly original within economic theory is its attempt to include institutions in macroeconomic theorizing and to build a frame in which institutions play an explicit and important role. A second insight comes from the open-endedness of the theory of Regulation. It is an open-ended institutionalism. This immediately creates a challenge: how to theorize without the closure associated with the more deterministic standard ways of thinking in the economic discipline? It is well known that this is the main criticism usually addressed to what has come to be called "old institutionalism" by proponents of the "new institutionalism".

Three concepts are at the basis of Regulation. They are the institutional forms, the regime of accumulation and the mode of regulation. The dynamics of regulation arises from their interplay.

The **institutional forms** set a bridge between observed regularities of socioeconomic life and agents' behaviour. Agents act within basic rules of interaction or institutional forms which consist in a codification of the main social relationships. It is consistent with the assumption of bounded rationality of agents. Five institutional forms are identified. First is the monetary and credit relationship. The configuration it takes depends on the type of monetary management, the kind of causality between money and credit, the structure and degree of development of national and international financial systems. The wage–labour nexus is the second institutional form. It has a key role since it is conceived as covering the main features of work organization and of the standard of living of wage-earners. Five components are distinguished by Boyer (1988a) : the organization of the work process; the stratification of skills; workers' mobility; direct and indirect wage formation and the use of wage income. Third are the forms of competition. A basic distinction is between traditional price competition and oligopolistic competition. Fourth is the configuration of the state. It is characterized in recent works as a mode of interaction between the state and the economy (Delorme 1995) in order to convey the idea of stabilized configurations over some periods of time with differences through history for a given country and also differences across countries. Finally, there is the relationship between an economy and other economies in the world. It is the mode of interaction with the international economy or, equivalently, the type of articulation with the international regime.

The logic of accumulation is a central feature of a capitalist system. History provides evidence that accumulation is not linear: there are cumulative growth patterns separated by crises. These patterns can be viewed as stabilized configurations of the economy over some periods of time. The concept of a **regime of accumulation** is aimed at depicting such patterns. It is defined by the set of regularities which allow a general compatibility between capital formation, production, the distribution of income and the genesis of demand. It expresses macroeconomic consistency. Given the evolution of technical coefficients, income shares, the composition of demand and time lags, it is possible to model these regularities in a dynamic setting.

Empirical investigations have revealed different regimes through history and across countries at a given period of time. A variety of regimes of accumulation exists, depending on the character and intensity of technical change and on the size and the structure of demand or, broadly speaking, on the norms of production and the norms of consumption.

The **mode of regulation** is a concept making it possible to pass from partial regularities involving numerous agents acting autonomously to the possibility of a consistent dynamic system. Several ways of adjusting production to demand, credit to money, income distribution to demand formation are possible. Institutional forms may, or may not, induce a coherent adjustment process for the economy as a whole. Institutions and forms of organization (markets, hierarchies – private firms and public units – and networks) jointly determine economic and social dynamics. Hence regulation depends on the behaviour of agents and of social groups in so far as it ensures the relative coherence and stability of the existing regime of accumulation. Then a more specificied definition of regulation can be given at this point. It is a conjunction of mechanisms of adjustment associated with a configuration of institutional forms. It provides an alternative to the notion of static equilibrium. A mode of regulation is a set of rules and individual and collective behaviours which render potentially conflicting decentralized decisions mutually compatible without the need for decision units of gathering the information necessary to understand the working of the entire system, and which regulate the regime of accumulation.

In the theory of Regulation, the long-run dynamics is seen as being discontinous. Periods of relative dynamic stability, during which basic regularities prevail, reach limits and leave room for phases of changes during which the consistency among previous components vanishes, with instability and disorder until a new consistency settles. These changes can be either structural or small. This is the reason why crises play such an important role in the theory of Regulation.

Tension and the potential for **crisis** are never absent from regulation. Indeed, although the term crisis is used in many ways, it is basic to the Regulation approach to distinguish two categories of crises: "small" and "large" crises. The former are of a rather cyclical nature. They are in the essence of **regulation**. They express the kind of self-equilibration process through which recurrent imbalances of accumulation occur within the system as a result of the necessary time-lags between the demand and capacity effects of investment, for instance. Variations in inventories, production, investment, employment and prices are part of these adjustments. These variations actually depict the usual business cycle. However, the institutional forms are likely to change only slowly, from cycle to cycle, leaving the character of regulation as a whole unaffected.

The latter are of a structural nature. In structural crises, the very process of accumulation becomes less and less compatible with the stability of institutional forms and the regulation which sustains it. In such a situation increasing doubts arise about the long-term viability of the system. It can no longer reproduce itself in the long run on the same institutional basis. Imbalances are such that within the given mode of regulation, former self correcting mechanisms become ineffective. Institutional forms become more and more questioned by the spreading of the misadjustments. Ultimately it is the whole combination of the mode of regulation, institutional forms and the regime of accumulation, which constitutes a mode of development (Figure 8.1) that may become questioned.

This distinction is central to the Regulation approach. Contemporaneous and lasting high levels of unemployment in many industrialized countries are for this approach the manifestation of a structural crisis which appeared in the early 1970s.

These notions render possible identification of varying institutional forms, regimes of accumulation and forms of regulation over time and across economies. Their combination constitutes a mode of development when some form of compatibility holds. Hence, post-Second World War growth is interpreted as the Fordist mode of development, combining intensive accumulation with mass consumption, modifications in the monetary regime with an increased place of credit-based money supply and primarily a shift in the wage–labour nexus (new wage norms involving the diffusion of productivity gains to wage earners on a nation-wide basis, extension of social security bringing a permanent improvement in consumption norms).

Figure 8.1: The basic notions of the theory of Regulation

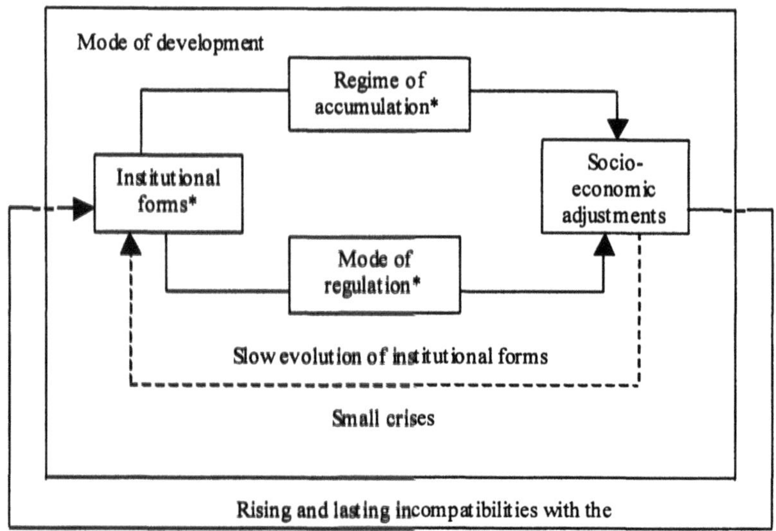

* Institutional forms: monetary regime
forms of competition
wage–labour nexus
place and role of the state
relationship with the international economy

* Regime of accumulation: dynamic compatibility between output, income distribution and demand.

* Mode of regulation: channelling of individual and group behaviours in accordance with the regime of accumulation; reproduction of institutional forms.

Source: Adapted from Boyer (Boyer/Saillard 1995)

A distinctive feature of the Regulation approach is to make it possible to account endogenously for both growth and crisis. Hence, the very development of Fordism as a social, economic and technical regime led to new conflicts and imbalances which, beyond some threshold, induced tendencies towards a lasting slowdown of growth, stagnation and pressures towards changes in institutional forms.

This much-schematized sketch of the theory of Regulation is only intended to present some basic information about this theory. The theory of Regulation developed in its founders' ignorance of *Ordnungstheorie*. Its origins are quite different. The central question arises from a macroeconomic level, although it converges with the classical basic interrogation about the coordination in a unified whole of a multiplicity of decentralized autonomous behaviours. This very question is also Eucken's starting-point, although he deals with it in a different way, emphasizing the micro-level of plans as a point of departure.

The necessary integration of economic history and economic theory is at the heart of the problematic of the theory of Regulation. The answer proposed by the theory of Regulation relies on an architecture of hierarchized intermediary notions of forms, regimes and processes identified through a historical and morphological investigation. This method renders it possible to account for their variability over time and across space and to articulate partial regularities with global regulation.

It provides a macroeconomic theory *with* institutions and an open-ended institutionalism *with* a theoretical core. It is worth pointing out these two features, since they are distinctive insights of the theory of Regulation into two traditional dilemmas of economics, namely systematically articulating institutions with macroeconomics in a unified setting and rendering compatible the open-endedness of history with theorizing.

The parallel with *Ordnungstheorie* is striking. Both theories attempt to answer the question of integrating the two traditionally irreducible features of history and theory in economics. They develop at a general level of theorizing. *Ordnungstheorie* relies on a microeconomic standpoint and provides a morphological theory highlighting form and process. The theory of Regulation starts from a macroeconomic standpoint and develops a morphological theory articulating form and process at the intermediary and macroeconomic levels.

III. A Common Salient Feature: Essential Complexity

1. The meaning of essential complexity

Complexity is increasingly evoked in the economic literature either to denote an otherwise undefined very important difficulty in solving some problem or to designate the growing body of work based on non-linear dynamics, especially chaos theory. Indeed, A. Kirman, in his reflection on the evolution of economics, refers only to the latter (Kirman 1997). This is not the place to develop this point, but it can be shown that it is a partial view (Delorme 1999). Non-linear dynamics is but one form of complexity. Complexity means the irreducibility to a satisfactory level of reduction. It can be applied to a phenomenon and to the knowledge we have of this phenomenon. In this case "knowing" means reducing our ignorance. Whether we perceive a phenomenon as complex or not will depend on our level of aspiration in terms of reduction of our ignorance. Taking again the example of chaos, chaos is complex because of the irreducibility of the deterministic unpredictability of the evolution it covers, to the extent that the goal is to obtain predictability. But things run differently here. The Great Antinomy presents *Ordnungstheorie* with a complex problem to the extent that history and theory appear irreducible to one another and that Eucken rejects considering them as dualistic notions, put side by side. A similar irreducibility arises in the theory of Regulation

Other theoretical systems are based on the recognition of a complexity intrinsic to the subject-matter of economics or political economy. This is the case for the Keynesian and the Hayekian approaches. It is also true of the so called "old institutionalism" in the tradition of Veblen.

An important feature of Eucken's argumentation and of the theory of Regulation, which does not seem to have attracted particular attention until now, lies in the reasons these authors find for departing from the established doctrines. In both cases, the morphological schemes appear as the common solution to the problem of irreducibility with which they are confronted. It is an irreducibility of such an importance that it leads these authors to reject unambiguously the available doctrines and the methods attached to them.

We already evoked Eucken writing that a "new start is necessary" and that a "completely new approach" must be made "because the established doctrines fail before the Great Antinomy". The Great Antinomy does not leave untouched the method for treating the subject-matter. It is different from the usual treatment of chaos for which it would appear difficult to find a declaration in the literature about the new methodological start it would involve. In another study we designate the former kind of complexity by reflexive complexity and the latter by object-based complexity. A synonym for reflexive complexity is essential complexity. We use it here since it has already been introduced in another context by F. A. Hayek (1989-1974; 1967). It is essential in the sense that it entails a profound change in scientific enquiry while non-essential complexity can be treated satisfactorily with available methods.

An example may help clarify this central difference. We borrow it from the article by W. Weaver (1948) which inspired F. A. Hayek. Weaver identified three degrees of complexity ranging from what he called simplicity to disorganized complexity and organized complexity. Organization plays an essential role in defining these notions.

Simplicity (or organized simplicity) occurs when there are a large number of insignificant factors and a small number of significant factors. Disorganized (or unorganized) complexity occurs when there is a large number of variables with a high degree of random behaviour. The behaviour of gas molecules is a common example. These two situations are tractable or reducible by the use of well-known analytical methods, the former by concentrating on few, specific elements, and the latter by statistical methods of calculation of the average properties of many variables. Organized complexity stands between these two polar cases and cannot be reduced either to few variables or to a large degree of randomness.

Essential complexity entails the same consequences as organized complexity. Unlike most fields of the physical sciences, social sciences have properties of essential complexity which render problematic importation of the methods of physical sciences into economics. The criticism of scientism by Hayek follows from this.

In our terms, Hayekian irreducibility in dealing with structures of essential complexity comes from the impossibility of depicting their significant properties by models having fewer variables than they do (Hayek 1967). It is a form of incompressibility. All in all, Hayek does not go as far as Eucken or the theory of Regulation in developing a theoretical scheme. He remains at a rather doctrinal, normative and methodological level. Essential complexity is simply said by him to put limits on what we can expect science to achieve, notably to its goal of prediction.

From these considerations we are entitled to ask what kind of change and what theoretical alternative Eucken and the theory of Regulation bring. An answer can be found by reference to the notion of a scientific research programme introduced by I. Lakatos (Latsis 1976). According to the definition presented below, it is not unreasonable to consider that *Ordnungstheorie* was, and theory of Regulation is, conceived of as a research programme.

A research programme is an organic unity which contains a rigid part, with essential components, and a flexible part. The rigid part includes a hard core and the heuristic. The hard core consists of the axioms and of the basic assumptions accepted by the scientists supporting the programme. They are not submitted to a test. The heuristic is divided into two parts. The positive heuristic consists of the "dos", the set of suggestions on how to construct hypotheses and testable variants of the research programme. It contains guidance as to how the programme should unfold, what falls within and what falls outside its scope. The negative heuristic comprises the "don'ts"; it indicates what should be avoided.

The flexible part or "protective belt" of a research programme is made of the non-essential, modifiable and replaceable components. They are the auxiliary assumptions and specific theories that can be submitted to the test of their validity and empirical evidence without compromising the hard core.

To take an image, a research programme can be pictured as a set of concentric circles, with the hard core at the centre and the protective belt schematized by the other circles. The protective belt is not uniform. The further we move from the centre towards the periphery the lesser the strength of the links with the core principles. The competition between research programmes in fact mainly occurs between peripheral, local theories and leaves untouched the hard cores. Whether a research programme is expanding, cumulative and fruitful or not depends essentially on what occurs within the protective belt. The kind of systematicity which is established in the research activity unfolding in the periphery is notably an important condition for cumulativeness and for assessing the consistency with the hard core. It is a central factor of clarification and for communication among the scientists adhering to the programme, and for communication with other scientists and other programmes. It serves both "internal" communication and "external" communication.

Two broad dimensions relevant for *Ordnungstheorie* and the theory of Regulation emerge at this point and offer a basis for comparison. They are the fruitfulness and the systematicity.

2. The challenge

Our focus on essential complexity helps understand the challenge facing the two theories. It directs attention to the way they are, or are not, not only scientific research programmes but, on top of this, whether they are alternative, competitive programmes for what they both perceive as a mainstream programme, namely the neoclassical programme. We do not need to insist on the fact that this is the common term of comparison and the actual test for any alleged alternative programme. It has to pass the test of comparison with what an important majority of economists regard as the legitimate way to do economic theory. On what terms can this comparison be made?

A distinguishing feature of *Ordnungstheorie* and of the theory of Regulation is that their founders discovered the irreducibility to one particular available theory of what they deemed especially relevant in their respective research paths. Consequently, they were driven to construct their own theories. The theories they obtain are naturally conceived at a high level of generality since they arise out of a rejection of basic principles of the available theories. This entails designing alternative, basic principles. In this sense they are general theories. If we extend to them the notion of a scientific research programme, we must consider the articulation between the general theory level, which furnishes the guidance, and the periphery or local theory level which constitutes the protective belt. Whether the general theory level is considered as a hard core or not, the question of an axiomatic, of the consistency assessed through the derivation from higher, explicit, first principles can hardly be eschewed. This is all the more true when a comparison with the neoclassical programme is pursued, since its axiomatic is often presented as its first strong point.

The other possibility is to emphasize the local theory side. If fruitfulness is to occur somewhere, it is mainly at the local theory level. Then whatever the ambiguities theories may contain, looking at them in terms of organic entities draws attention to their fruitfulness and to what we may call their systematicity. Systematicity pertains to the hard core, and to the strength of the link between it and what is practised. Paraphrasing M. Blaug who asks "Do economists practise what they preach?" (Blaug 1976 p. 171) one may consider that the more what is preached is practised, the more systematic it is. This pragmatic systematicity comes in addition to the systematicity contained in the hard core and ensuring the consistency of the declared principles of the theory. How do the two theories fare on this account? The theory of Regulation, in its greater part, follows clearly a strategy insisting on the fruitfulness side. The systematicity side consists essentially in basic concepts (mode of regulation, set of institutional forms, regime of accumulation). The reference to any derivation from an axiomatic is rejected on the ground that it would entail a too-restricted view which would be at any rate incompatible with preserving the open-endedness of the theory. Moreover, it can be claimed that the neoclassical axiomatic is accompanied by numerous *ad hoc* assumptions thanks to which an overall consistency may still be proclaimed. But "ad hocness" plays an important role in it. This is the reason why some authors in the theory of Regulation defend a "well-tempered 'ad-hocness'" in the way of a "specific scientificity" (Amable *et al.* 1997). In this case *ad hoc* is understood in its common sense of methods appropriate to the subject studied, not as an assumption either insufficiently derived from the axiomatic or intended to produce the result sought by the theoretician and indispensable to that production (ibid. pp. 253-254). A similar claim is made by G. Duménil and D. Lévy (1997). They emphasize the need to articulate local theories submitted to direct empirical tests and reciprocal control as a test of their compatibility and to combine their explanatory powers at a global level, or at a general theory level in our own terms.

This means accepting the coexistence of various theories and models whose consistency depends heavily on a virtuous circle of mutual interplay, communication and criticism in the scientific community, a situation which seems to be still highly problematic in economics. Relying to such a degree on the fruitfulness side even at the price of floating, methodologically uncontrolled eclecticism, finds some support in D. Hausman's advice, "...I would urge economists to be more eclectic, more opportunistic, more willing to gather data, more willing to work with generalizations with narrow scope, and more willing to collaborate with other social scientists" (Hausman 1992 p. 280).

Eucken's presentation of *Ordnungstheorie* emphasizes the method. Its originality comes primarily from it. In a first step, Eucken claims to solve the complexity of the Great Antinomy by reducing it through a morphological analysis grounded in isolating abstraction and thinking in terms of orders. This leads to defining ideal-types of orders. In a second step, the morphological moment provides the particular, historical frame to which theory is applied. Theory here seems to mean mainly established microeconomic theory. It is only in a third step, not analysed here, that the consequences of the superiority attributed to competition in curbing economic power are fully developed in terms of economic policy principles and orientations. It constitutes *Ordnungstheorie* proper. All in all *Ordnungstheorie* seems to have been more fruitful for reflecting on economic policy than in the strict theoretical field. It seems that *Ordnungstheorie* did not give rise to significant cumulative local theorizing.

Finally, one has the impression that essential complexity creates a special difficulty for theorizing when it is considered, as is done here, in terms of articulating general and local theory. The two theories studied here are caught in a kind of unresolved trade-off between systematicity and fruitfulness. The theory of Regulation emphasizes the fruitfulness side and is expanding on this basis. *Ordnungstheorie* relied on an attempt at systematicity and almost vanished. Can the expansion of the former survive its founders and retain some durability? Can the latter be revived or prolonged with changes, as is envisaged (below, chap.7) by M. Wohlgemuth? These questions arise directly from the competition between research programmes in which these theories are taking part. Their ambition is to present alternatives to what they consider as a neoclassical mainstream. Yet, by comparison with it, they suffer from a weak articulation of systematicity and fruitfulness. Systematicity without fruitfulness is useless. And fruitfulness without systematicity can succeed in the short run but at the risk of rising confusion and of being swallowed up in the long run by the expanding character of the mainstream. Neoclassical economics can claim consistency obtained by its reliance on a chain of reasoning derived from axiomatic principles. It leaves unanswered the question of the validity of the predictions. But it relies on an explicit articulation between its foundations and the way it is applied. This is not saying whether this articulation is satisfactory or not. Simply, it is firmly claimed. No such firmly established linkage appears to exist in the two theories studied here.

IV. Conclusion. The Need for a Theory of Essential Complexity

The goal pursued here has been rather modest. It was to show how *Ordnungstheorie* and the theory of Regulation are facing a common challenge as general theories. This challenge follows from their attempt at integrating essential complexity. Difficulties follow from it in their cumulativeness (or fruitfulness) and in their capacity to become more systematized programmes able to compete with the already systematized neoclassical programme. I simply focused on a consequence of essential complexity for theorizing. I did not attempt a thorough investigation as to whether these theories succeed in this task. I wish only to point to the available strategies that can be followed. I take the liberty of repeating that the theory of Regulation puts the emphasis on the fruitfulness aspect more than on the systematicity side. With today's hindsight, *Ordnungstheorie* appears to have relied more on a detailed construction of a method than on its theoretical cumulativeness. I would claim that it has been more fruitful in its inferences for economic policy, through *Ordnungspolitik* and the social market economy orientation, however debated this success may be today. Research in progress by the author derived from his own experience suggests that a third strategy might be explored. It would be based on theorizing about essential complexity in its own right, since it is merely taken in these theories as a consequence of a profound problem of irreducibility which creates the need for a change of research programme. This change is dealt with in broad methodological terms. But nowhere can one find an attempt at exploring what essential complexity would mean if it were informed by a theory of essential complexity.

Given the pervasiveness of complexity and the important role it plays in these and other theories, one might think that it would be worthwhile to explore the connections between the various definitions and ways in which this notion is called for and to identify the possible regularities and properties which emerge. Such an investigation would at least be a starting-point for building a theoretical argumentation about what can be done with complexity. This is needed if we want to reduce the confusion that surrounds this notion. Theorizing might help clarify and systematize the knowledge of complexity and of its scope in economics.

In the two theories studied here, complexity is considered so important as to drive their authors to reject conventional theory and to design alternatives to it. The same situation and the same consequences can be observed in other general unorthodox theories. Among them are Veblen's plea for an evolutionary economics because of the historical and open-ended character of the subject-matter of economics (Veblen 1919), Keynes's rejection of atomism in the name of the organic interdependence characteristic of social phenomena (Keynes 1921; Carabelli 1988), and Hayek's view of complex phenomena calling for a specific scientific practice. However, considering the diversity of these theories, it would be interesting to know if there is some unity in the essential complexity that is operating and, if such were the case, how it remains compatible with such diversity. Or, if there is no unity, a systematic enquiry into the differences among essential complexities would be useful. Another issue is understanding when complexity becomes essential, or what is non-essential complexity. What are the kinds of complexity that do not entail the same consequences as essential complexity? Do they imply specific behaviours or not? We introduced above the notion of organization when we evoked Weaver's classification. Is it the only way to differentiate within complexity?

Examination of the theories mentioned here suggests strongly that morphological analysis is one theoretical strategy among others in order supposedly to overcome essential complexity. We are left again with a feeling of diversity about the implications of complexity which cannot be clarified in the absence of a firm way of connecting these differences. Why not suppose that a part of what is perceived as a lack of clarity or even as an ambiguity is indeed an intrinsic ambivalence reflecting the open-endedness associated with complexity? But such a property can hardly be established without a firm grip on the mechanism which generates it and renders it legitimate. This example illustrates how the rather confused atmosphere enveloping complexity can hardly be clarified without a much deeper investigation susceptible of leading to a theory of complexity.

If the above development has some relevance, then we are entitled to wonder whether essential complexity can remain so centrally present in these theories without being based on a theory of essential complexity in its own right. Such a theory would aim at identifying essential complexity, at comparing it with other notions of complexity, at establishing in what sense it is original or not, at surveying the theories which include it, without using the name, and at discussing systematically the solutions proposed and the obstacles remaining. Some reflection on this issue has already started in the philosophy of science and in methodology.[4] However, there seems to be no easy and straightforward way to apply it to theorizing on a substantial matter like the economy. It requires specific research. Work done in this direction has begun to be published.[5] Like any other theoretical work, it will have to pass the test of its fruitfulness.

References

AGLIETTA, M.: *Régulation et crises du capitalisme. L'expérience des Etats-Unis* Paris (Calmann-Lévy) 1976. Published in English as: *Regulation and Crisis of Capitalism*, New York (Monthly Review Press) 1982.

AMABLE, B., BOYER, R., LORDON, F.: "The *ad hoc* in Economics: The Pot Calling the Kettle Black ", in: A. D'AUTUME, J. CARTELIER (Eds.): *Is Economics Becoming a Hard Science?* Cheltenham (Edward Elgar) 1997, chap. 18.

BLAUG, M.: "Kuhn versus Lakatos or Paradigms versus Research Programmes in the History of Economics", in: S. J. LATSIS (Ed.): *Method and Appraisal in Economics*, Cambridge (Cambridge University Press) 1976.

BOYER, R., MISTRAL, J.: *Accumulation, inflation, crises*, Paris (PUF) 1978.

BOYER, R.: *Théorie de la régulation: une analyse critique*, Paris (La Découverte) 1987.

BOYER, R. (1988a): "Technical Change and the Theory of 'Régulation'", in: G. DOSI, C. FREEMAN, R. NELSON, G. SILVERBERG, L. SOETE (Eds.): *Technical Change and Economic Theory*, London (Pinter) 1988, chap. 4.

BOYER, R. (1988b): "Formalizing Growth Regimes ", in: G. DOSI, C. FREEMAN, R. NELSON, G. SILVERBERG, L. SOETE (Eds.): *Technical Change and Economic Theory*, London (Pinter) 1988, chap. 27.

BOYER, R.: "La crise de la macroéconomie, une conséquence de la méconnaissance des institutions?", *L'Actualité économique. Revue d'analyse économique*, 68/1-2 (1992), pp. 43-68.

BOYER, R., SAILLARD, Y. (Eds.) *Théorie de la régulation. L'état des savoirs*, Paris (La Découverte) 1995.

CARABELLI, A. M.: *On Keynes's Method*, London (Macmillan) 1988.

D'AUTUME, A., CARTELIER, J. (Eds.): *Is Economics Becoming a Hard Science?* Cheltenham (Edward Elgar) 1997.

4 See the references in Morin (1977-1991) and Le Moigne (1990).

5 Delorme (1997; 1999).

DELORME, R.: "An Alternative Theoretical Framework for State-Economy Interactions in Transforming Economies", *Emergo*, 2/4 (1995) pp. 5-24.

DELORME, R.: "The Foundational Bearing of Complexity", in: A. AMIN, J. HAUSNER (Eds.): *Beyond Market and Hierarchy. Interactive Governance and Social Complexity*, Cheltenham (Edward Elgar) 1997, chap. 2.

DELORME, R.: *Complexity and Evolutionary Theorising in Economics*, Paper presented at the International Workshop on the Evolution and Development of Evolutionary Economics, The University of Queensland, Brisbane, July 1999.

DELORME, R.: "Regulation as an Analytical Perspective. The French Approach", in: A. MIDTTUN, E. SVINDLAND (Eds.): *Approaches to and Dilemmas in Economic Regulation*. London (Macmillan) forthcoming 2000.

DUMENIL, G., LEVY, D.: "Should Economics be a Hard Science?", in: A. D'AUTUME,, J. CARTELIER (Eds.): *Is Economics Becoming a Hard Science?* Cheltenham (Edward Elgar) 1997.

EUCKEN, W.: *The Foundations of Economics*. London/Edinburgh/Glasgow (William Hodge) 1950.

HAUSMAN, D. M.: *The Inexact and Separate Science of Economics*. Cambridge (Cambridge University Press) 1992.

HAYEK, F. A.: "The Pretence of Knowledge", *Nobel Memorial Lecture*, 11 December 1974, *American Economic Review*, 79/6 (December 1989), pp. 3-7.

HAYEK, F. A.: "The Theory of Complex Phenomena", in: F. A. HAYEK: *Studies in Philosophy, Politics and Economics*. Chicago (University of Chicago Press) 1967, chap. 2.

KEYNES, J. M.: *A Treatise on Probability*, London (Macmillan) 1973 (= Collected Writings, Vol. 8). First edition: 1921.

KIRMAN, A. "The Evolution of Economic Theory", in: A. D'AUTUME, J. CARTELIER (Eds.), *Is Economics Becoming a Hard Science?* Cheltenham (Edward Elgar) 1997, chap. 8.

LATSIS, S. J.: "A Research Programme in Economics", in: S. J. LATSIS (Ed.): *Method and Appraisal in Economics*, Cambridge (Cambridge University Press) 1976.

LE MOIGNE, J.-L.: *La modélisation des systèmes complexes*, Paris (Dunod) 1990.

LE MOIGNE, J.-L.: *Le constructivisme. Tome 1: des fondements, Tome 2: des épistémologies*, Paris (ESF) 1994.

MORIN, E.: *La méthode*, 4 Vols., Paris (Seuil) 1977-1991.

VEBLEN, T.: *The Place of Science in Modern Civilization and Other Essays*, New Brunswick (Transaction Publishers) 1990. First edition: 1919.

WEAVER, W.: "Science and Complexity", *American Scientist* 36/4 (1948), pp. 536-544.

Chapter 9

How to Research Complex Systems: A Methodological Comparison of Ordoliberalism and Regulation Theory

CARSTEN HERRMANN-PILLATH*

I. Complex Systems and Current Economic Methodology

The economy is without doubt a highly complex system. However, this simple fact rarely gives economists cause to reflect on the methods appropriate to this special property of complexity. This is no trivial matter: British philosopher Roy Bhaskar (1989 pp. 44-5) has clearly shown that even widespread explanatory schemas, such as the deductive–nomological (D–N) model, are inadequate in analysing complex, "open" systems in which conditions are constantly changing and give rise to endogenously engendered novelty.[1] Yet economists assume a methodological position which gives the discipline the appearance of being a strict, empirical science. This entails:

- an adoption (where possible) of the D–N scheme which demands that a theory be presented as follows: the statement of "initial conditions", the formulation of hypotheses, the deduction of testable statements and their quantitative empirical testing;
- an attempt to account for deviations between test statements and results by appeal to differences between the ideal conditions in which the theory is

---

* I am very grateful to Mark Peacock, who has not only produced a fine translation of the German original of this paper, but who also raised many questions and put forward important comments. He helped me to clarify my argument considerably. Of course, the usual disclaimer applies.

1 One of the first economists to note the methodological challenge posed by complex systems was HAYEK (1972).

applicable and those which actually pertain;[2]

- a *de facto* instrumentalist stance whereby the accuracy of predictions is given more weight than the realisticness of the theory and the validity of the ideal conditions of applicability.[3]

The dominance of this methodology in contemporary economics should not be taken for granted, especially when one considers other possibilities. However, those alternative methodological approaches are, as a rule, taken to be the preparatory or complementary stage in the generation of hypotheses which are to be subsequently tested according to the above schema. Other methods of testing are, for example:

- Simulation, whereby patterns are generated which are then compared to data and structures according to the principle of similarity.[4]

- Experiment, through which artificially controlled conditions can be created which correspond to the assumptions of the theory.[5]

- Qualitative description and comparison of historically and geographically situated structures, especially in the analysis of institutions.[6]

In the light of Bhaskar's views, one must ask why economics uses a method which is singularly inappropriate in dealing with complex systems. This is without doubt a result of mimicking the natural sciences but is also related to the adoption of Popper's highly acclaimed falsification principle as a rhetorical device giving methodological blessing to the argument. For it is precisely the Popperians, such as Blaug, who sharply criticize the practice of empirical testing in economics: the preceding decades have shown that falsifications (in

---

2 HAUSMAN (1992) has shown that this scheme is already a deviation from that of Popper. BACKHOUSE (1997) nevertheless sees Hausman's position as still very much in the Popperian (and Lakatosian) mould.

3 As the discussion of Friedman's (1953) instrumentalism shows, economists – often precisely because of their recognition of falsification as the method of testing hypotheses – are, in the main, opposed to instrumentalism, although their practice mostly accords with this principle. One could say, paraphrasing Friedman, that economists practice their discipline *as if* instrumentalism were true. On the issue of instrumentalism see CALDWELL (1994 chap. 8).

4 One example is to be found in the "New Economic Geography", which tries to bring certain heterodox approaches back into the mainstream economic thought (see KRUGMAN 1996).

5 Experiments do play a role in unorthodox approaches and have cast doubt on the rational choice model. However, they are also applied in examinations of e.g. the neoclassical theory of international trade (see NOUSSAIR *et al.* 1995).

6 A classic example is the analysis of patterns of land tenure in neoinstitutional economics (see EGGERTSON 1990 pp. 213-230, for an overview).

the sense of the D-N model) do not lead to the rejection of theories (see Redman 1993 pp. 114-115; Caldwell 1994 pp. 124-125). It is here that instrumentalism makes its presence felt: if experiments have shown that people do not act rationally (according to the strict standards of economic theory at least), then this represents a refutation of the *homo economicus* model. But this fact has been of no consequence to macroeconomic modelling so long as the econometric tests manifest a close enough fit (cf. Mayer 1993). Much-cherished theories remain in place in spite of refutations of their conditions of applicability which are part and parcel of their conceptual system.

The usual methodological policy in justifying such an approach is to appeal to Lakatos's concept of a "research programme" (cf. Redman 1993 pp. 144 ff.; Hands 1993). This is understandable given that Lakatos weakened Popper's criteria by distinguishing between a "core" of central hypotheses and a "protective belt" of peripheral hypotheses. Of even greater importance in the context of economics, however, is Lakatos's point regarding the "progress-siveness" of a research programme, particularly if rivals with competing explanatory power do not exist (Bensel/Elmslie 1992). If we take a look at the recent history of economics, we find the paradoxical situation in which a research programme is progressive despite the fact that its central hypotheses have been falsified; and the programme even spreads its tentacles into other disciplines (so-called "economic imperialism"). In the meantime, unorthodox approaches remain badly organized and unfocussed, encompassing at most only small fields of research. They are easily neutralized by "compartmentalizing" the field. Furthermore, orthodox theory has been able to integrate particular non-orthodox hypotheses as peripheral assumptions without thereby calling its central core into question. Hence, from the Lakatosian perspective the orthodox approach is recurrently being validated by the progressiveness of its research programme and the lack of a powerful rival.

This development is all too predictable when complex systems are under investigation. Bhaskar's somewhat formal critique notwithstanding, it is obvious that the gap between theory and reality is very large in the case of complex systems because the choice of observational language, which influences the kinds of data being regarded as relevant to the theory, is already in itself extremely difficult. Many economic theories effect a qualitative leap between theory and observation in that they lack a half-way house consisting of phenomenological laws. This excludes the possibility of generalizing from certain observed regularities which should be theoretically explained only in the second step of the argument (Lind 1993). Sometimes such regularities even play an odd part in economics, for example, the Phillips curve which only came to be explained away as a result of endeavours to demonstrate its irrelevance for economic policy. There are many such regularities which the orthodox theory looks upon almost as "curiosities", for instance attempts to identify "long waves" or ordered patterns in the spatial distribution of cities.

There are many *observational* problems involved in complex systems and the identification of causal relations is difficult. This is because complex systems:

- engender new phenomena endogenously, making particular states of the system singular, historically specific and perhaps irreversible;
- are characterized in the broadest sense by non-linear and frequency-dependent processes. Consequently, causal relations exist where the strength and direction of both causes and effects can be highly divergent in terms of magnitude and power (cf. Mainzer 1997 pp. 270 ff.). For this reason, even the unambiguous observation of systems according to deterministic causality is not possible;
- have no fixed boundaries because the process of engendering new phenomena endogenously changes the meaning of exogenous influences on the system;[7]
- consist of agents whose actions are not fully determined by the system but rather stem from autonomous cognitive motives. This means that events in the system are not completely dependent on the environment and are therefore "coincidental";
- are heterarchically ordered. This means there are causal relations between elements and particular emergent properties of the system which run in both directions. There is thus no possibility of ontological reduction in either direction.

This abstract characterization of complex systems can be concisely expressed by two peculiarities of such systems. The first point concerns the actors. Unorthodox theories typically propound alternative theories of action by way of critique of neoclassical economics. Factors such as creativity and the orientation to the future (expectation formation) become important here. Compare the analysis of such factors to the orthodox "rational expectations" approach in which stochastically deterministic causal relations between system and action are posited, thus minimizing the disturbing effect of agent-autonomy as well as creating the possibility of reduction.

7 The cognitive concept of meaning is always related to a knowing subject, whilst its information-theoretical counterpart is more general. According to the latter, every form of order contains informational content which can, in Shannon's sense, be quantitatively determined and also manifests semantic reference, e.g. genetic information (see Ayres 1993 chap. 2).

The second point is not as transparent and has yet to receive sufficient attention from unorthodox schools. Consider "singularity", referred to above. This property can only be meaningfully described from the standpoint of an observer (Ayres 1993 p. 44).[8] Hence, any explanation of a singular phenomenon ultimately cannot but include a description of the observer unless the subject attempting such an explanation were endowed with perfect and complete information about the system. Yet, the latter presupposes, of course, just such an explanation as the one being sought, so a contradiction in terms would be involved. Complex systems can only be analysed if the observer, too, is observed. Singularity can, for example, be traced back to a non-linear dynamic in the economy: small changes in systemic characteristics can have large-scale implications at the macro level. If the observer were in a position to describe the deterministic relations completely, the system would no longer be singular. However, the fundamental problem of all economic analysis is that, although macroeconomic phenomena are observable, frequently their causes are not (because, for instance, an exhaustive microeconomic description of the system is impossible, available data being too crude). As a result of this asymmetry, the non-linearity of the process involved must be reflected on the level of the *observation language* as "singularity" of the system proper. The problem would be soluble in the long run only on the assumption that the system be deterministic and that full information about its workings be attainable. But even this turns out to be unreachable because economic systems contain agents with the same cognitive potential as observers, so that a complete systemic description of the kind imagined is in principle impossible.[9] As we see, the ideas of "singularity" and "creativity" appear to be very closely related.

8 As the classical Aristotelian *principium individuationis* already states, singularity must be defined by reference to the lack of subsumability of certain properties under a general term so that the singular phenomenon (the "individual") can only be referred to by a proper name or by ostension. Concepts and names are lingustic categories and thus by necessity linked with the presumption of a cognitive relation between observer and observed.

9 HAYEK (1952) referred to this fact in his *Sensory Order* when he argued that a full explanation of nature would, in the end, have to include an explanation of the human mind explaining nature which in turn presupposes that mind be conceived to be capable of higher complexity than it in fact is. This results in a paradox. Hence, a full explanation of mind and nature is impossible and will never be achieved. For an extensive discussion of Hayek's argument, see HERRMANN-PILLATH (1992).

Thus, the description of complex systems is plagued by difficulties which the orthodox method simply ignores. But paradoxically, the persistence of this method is to be explained by just those difficulties, because empirical tests are hardly practicable and thus falsification can always be avoided by a new specification of hypotheses. Moreover, given the ambiguities in fixing the meaning of complex systems in the context of the relation between observer and system, it is easy to construct new specifications willy-nilly which, for example, define the boundaries of the system anew as in models with and without lobbying. With such a high degree of freedom in theory construction, one must presume that theory development is governed by exogenous factors such as current trends in the academic market and aesthetic considerations regarding formality.[10]

From the foregoing it is clear that the formulation of alternative theories must begin with the simple tasks of observation and description. The challenge consists in the following:

- finding methods for dealing with singularity and historicity, which simultaneously allow for the formulation of phenomenological laws (namely, empirical regularities, stylized facts and so on);
- taking explicit account of the system's boundaries;
- reflecting the cognitive and creative potential of actors in the system as well as those of the observer.

With this point of departure it is now time to consider two European *Sonderwege* in economics:[11] the theory of Regulation (or simply Regulation) and the theory of *order* (or "Ordoliberalism"). To this end I will deal only with parts of each theory, which defy simple categorization as a result of their many points of contact with other theories.[12]

---

10 Cf. my recent papers on growth theory, the theory of international trade and the economics of economics (HERRMANN-PILLATH 1998a, 1998b; 1999). A somewhat similar point has been made by MÄKI (1993b).

11 The term *Sonderweg* (literally "special way") is normally used to refer to Germany's special course of social development in the nineteenth-century in comparison to that of other European nations. Nowadays, there are "Russian" and "Chinese" *Sonderwege* as well. Obviously, we encounter the problem of singularity yet again. Note that the idea of economic *Sonderwege* is indeed closely linked to the claim that theoretical *Sonderwege* need to be discovered in order to deal appropriately with the economic *Sonderwege*. This Listian thinking is implicit, for example, in the contemporary Russian critique of the so-called mainstream "Chicago boys" approach.

12 The textual basis of Regulation Theory will be drawn above all from the collection BOYER/SAILLARD (1995b); for Ordoliberalism, the emphasis will be on Eucken's

II. Descriptive and Normative Statements about the Economy

Exploring the intellectual roots of these two theories, it becomes apparent that they proceed from separate junctures. The Regulationists take their cue from Marx and stress the regularity in the temporal transformation of the economy; the ordoliberals aim to describe and classify economic systems according to particular taxonomies (so-called "types") (Grossekettler 1994). Both schools react against peculiar methodological dilemmas of their intellectual forebears, thus necessitating the development of an alternative approach. In the case of the Regulation school, the reaction is against Marxist historical determinism; in that of the ordoliberals, the inadequate theoretical justification of previous taxonomies, for example, the theory of historical stages in the development of the economy. One might think that the one could complement the other as static and dynamic counterparts;[13] but there seems to be a fundamental problem with synthesis, one which lies in the theories' respective normative components.

It is well known that economic orthodoxy contains a strong normative element (see Rosenberg 1994) which arises in the intellectual quest for the optimal form of economic regulation. Central concepts, such as "equilibrium", thus play a double role, first as analytical assumptions (in that the economy is analysed "as if" it were always in equilibrium) and, secondly, as normative statements (it being desirable 'that' the economy be in equilibrium). Although this mishmash of positive and normative analysis is often looked upon critically, one wonders whether complex systems and their development must be considered normatively in order that different states of the system may be compared. In this sense, the normative dimension represents the possibility of dealing with the *descriptive* problem of singularity. The difference between normative and positive criteria in the case of complex systems is presumably then one of degree and reflects the ignorance of the observer *vis-à-vis* the system. Normative criteria are a substitute for positive description and measurement in the face of fundamental ignorance.[14]

---

work and its subsequent interpretation (1939; 1952; 1954). More attention will be given to Regulation Theory.

13 The ordoliberals have indeed been criticized for being unable to explain economic change (see e.g. KLEINEWEFERS 1988).

14 The theory of biological evolution – a theory of complex open systems *par excellence* – likewise suffers great difficulties in gauging "progress" in the sense of improved adaptability or anagenesis (see AYALA 1979). This indeterminateness is

A decisive difference between Ordoliberalism and the theory of Regulation is that the latter, in eschewing Marxist historical determinism, has relinquished the normative element in Marxism without seeking a substitute.[15] Ordoliberalism, on the other hand, cannot be understood without reference to the socio-political debate with Marxism and socialism. In some respects, one could say that Ordoliberalism stands closer to the concerns of Marxism than the theory of Regulation does, in that the former conceives the problem of *power* as central. As with the young Marx above all, Ordoliberalism actually starts out from a philosophical anthropology which serves as a normative yardstick in assessing states of the economy, namely the extent to which such states promote human freedom.[16]

Within the Ordoliberal school, there are, of course, various ideas regarding the evaluation and assessment of economic systems. The problem of power plays an important role in the analysis of the dualism, market–plan; and even more so in the context of justifying economic policy on competition.[17] As well as power, the problems of *knowledge* and *allocative efficiency* play important

---

connected to the fact that although plausible assumptions regarding partial positive measures can be made, eventually more encompassing measures result in a qualititative leap to overcome our ignorance by shifting onto the normative level. For example, it is almost impossible to explain completely the interaction between proximate measures of fitness like the engineering optimality of anatomical features and the ultimate evolutionary currency of reproductive success (including fitness). Even in biology we observe recurrent intellectual movements that link normative theories with the idea of a evolutionary teleology (for a brief description of some of those theories, see BLITZ 1992). I mention this for the reason that there are efforts to find integrated measures for biological and economic systems (see Ayres's 1993 information-theoretical approach).

15 One must note, it is the normative dimension in Marxism which grounds its historical teleology. Nadel (1995 p. 44) writes: “Si la régulation rompt avec l'eschatologie marxiste, elle développe la dimension institutionelle du projet marxien”. Paradoxically, any attempt at building a pure explicatory variant of Marxism cannot but reject its normative claims.

16 This anthropological dimension is salient, for example, in Röpke's (1948) discussion of small- and medium-scale enterprises in the economy, where he emphasizes the interaction between technological change, market structures and the cultural crisis of contemporary civilization as he sees it.

17 The roots of the old Franco-German divergence *vis-à-vis* the separation of industrial policy and competition policy lie here. Ordoliberals justify their critique of industrial policy by appeal to potential misuses of power which could arise from it, even in cases where there are good allocation-theoretical arguments in favour of industrial policy based on the theory of market failure.

roles in Ordoliberalism.

The problem of knowledge is a major concern of the "Freiburg school" (especially Hayek) from which Ordoliberalism is nevertheless distinct; the allocative problem meanwhile has been addressed by the Marburg tradition of comparative economics (in particular Hensel). This distinction is important because the ordoliberal stress on allocative efficiency places the school near to the concerns of orthodoxy, the concept being used in the sense of neoclassical equilibrium theory. Fundamental contradictions do not arise here with the criterion of power because the theory of perfect competition dissolves the power problematic by defining perfect competition in terms of the powerlessness of agents. That is, economic agents are taken to be "price-takers" in conditions of perfect competition, and cannot, therefore, influence market processes: they have no market power. Thus, Eucken's critique of Marxism stresses that the defects of early industrial labour markets were the result of monopolistic regional market structures which bestowed disproportionate power on employers.

On the other hand, certain tensions arise between the allocative and power criteria in the case of planned economies; such economies can be considered purely from an allocative point of view and as theoretically functional systems in the sense of a purely deductive, decision-theoretical equilibrium solution (*à la* Lange).[18] This equilibrium approach is closely related to the dualistic thesis that only the two poles – market and planned economy – are stable whilst all mixed systems would be inherently unstable. This can be seen from the fact that this thesis is based on the idea of "consistency of decisions". All mixed systems, it is alleged, lead to internal contradictions which can be resolved only by steering the system in the direction of one of the two "consistent" poles – market or plan. In comparing the desirability of these two poles, ordoliberals look beyond this purely decision-theoretical criterion and ground their preference for markets by appeal to both the power and knowledge criteria. As soon as the latter is brought into play, the market is valued positively: as "Austrian" economists showed in the "socialist calculation debate", a planned economy cannot overcome the problem of knowledge because not all socially distributed knowledge can be centralized (see Nienhaus 1984).

With this stress on the problem of knowledge, the ordoliberal connection with the mainstream is broken, thus constituting the former as an "unorthodox"

---

18 In the parlance of the time, the solution of the economic problem of calculation involved determining the scarcity of goods (of the order 1...n, namely, including capital goods) through material balances (see HENSEL 1954 pp. 128-139).

school of thought. The knowledge problem could be methodologically side-stepped only by focusing exclusively on end-states of a system in which all possibilities for arbitrage over all available information were considered. This is the case, for instance, in rational expectations models.[19] All mainstream models assume - explicitly or implicitly - a complete integration and diffusion of information in the economy. They perceive states in which individual or local divergences exist in the last instance as disturbances (see Lucas's "islands paradigm" which explains the real (transitory) effects of monetary policy).[20] Ordoliberalism, on the other hand, conceives the states of economic systems as constitutive of the economic process which is characterized by a lack of informational integration and diffusion. The economic process is represented as a diffusion process which never comes to an end because new events and information are constantly appearing. From this fact, a fundamental difference between "competition" and "allocation" is apparent.

In comparison with the orthodox point of view, Ordoliberalism posits a detailed and differentiated approach to the evaluation and assessment of economic systems; and when the Ordo school's is compared to that of the Regulationists, it becomes clear that the latter has no criterion of judgement except *adaptive efficiency* or *viability*. This is completely legitimate from a methodological viewpoint and finds echoes in developments on the fringes of the mainstream.[21] In the last instance, this stance can be traced back to Marx,

---

19 It is often overlooked that the original *raison d'être* of the rational expectations hypothesis consisted in the fact that better informed economic subjects would diffuse their knowledge through arbitrage actions, whilst those who did not have the "correct" model at their disposal would be "de-selected" from the market as a result of their false prognoses. This means that rational expectations themselves are the outcome of an arbitrage equilibrium and do not represent a "state of nature" (see HAUSMAN 1989).

20 I would go as far as to say that this is a central characteristic of mainstream economics which distinguishes it from unorthodox approaches.The adoption of unorthodox arguments, e.g. multiple equilibria and suchlike, in "New Economic Geography" was noted above (note 4). In his critique of Krugman, SCHWEITZER (1998 p. 110) stresses precisely the assumption of the complete diffusion of information.

21 North (1990 pp. 80 ff.) is without doubt the most important proponent of such an efficiency concept. Others, e.g. DIXIT (1996), stand closer to the mainstream and try to endogenize policy using the "transactions costs" concept. He reaches the conclusion, however, that there is no exogenous measure of efficiency with which to judge political intervention aside from evolutionary criteria. *Viability*, that is adaptive efficiency, is discussed in detail by PENZ (1998).

whose deterministic theory of development assumed that the capitalist mode of production was moving towards a crisis and subsequently to a Communist revolution. For this reason, the adaptability of capitalism and the nature of the crises which beset it must be analysed. It is no surprise, therefore, that the Regulationists characterize themselves as crisis theorists.[22] It is also clear that a theory of crisis must also be a theory of its absence, or in other words, a theory of stability.

Viability and *adaptive efficiency* thus form the assessment criteria of the Regulation school. Here, as in the case of the ordoliberals, there is a connection with economic orthodoxy, namely the macroeconomic equilibrium of aggregate supply and aggregate demand (not the criterion of microeconomic equilibrium, which is, of necessity, subject to the justification of allocative efficiency in ordoliberal thought). This plays a decisive role in analyses of the long-term viability of Fordism, the crisis which goes hand in hand with macroeconomic disequilibrium. Independently of the specific regulationist structural model developed so far (which is Kaleckian in nature), the equilibrium concept in general represents an important empirical criterion in assessing viability.[23]

Nevertheless, the regulationist connection with orthodoxy is more tenuous than it is in the ordoliberal case;[24] the regulationist theory of wages is utterly different to, say, the new classical macroeconomic theory, to take an extreme example. The ordo theory, by contrast, does identify labour market irregularities which necessitate state intervention (Eucken 1952 p. 303); but beyond this, there is no disagreement between the ordo and orthodox theories of distribution, thus giving the criterion of allocative efficiency its purchase. According to the Fordist paradigm of the theory of Regulation, however, macroeconomic equilibrium is merely an epiphenomenon of the institutional regulation of the labour market (cf. Billaudot 1995).

The problem for the Regulationists is how to assess and evaluate the institutionalized principles of distribution in the labour market. The ordoliberals have, to repeat, three criteria – allocation, power and knowledge – from which

22 The heading, "Une théorie pour temps troublé", is to be found in the preface to BOYER/SAILLARD (1995b p. 12). It is debatable whether this type of "theory-marketing" is of much use.

23 North simply suggests that positive economic growth be used as an indicator of efficiency. FURUBOTN (1998), on the other hand, proposes a "positive-profit-criterion". Compared to this, the regulationist concept is by far and away more sophisticated because it refers to structural and dynamic criteria.

24 I am in no way suggesting that such a connection is desirable.

to choose. If the latter two are taken together, then the labour market will be considered according to the degree of absence of power as well as the special role of residual profits in attracting and stimulating entrepreneurs. Encompassing institutions, such as personal liability, should limit the misuse of entrepreneurial power; and competition will lead to the dissolution of monopoly profits. This allows the ordoliberals to assess institutional structures even when labour markets do not correspond to the orthodox ideal: every economic agent must, via her or his own competitive actions, have the opportunity to share in the temporary monopoly profits accruing to particular entrepreneurs. Here one can see how important the concept of freedom is.

The Regulationists, as I have already stated, have recourse to the endogenous criterion of *viability* in such cases, whereby conflicts are interpreted as indicators of instability. The concept 'conflict' is a highly general anthropological one and is comparable with that of "freedom". To my knowledge, this fact has not been addressed by the Regulation school. It nevertheless seems that valuable judgements can be gleaned from it, at least on the abstract level: the labour market can be judged both negatively – according to the intensity of conflicts – and positively – according to the degree of consensus surrounding the institutional rules of the labour market. This would, interestingly enough, establish a connection between the Regulationists and a modern sibling of Ordoliberalism, *constitutional political economy*. What *really* underlies the Regulationists' concept of macroeconomic equilibrium is a notion of labour market regulation which is both viable and adaptively efficient (when coupled with the condition of macroeconomic equilibrium) and, furthermore, which is agreed to *consensually* by all participants.[25] However, how such a consensus might come about, and how it is to be assessed, are not matters upon which much light is shed by the Regulation school. The evaluation of states of affairs in which some of those affected are excluded from the debate is something which needs to be considered. Were this to become part of the theory of Regulation it would bring the latter into line with normative political theory, and thus make reference to the claims of Marxism to be a comprehensive theory of economic, social and political development.[26]

---

25 PENZ (1998 pp. 73-98), in his discussion of the relationship between consensus and efficiency, does not consider Regulation theory. He addresses the fact that "freedom" and "consensus", in the work of Habermas, do not stand in a relationship of tension. It is possible that the real point of connection between Regulation theory and Ordoliberalism lies in this fact.

26 To recapitulate, one must distinguish between two things: first, the *positive* analysis of states of affairs, institutional arrangements, changes in these and macroeconomic

III. The Theoretical Taxonomy and Singularity of Systems in Time and Space

The regulationist theory of crisis characterizes a singular phenomenon *par excellence*. Methodological problems arise from this fact – as is the case with all crisis theories – when a crisis develops in a manner contrary to expectation. In face of this eventuality, theorists can cling on to their theory, but are accused of immunizing the theory with *ad hoc* hypotheses. This argument gains in strength when, for example, the Regulationists distinguish between types of crisis which, although superficially similar, differ at a structural level because they involve different levels of regulation.[27] In actual fact, neither participant nor observer can know, during the occurrence of a crisis, whether it is a serious one or not.[28] Note that each predictive failure can be explained away by claiming that the crisis is only "within the system" and not "against the system". There is no comprehensive one-to-one mapping of kinds of crisis phenomena and the plethora of possible deficiencies of complex systems of regulation.

---

aggregates; and, secondly, *normative* evaluation. LEROY (1995 p. 122) regards the problems of positive analysis as responsible for the overemphasis on historical-descriptive studies by the regulation school; normative judgements are indispensable, especially if policy prescriptions are to be made. DELORME (1998) describes this as a "fundamental deficiency" in Regulation; compared to the eschatological proclivities of Marx, Regulation does, indeed, look anaemic. In Nadel's (1995 p. 45) words: "les hommes instituent des formes sociales structurantes qui, à travers les tâtonnements complexes, approximatifs de leurs luttes d'intérêts, s'imposent finalement comme autant de compromis". This really says very little indeed.

27 For example, a crisis may be restricted to a certain phase of techno-economic dynamics without affecting the fundamental institutions of the labour market. Both kinds of crisis can be reflected in a loss of macroeconomic equilibrium, but the former, *qua* "creative destruction" may contribute to the progress of the system, whereas the latter ends up in a transformation of the core institutions of the economy. For example, the Asian crisis of 1998 cannot be assessed in final terms, because it is still difficult to distinguish neatly between the financial crisis and a possible fundamental crisis of the so-called "Asian model". On the typology of crises, see BOYER (1990 pp. 48-60).

28 BOYER/SAILLARD (1995b p. 63) write: "En effet, seule l'experience permet de juger *ex post* de la viabilité d'un mode de régulation".

Now it could be argued that a "soft" approach is appropriate to the phenomenon of singularity and that the immunization allegation is a Popperian one which, as we have already seen, is inadequate to the task of analysing complex systems anyway. The problem of immunization is common to all theoretical "just so stories" which operate with the notion of "adaptive efficiency", in particular evolutionary theories. In such cases, singularity and historicity are characteristic properties of the complex systems observed. Prognoses are either difficult or impossible to make, and testing hypotheses must be carried out through retrospective reconstructions.[29]

The problem of singularity is the most central here. That the solution to precisely this problem has been a prime concern of Ordoliberalism has been forgotten (Herrmann-Pillath 1991; 1994). One could say that the ordoliberals' interest was primarily in providing systematic descriptions of different economic orders to which theoretical explanations could be attached. The conceptual tools involved in this endeavour are, first, a distinction between "system" and "order", and secondly, between "real-" and "ideal-types".

With these distinctions, the difference between observer and object is explicitly addressed, for "system" is a purely cognitive, conceptual category used for the purpose of building statements about reality; and "order" as a result of this application of concepts is a set of statements or hypotheses. In this sense, since the set of statements is contingent, "orders" are always singular, whereas "system" is an abstract, theoretically founded category which has no relation to reality whatsoever. More recently, this solution to the "Great Antinomy" between theory and the uniqueness of historical realities has been poorly understood as a result of unfamiliarity with the philosophical context (Husserl's phenomenology). Similarly with the concepts of "type", which were used and analysed by Max Weber: "real-types" are basically classifications of different types of order, *qua* singular historical phenomena; "ideal-types", on the other hand, represent a systematic corpus of theoretical hypotheses pertaining to an abstract model of order, like the *Verkehrswirtschaft* (exchange economy).[30]

---

29 Popper was, of course, the one who held Darwinism to be unscientific because unfalsifiable. He changed his mind very late. Regulation theory is thus in good company here. On the idea of a 'explanatory, non-predictive' model of social science, see BHASKAR (1989 pp. 43 ff.). Hayek (1972) proposed a special method of "pattern prediction" for very similar reasons; cf. also Mäki's (1993a) comments on "story-telling" in institutional economics.

30 These methodological points are to be found above all in the extensive footnotes to EUCKEN (1939), where he criticizes Weber's use of the terms. Ordoliberal methodology has been neglected as a result of the almost exclusive attention given

If we consider the theory of Regulation from the viewpoint of ordoliberal categories, we find that "Fordism" is a real-type pertaining to a number of concrete economies (see Boyer 1995). The regulationist use of the term was for a long time somewhat diffuse, because in the beginning few authors like Aglietta applied the concept to the case of America. It was thus unclear to what extent the term could be generalized. Eventually the problem of singularity came to the fore here: international comparisons made it clear that there were significant deviations from the Fordist model attributable to different stages of development and historical contexts. The same issue arose in the case of "post-Fordism". These concepts are thus comparable to Max Weber's attempt to classify historical systems of domination as types without deriving them deductively from a general theory.

Hence, from a methodological perspective we hit upon the surprising insight that the archetypal Fordism concept is at odds with the core theoretical categories of the theory of Regulation. In its own way, it is just as much of a hindrance to the application of the theory as the realist interpretation of the dualistic system concept (market–plan) has been to Ordoliberalism. Regarding the latter, the statement, "there are only two systems", was understood to mean that in reality there could only be either market economies and planned economies and nothing else besides, an interpretation which quickly took hold in a divided Germany. This is to misunderstand Ordoliberalism in two ways:

- the "systems" of Ordoliberalism are ideal-typical and do not characterize a real situation; "orders", *qua* real phenomena, can take on very different characteristics which none the less can be grasped analytically by using the "systems" concepts;
- presumptions about the stability and direction of change in particular orders are made, which are grounded in the polarity of the system concepts.

With respect to the latter point, there is a general suspicion of state intervention in the market, as it is feared that such interventions generate the case for further intervention until the market economy is transformed through an intervention spiral into a planned system.[31] This fear is grounded on

by German liberal economists to Popper (particularly in the debate with the Frankfurt school). This resulted in a somewhat strained relationship of many liberal economists with Eucken's work because it could not be fitted into the Popperian methodological framework. On the relation between Weber's and Eucken's approach, see MEYER (1989).

31 This idea is most apparent in the context of foreign trade policy (in his penetrating critique of ordoliberalism KLEINEWEFERS (1988 pp. 66f.) even argues that Eucken seems to believe paradoxically that the centrally planned economy is the only stable

arguments of consistency as well as on the ordoliberals' sceptical attitude towards political power. Obversely expressed, the market is stable only if its institutions are legally protected in order to prevent a creeping concentration of power from occurring which might ultimately lead towards collusion between economic and political power.

This conceptual jumble is also to be found, *mutatis mutandis*, in the regulationists' Fordism concept. This concept is itself devoid of theoretical content because the only explanatory concepts referred to are *institutional forms, accumulation regime* and *mode of regulation*, all of which are applicable to concrete situations quite independently of "Fordism". The question to be addressed in this context is: What is to be observed as singularity? The Fordist concept suggests that it involves a spatio-temporally specific phenomenon. Problems arise because Fordism is conceived as a phase of development of the capitalist mode of production; but spatio-temporal specificities threaten the validity of applying the term to other situations. In practice, the term "mode of development" was introduced to characterize singular developmental trajectories. However, this "mode" is independent of the other regulationist categories and, on closer inspection, one ascertains that it encompasses all spatio-temporal specificities, for example the degree of acceptance of particular wage regimes. The concept is thus merely a substitute for "historical specificities"; renaming the phenomenon does not remove the problems it poses.[32]

---

institutional regime because every kind of state intervention as well as the autonomous anti-competitive forces in the market economy would ultimately lead towards the destruction of the liberal economic order. This presumption, of course, seems to fly in the face of the firm conviction that the planned economy is by no means a workable economic order).

32 More critically: the concept is really only a reaction to the dogmatic use of the Fordism concept, but it thereby becomes somewhat arbitrary. A possible solution would be to conceive the various "modes of development" as integral components of a single, unitary international economic regulation regime. This problem has not been sufficiently addressed by regulationists for whom national systems were originally taken *ex hypothesi* to be the unit of investigation. BOYER (1995 pp. 375-376) does not transcend the level of the fracas: *régulation contra régulation* in the global economy. Also of note is the fact that the regulationists did not appropriate the Anglo-Saxon "post-Fordism" paradigm of flexible production because it was regarded as insufficiently differentiated (BENKO/LIPIETZ 1995). On the other hand, one of the distinctive features of classical Fordism is autocentric development and hence, relative autonomy from the global economy.

The original version of ordoliberal theory is more radical in this respect because it reduces singularity to particular points in time, partly as a reaction to the older, time-space-orientated stage theories. It is here that one finds the justification of the static character of the approach: particular institutional phenomena of an economy at particular points in time are analysed according to their implications for the planning/decision process and their allocative consequences. "Development" means nothing other than the chronological series of such time snippets, whereas such a series is captured by the regulationists' "mode of development" as being one coherent singularity. One must, however, question the validity of the latter because the associated theoretical categories yield no additional insights. To be precise, the validity of the accumulation regime, and this alone, is decisive for the stability of the mode of development.[33]

Thus, it is revealed that the decisive difference between the ordo and Regulation theories consists in the methodological role of the concept "accumulation regime". This is clear from the structure of the models, the ordoliberals' being microeconomic, the regulationists' macroeconomic. For the latter school, the concept of "crisis" is also decisive in differentiating between modes of regulation; for "great crises" are indicative of fundamental systemic change. As long as no crisis rocks the basic stability of the system, one must assume that the accumulation regime is stable. Such a regime can only be analysed, to recapitulate, with the aid of the concepts of *institutional forms* and *mode of regulation.* These concepts correspond therefore, in their methodological role, to the "systems" of Ordoliberalism as being the core theoretical concepts around which general hypotheses would have to be arranged.[34] In the ordo approach the concepts are introduced as ideal-types consisting of a set of deductive, systematically interrelated hypotheses (for example about the laws of perfect competition).

---

33 This point becomes clear in analyses of planned economies and their collapse. The concept "mode of development" is clearly redundant if one is dealing with fundamental systemic characteristics and causes of "great crises" in the mode of regulation (see CHAVANCE 1995).

34 Here it is of great import that the descriptive and normative versions of Ordoliberalism be distinguished; for the equivalents to the regulationists' "institutional forms" are to be found in the ordo theory of economic policy, e.g. the form of money (Regulation) and the "currency stabilizer" (Ordoliberalism). It is undeniable that the normative arguments of the ordoliberals are made far more strongly than those of the Regulationists. Hence a direct comparison may end up with a misleading mix of categories on different conceptual levels.

Consequently, the next question to address is how the Regulationists combine their core concepts with law-like statements. The ordoliberals are quite clear on this point, for example, in oligopoly theory where the empirical generalizability of the deductions depends on whether the conditions of applicability pertain.[35] The Regulationists, on the other hand, provide no systematic justification for their taxonomies in the sense of a comprehensive list of all possible institutional forms and a description of the corresponding hypotheses. This is connected to the fact that, unlike Ordoliberalism (in its classical Euckenian form), the Regulationists renounce the use of a single criterion of analysis. (For the ordoliberals, this criterion concerns decisional authority and market structures).

It is for this reason that the Regulationists are accused of eclecticism. But this is to misunderstand their method in a way similar to that in which the accusation of dualism against Ordoliberalism misses the point. There is indeed an order to the hypotheses formed by the Regulationists: "mode of regulation" has the status of the central theoretical term; "institutional forms" that of observational statements; and "accumulation regime" the role of mediator between the two. The hope, yet to be fulfilled, has always been that empirical data would elucidate the connection between "mode of regulation" and "institutional forms", the problem being that the number of hypotheses linked to the list of institutional forms is subject to no limit. The observer, for example, describes particular organizational forms of the labour market and can illuminate each possibility under the aspect of diverse analytical categories. In this way, the Regulationists can accommodate virtually any imaginable labour market theory, taking these as special cases if need be.[36] On the level of direct observation, eclecticism becomes method. The question, though, is how these diverse institutional factors operate to engender a viable system. This system is the mode of regulation, a category which is inaccessible to direct observation.

It is precisely here that the Fordism concept poses problems because it is directly referred to the observed institutional forms; but Fordism does not have the status of an observation statement in the theoretical structure. Thus, the mode of regulation is not comparable to the ordo theory's derivation of "systems" from basic analytical principles. Rather, the mode of regulation is

35 In this respect, the methodology of ordoliberal theory corresponds closely to that of HAUSMAN (1992).

36 BOYER/SAILLARD (1995 p. 81) offer a highly instructive overview of available analytical structures regarding Fordism. BOYER (1990 p. 105) states explicitly: "This leads to a monumental research agenda: examining the different theories of institutions to find those that best fit with the regulation approach".

grounded in the observable self-reproduction of a particular pattern so that any pattern can lead to the identification of a corresponding mode of regulation. Accordingly, the list of modes of regulation is an open-ended one. The main indicators of self-reproduction are crises, or rather the systems' ability to weather these. Such crises lead to macroeconomic disequilibrium and conflicts over institutions.

It also becomes clear that both schools of thought actually represent procedures for attaining pattern statements about reality, a type of guideline for the construction of classifications. One is tempted to depict the difference between the two using the synchronic–diachronic distinction, Ordoliberalism occupying the synchronic, the theory of Regulation the diachronic pole. As a synchronic taxonomy, Ordoliberalism need not concern itself with transitions between states, a question which a diachronic approach cannot dodge. This represents something of a challenge. A synchronic theory takes certain universal building blocks and principles and assumes that the corresponding taxonomy is a closed one. Yet, this approach means necessarily that the theory renounces explanation of historical change. A diachronic theory, conversely, must assume that new taxonomic patterns can arise in varying historical circumstances. The theoretical problem concerns the novelty and emergence in modes of regulation. However, this problem is hidden when the taxonomy with concepts like Fordism is directly related to diachrony and thus time-space segments conceived as types in a closed manner.

IV. The Challenge of Laws of Transition

Ordoliberalism characterizes itself as a radical critique of all attempts to discover historical laws and regularities. This explains its intellectual affinities with Popper, even though the epistemological foundations of Ordoliberalism do not lie in critical rationalism. However, Ordoliberalism does not always adhere to the stipulations of this self-imposed limitation: as shown above, the theory conceives an institutional logic according to which, for example, state intervention necessarily leads to more state intervention, and so on. On the other hand, ordoliberals see the main cause of institutional change not in any internal logic inherent in the *economic* system, but in extraneous, non-economic factors, namely power struggles (Eucken 1952 pp. 16 ff.).

It has been typical of the intellectual history of Germany since the end of the nineteenth-century to seek "irrational" and theoretically non-systemizable elements of socio-economic change in the concept of *power*. In this light, it is wrong to attribute a primarily normative approach to Ordoliberalism, for these normative considerations only reflect the assumption that socio-economic change is caused by power struggles; "order" is counterposed to "power". The idiosyncratic solution of the tension between order and power is the source of the singularity of different orders. To this extent, there can be no *laws* of transition, merely contingent successions of power struggles.

What this negative statement amounts to is that there are no specifically *economic* laws of transition. This invokes an explicit imposition of boundaries around the "economic system" meaning that attempts (of which there have been many) to widen the scope of Ordoliberalism, so that it might encompass modern forms of the economic theory of politics, cannot work: methods of analysis appropriate to the economy cannot be extended beyond the boundaries of the system. This means that laws of transition could not be economic in nature and would have to be derived from a theory of power (cf. Herrmann-Pillath 1991). To hold that such a theory is, in principle, impossible is the ultimate expression of irrationalism in classical (not modern) thought.

Presumably, the much-cited principle of the "interdependence of orders" could fill this void. One is tempted to relate it to the French term, *imbrication*, used when describing the complex institutional patterns of modes of production (Boyer/Saillard 1995a p. 64). But it is apparent that such concepts fail to reach beyond the level of analysis already attained without them; further levels could only be added with the identification of transition *potentials*. For instance, Ordoliberalism assumes that political totalitarianism is incompatible with the freedom of the market and that corresponding powers of change exist which would make coexistence of the two unsustainable. In a similar way, traditional ties of community in a particular territory are a hindrance to the functioning of a modern labour market. Such statements, however, say nothing about the strength or speed of such transitions. Thus, the idea of interdependence is still a far cry from a set of laws of transition.

As I have already mentioned, the (explicitly diachronic) theory of Regulation cannot avoid the issue of laws of transition. The crisis concept is an indicator of transition, but a clear theory of the genesis of crisis is required. That we are dealing with meta-theories *vis-à-vis* conventional macroeconomics can be seen from the *imbrication* of macroeconomic and institutional regularities. Consequently Billaudot says (1995 p. 209): "La dynamique macroéconomique n'est soumise à aucune loi générale: celle qui est observable, dans tel pays à tel moment, est relative aux institutions en place". Ordoliberals share this view: Eucken (1939 pp. 180 ff.) rejects the possibility of a general theory of business cycles which is directly applicable to observable data because cycles are singular phenomena. The Regulationists, on the other hand, capture the phenomenon of cycles (in the sense of a "system crisis") with the aid of the term "accumulation regime". The search for laws of transition cannot be avoided here.

In this context, it is interesting to note that the theory of Regulation has incorporated modern techniques of formal modelling for complex systems.[37] This is doubtless connected to the fact that, in non-linear models, very small changes in parameters can cause very large systemic changes. It is therefore *ex ante* impossible to predict the magnitude of a crisis from a parameter change. Non-linear and synergetic models also allow one to explain the genesis of stable patterns which, from the viewpoint of the observer, can change without apparent (meaning, observable) cause. A change from one pattern to the next can occur, separated only by short periods of disturbance. The problem in transposing non-linear models across various situations consists in their empirical unspecifiability; that is, although it can be proved that endogenizing the structural determinants of economic dynamics leads to phenomena such as bifurcation, simultaneously, no direct correspondence to economic data can be made. I have already shown in the first part of this essay how complex systems of this sort set limits to this approach in empirical analysis so that alternative

37 One should note that every type of frequency-dependent behaviour implies non-linearity, so that this formal character is highly general, e.g. in the "mass phenomena" (BILLAUDOT 1995 p. 214).

methods like simulation have to be adopted (cf. Allen 1998).

This negative conclusion is at the same time a positive confirmation of the necessary role of other forms of observation. If the dynamic process of growth and institutional change is shown to be non-linear, then one must ask whether special methods of pattern recognition are required in formulating laws of transition. Real-world non-linear systems must be analysed according to the emergence of a "*gute Gestalt*" (well-behaved pattern) without this necessarily being fully describable.[38] Observation is thus a necessarily hermeneutic process of endowing an object with meaning. The observer can either be within or without the system, the question then being how far a change in observation *within* the system leads to changes in that very same system. Concretely, this means that the perception of agents in the system has an impact on the system itself (for example, if agents perceived themselves as part of Germany's "economic miracle"). An outside observer will not necessarily form the same picture of the system. This two-way influence corresponds exactly to the regulationist assumption that modes of regulation are anchored in particular cognitive patterns and routines of agents, hence cannot be fully understood via the analysis of external institutions (compare Denzau/North 1994).

This leads to the view that those observing complex systems must construe particular states of the system as "time pictures". Laws of transition would thus have to include the self-observation of the theorist as well as the cognitive processes of economic agents. One would have to refer to the change from one mode of regulation to another, thereby making explicit claims about the cognitive (and perhaps emotional) processes of individuals which accompany such changes (and may contribute to them). As far as one can tell, however, the Regulationists seek the determinants of change in "real" factors, for example, the relationship between distribution and growth. This does not clear up the nature of the connection between historical analysis and model-building. This can be seen from the fact that the "time pictures" correspond to the "*fresques historiques*" of the *Annales* school (Clio 1995). And it would be utterly impossible to reduce transitions between such *fresques* to the components of macroeconomic models. This means that laws of transition cannot be directly connected to the analysis of accumulation regimes because such laws are quite different from conventional "economic laws" in that they have an essentially hermeneutic moment. The Regulationists have not taken sufficient account of this, something which explains why terms like "Fordism" are so difficult to replace despite their becoming more and more problematic.

38 The possible use of gestalt theory in analysing institutions has been innovatively discussed by KUBON-GILKE (1996).

V. Complex Systems and the Missing Links in Ordoliberalism and Regulation

Returning to the remarks on complex systems in the first part of this chapter, it is apparent that both theories have strong and weak points. Both are, in their different ways, attempts to confront the problem posed by the singularity of complex systems. They tackle the problem by constructing theoretical taxonomies. There are good reasons for comparing the regulationists' taxonomies to "phylogenetic" classification, in that "mode of regulation" is a fundamental "principle of organization" which makes its appearance at a particular historical juncture and manifests itself in subsequent phylogenetic branches of this "species" (the temporal sequence being identified as "mode of development"). "Catastrophes" can bring development along such a branch to an end inasmuch as they are an expression of insufficient reproductive potential (meaning that the coexistence of different modes of reproduction is not sustainable, for instance nineteenth-century slave and wage labour in America). In contrast, the ordo theory is better described as "physiological" (and, in a certain way, "sociotechnological"), analysing the time-independent, law-like functioning of economic systems.

Methodologically, ordoliberals are clearer on the role of the observer in the construction of orders; different types of order, that is, are conceived as complex phenomena perceived with the help of a conceptual apparatus. The epistemological stance of the theory of Regulation is not so clear because the role of the observer of crises is not well defined. Crises form the decisive criterion in differentiating *ex post* between different modes of regulation; but the magnitude of a crisis depends, in the last instance, on the perception of the situation by the agents involved in and affected by it, hence the lack of epistemological clarity. For it is these agents who affect the situation through their capacity and readiness to engage in conflict. This fact is, however, somewhat obscured by the regulationists' conception of accumulation regimes as objective, quantifiable phenomena. But one cannot deny that macroeconomic disequilibria (for example, high unemployment rates) do indeed lead to conflicts of a particular sort and intensity and only through the mediation of agents' perception of the situation.

This aspect seems explicitly to address the creativity of agents, one of the central characteristics of complex systems which I discussed above. It can only be described as astonishing that neither theory takes up this point; there is an implicit theory of action in the theory of Regulation but it is not elaborated into an explicit model and, in a way, plays the merely negative role of criticizing orthodox decision theory: the only good models are refuted models. With this, however, the baby is thrown out with the bath water. The regulationists work with concepts of compromise and conflict and they regard the internalization of certain social norms (*qua habitus*) as an important element in the viability of a mode of regulation. This should have given them cause to construct a theory of *habitus*-formation, simultaneously allowing for a critique of neoclassical rational choice theory. But such has not been forthcoming in the theory of Regulation thus allowing structural determinism entry by the back door. This is surprising: was it not the Regulationists who reacted *against* structural Marxism (Vidal, above, chap. 1)? Yet it is clear that there is little explanation which goes beyond the old Marxist categories. Here, one might think of examples of real wage collapse in Brazil and Mexico; why did such episodes not lead to greater social conflict (Boyer 1995 p. 372)? What is missing in the theory of Regulation is a theory of conflict in which the *individual* plays an explicit role.

The above is connected to the regulationist theory of the state, a very late addition to their theoretical canon. To the extent that the state acts autonomously it, too, can be conceived as a creative element.[39] This leads one to question the role of ideologies representing particular interest groups which lurk behind the state, a role which is neglected by the Regulationists because it is obscured by the facade of a value consensus. It is well known that nearly all streams of economic thought rediscovered the state in the 1980s. Before this, authors from "left" and "right" were in tacit agreement that the state possessed no autonomy of its own but was an instrument of particular interests. The latter idea has been shown again and again to be at best a partial truth, at worst false; the political sphere can create economic disturbances and even radical structural changes in the system as a whole, for example, the introduction of central planning. As far as can be ascertained, recent attempts by the Regulation school to "bring the state back in" have been purely classificatory in the sense of differentiating the functions of the state. It is interesting that the boundary between state and market thereby becomes blurred: "l'état relationnel intégré complexe" is functionally omnipresent (Delorme 1995).

In lacking a theory of the acting subject, the Regulationists find themselves in good company with the ordoliberals. With the latter, one could say that power struggles represent a type of action-theoretical irrationalism. This

39 CHARTRES (1995) stresses that political processes – of great significance in cases of war and invasion – are important exogenous disturbances to development and can thus lead to fundamental changes in the mode of regulation.

"theorylessness" is also apparent in Schumpeter's conception of the entrepreneur (which is of the same intellectual vintage): the entrepreneur is orientated to non-economic, irrational goals. This stance is an explicit attempt to capture creativity and is connected to the growth of elite theories in the early twentieth century. But this cannot be described as an explanation of the phenomenon. The ordoliberal theory, too, neglects the state as a theoretical category, although much attention is paid to it at a normative level where the state (and only the state) can institute and guarantee a competitive order. Nevertheless, the state plays no role as an object of investigation. Analytical attention to the state has been a consequence of reimporting political economy along Anglo-American lines (see Broyer, above, chap.3). Contemporary *Ordnungsökonomik* thus appears to be an amalgam of Eucken, North and Buchanan (e.g. Streit, 1995). But the analytical potential of Ordoliberalism proper is being neglected, especially regarding power-theoretical analyses of the state and the drawing of boundaries between economy and polity. It was not for nothing that German political science after the Nazi period felt obliged to deal with the irrationalist theories of the state of the 1920s and 30s. Current literature on this theme took shape as the individual disciplines of economics, social science and politics became institutionally separated. Today the autonomy of the state does not imply its irrationality. It is noteworthy that now even the Regulationists recognize its autonomy (Boyer/Saillard 1995a p. 63). It might therefore be worth reconsidering a fresh synthesis of economic theory, political science and sociology.[40]

Although there are other missing links in the two theories, I have covered the most important: the theory of action, the theory of the state and the problem of the system–observer dyad. Most of the other theoretical problems of the theory of Regulation and Ordoliberalism stem from these three points. Despite their quite different starting points, the two schools of thought manifest a surprising degree of convergence in the blanks on their intellectual map.

40 To quote one of the rare examples of this line of thinking in mainstream economics: BALDWIN (1996).

VI. Conclusion

In conclusion, it should be observed that methodological reflection must play an important regulative function in disputes between unorthodox and mainstream economics. Both the theory of Regulation and Ordoliberalism would be at a disadvantage were they required to demonstrate their potential *vis-à-vis* the mainstream by comparing their respective findings. Neoclassical theory presents itself, of course, as a dynamic, fruitful research programme of great intellectual value which is employed by economists in myriad ways. The kind of comparison imagined would lead to a strengthening of that theory in which most time and human capital is invested. But such comparison does not necessarily reveal the truth: the decisive question must refer to the way in which the object of research is constituted and the methods appropriate to it. Anyone cultivating a garden can obtain high yields with sufficient dedication and energy, even if her or his methods are not those most fitting for the conditions. A better approach, though, would be to assess the conditions first and then cultivate the garden.

The starting-point must consist in an ontological investigation into the properties of complex systems in the sense of positing a priori assumptions about the structures of what has hitherto been the "mystery object". From such assumptions, such as non-linearity, considerable methodological conclusions can be drawn. Hence it is mistaken to advocate a priori the use of the D–N schema. Complex systems may demand quite different, for example, hermeneutic, methods. In this respect, the approaches of Ordoliberalism and the theory of Regulation correspond well.

The question is, of course, which criteria for the selection of methods should be used. In view of the obviously complex relationship between observer and object, the idea of reaching a correspondence between theory and reality is totally inappropriate. An alternative would be criteria allowing for creativity, normative considerations and pragmatic value (cf. Rescher 1973, chaps. 10, 13). In this sense, economic theory would be possible only as a reflection of economic policy. It is precisely this side which has been developed by Ordoliberalism. The Regulationists, however, have paid too little attention to whether the normative element in Marxism can be reinvigorated. Factors such as "distributional justice" could play a role here. This would create a point of contact with other concepts, for example "social market economy" in Germany which is the intellectual fruit of Ordoliberalism (hanging, it must be added, at some distance from the tree which bore it: see Wohlgemuth, above, chap. 7).

On the other hand, the Marxist tradition in Germany has all but died out, a fact which does not bode well for the prospect of a productive discourse. Hence a Franco-German dialogue between Ordoliberalism and the theory of Regulation might bear fruit in the future.

References

ALLEN, P. M.: "Modelling Complex Economic Evolution", *Selbstorganisation. Jahrbuch für Komplexität in den Natur-, Sozial- und Geisteswissenschaften*, 9 (1998), pp. 47-76.

AYALA, F.: "The Concept of Biological Progress", in: F. AYALA, T. DOBZHANSKY, (Eds.): *Studies in the Philosophy of Biology. Reduction and Related Problems*, Berkeley/Los Angeles (California University Press) 1974, pp. 339-356.

AYALA, F., DOBZHANSKY, T. (Eds.): *Studies in the Philosophy of Biology. Reduction and Related Problems*, Berkeley/Los Angeles (California University Press) 1974.

AYRES, R. U.: *Information, Entropy, and Progress. A New Evolutionary Paradigm*, New York (AIP Press) 1993.

BACKHOUSE, R. E. (Ed.): *New Directions in Economic Methodology*, London/New York (Routledge) 1994.

BACKHOUSE, R. E.: "An 'Inexact' Philosophy of Economics?", *Economics and Philosophy*, 13 (1997), pp. 25-37.

BALDWIN, R. E.: "The Political Economy of Trade Policy: Integrating the Perspectives of Economists and Political Scientists", in: R. C. FEENSTRA *et al.* (Eds.): *The Political Economy of Trade Policy*, Cambridge MA/London (MIT Press) 1996, pp. 147-174.

BENKO, G., LIPIETZ, A.: "De la régulation des espaces aux espaces de régulation", in: R. BOYER., Y. SAILLARD (Eds.): *Théorie de la régulation. L'état des savoirs*, Paris (La Découverte) 1995, pp. 293-303.

BENSEL, T., ELMSLIE, B. T.: "Rethinking International Trade Theory: A Methodological Appraisal", *Weltwirtschaftliches Archiv*, 128 (1992) pp. 249-265.

BHASKAR, R.: *The Possibility of Naturalism*, 2nd edition, Hemel Hempstead (Harvester) 1989.

BILLAUDOT, B.: "Formes institutionelles et macroéconomie", in: R. BOYER., Y. SAILLARD (Eds.): *Théorie de la régulation. L'état des savoirs*, Paris (La Découverte) 1995, pp. 209-214.

BLITZ, D.: *Emergent Evolution. Qualitative Novelty and the Levels of Reality*, Dordrecht/Boston/London (Kluwer) 1992.

BOYER, R.: *The Regulation School: A Critical Introduction*, New York (Columbia University Press) 1990.

BOYER R.: "Du fordisme canonique à une variété de modes de développement", in: R. BOYER, Y. SAILLARD (Eds.): *Théorie de la régulation. L'état des savoirs*, Paris (La Découverte) 1995, pp. 369-377.

BOYER, R., SAILLARD, Y. (Eds.): *Théorie de la régulation. L'état des savoirs*, Paris (La Découverte) 1995.

CALDWELL, B. J.: "Clarifying Popper", *Journal of Economic Literature*, 29 (1991), pp. 1-33.

CALDWELL, B. J.: *Beyond Positivism. Economic Methodology in the Twentieth Century*, revised edition, London (Routledge) 1994.

CHARTRES, J.: "Le changement de modes de régulation. Apports et limites de la formalisation", in: R. BOYER., Y. SAILLARD (Eds.): *Théorie de la régulation. L'état des savoirs*, Paris (La Découverte) 1995, pp. 273-281.

CHAVANCE, B.: "Institutions, régulation et crise dans les économies socialistes", in: R. Boyer., Y. Saillard (Eds.): *Théorie de la régulation. L'état des savoirs*, Paris (La Découverte) 1995, pp. 417-426.

CLIO, J.: "Régulation et histoire: Je t'aime, moi non plus", in: R. BOYER., Y. SAILLARD (Eds.): *Théorie de la régulation. L'état des savoirs*, Paris (La Découverte) 1995, pp. 49-57.

DELORME, R.: "L'état relationnel intégré complexe (ERIC)", in: R. Boyer., Y. Saillard (Eds.): *Théorie de la régulation. L'état des savoirs*, Paris (La Découverte) 1995, pp. 180-188.

DELORME, R.: "Regulation as an Analytical Approach: The French Approach", in: A. Midttun, E. Svinland (Eds.), London (Macmillan), forthcoming.

DENZAU, A. T., NORTH, D. C.: "Shared Mental Models: Ideologies and Institutions", *Kyklos*, 47(1994), pp. 3-32.

DIXIT, A.: *The Making of Economic Policy: A Transaction-Cost Politics Perspective*, Cambridge MA/London (MIT Press) 1996.

EGGERSTON, T.: *Economic Behavior and Institutions*, Cambridge (Cambridge University Press) 1990.

EUCKEN, W.: *Grundlagen der Nationalökonomie*, Berlin (Springer) 1939.

EUCKEN, W.: *Grundsätze der Wirtschaftspolitik*, Tübingen (Mohr) 1952.

EUCKEN, W.: *Kapitaltheoretische Untersuchungen*, Tübingen (Mohr) 1954.

FEENSTRA, R.C. *et al.* (Eds.): *The Political Economy of Trade Policy*, Cambridge MA/London (MIT Press) 1996.

FRIEDMAN, M.: *Essays in Positive Economics*, Chicago (Chicago University Press) 1953.

FURUBOTN, E.G.: "Economic Efficiency in a World of Frictions", Diskussionsbeitrag 09-98, Jena, (Max-Planck-Institutes zur Erforschung von Wirtschaftssystemen) 1998.

GROSSEKETTLER, H.: "On Designing an Institutional Infrastructure for Economies. The Freiburg Legacy after 50 Years", *Journal of Economic Studies*, 21 (1994), pp. 9-24.

GUTMANN, G.: "Euckens Ansätze zur Theorie der Zentralverwaltungswirtschaft und die Weiterentwicklung durch Hensel", *ORDO*, 40 (1989), pp. 55-69.

HANDS, D. W.: "Popper and Lakatos in Economic Methodology", in: U. MÄKI *et al.* (Eds.): *Rationality, Institutions & Economic Methodology*, London (Routledge) 1993, pp. 61-75.

HAUSMAN, D. M.: "Arbitrage Arguments", *Erkenntnis*, 30 (1989), pp. 5-22.

HAUSMAN, D. M.: *The Inexact and Separate Science of Economics*, Cambridge (Cambridge University Press) 1992.

HAYEK, F. A.: *Die Theorie komplexer Phänomene*, Tübingen (Mohr) 1972.

HENSEL, K. P.: *Einführung in die Theorie der Zentralverwaltungswirtschaft. Eine vergleichende Analyse idealtypischer wirtschaftlicher Lenkungssysteme an Hand des Problems der Wirtschaftsrechnung*, Stuttgart (G. Fischer) 1954.

HERRMANN-PILLATH, C.: "Der Vergleich von Wirtschafts- und Gesellschaftssystemen: Wissenschaftsphilosophische und methodologische Betrachtungen zur Zukunft eines ordnungstheoretischen Forschungsprogramms", *ORDO*, 42 (1991), pp. 15-68.

HERRMANN-PILLATH, C.: "The Brain, Its Sensory Order and the Evolutionary Concept of Mind, On Hayek's Contribution to Evolutionary Epistemology", *Journal of Social and Biological Structures*, 15 (1992), pp. 145-187.

HERRMANN-PILLATH, C.: "Methodological Aspects of Eucken's Work", *Journal of Economic Studies*, 21(1994), pp. 46-60.

HERRMANN-PILLATH, C. (1998a): "Das Dilemma unrealistischer empirischer Theorie – Wissenschaftstheoretische Aspekte der Neoklassik und einige didaktische Folgerungen", *Wittener Jahrbuch für ökonomische Literatur* (1998), Marburg (Metropolis), pp. 41-72.

HERRMANN-PILLATH, C. (1998b): "Theorie und Beobachtung in einem neoklassischen Modell – eine intuitive strukturalistische Rekonstruktion und Implikationen für die evolutorische Ökonomik", Diskussionsbeiträge der Fakultät für Wirtschaftswissenschaft der Universität Witten/Herdecke, Heft 8 (1998).

HERRMANN-PILLATH, C.: "Ein Versuch über die Anwendung der Ökonomik auf die Ökonomik, oder: Warum wir nicht wissen können, daß und was wir wissen", Diskussionsbeiträge der Fakultät für Wirtschaftswissenschaft der Universität Witten/Herdecke, 1999.

KLEINEWEFERS, H.: *Grundzüge einer verallgemeinerten Wirtschaftsordnungstheorie*, Tübingen (Mohr) 1988.

KRUGMAN, P.: *The Self-Organizing Economy*, Oxford (Blackwell) 1996.

KUBON-GILKE, G.: *Verhaltensbindung und die Evolution ökonomischer Institutionen*, Marburg (Metropolis) 1997.

LEROY, C.: "Les salaires en longue periode", in: R. BOYER., Y. SAILLARD (Eds.): *Théorie de la régulation. L'état des savoirs*, Paris (La Découverte) 1995, pp. 115-123.

LIND, H.: "A Note on Fundamental Theory and Idealizations in Economics and Physics", *British Journal for the Philosophy of Science*, 44 (1993), pp. 493-503.

LORDON, F.: "Formaliser la dynamique et les crises régulationnistes", in: R. BOYER., Y. SAILLARD (Eds.): *Théorie de la régulation. L'état des savoirs*, Paris (La Découverte) 1995, pp. 264-272.

MÄKI, U. *et al.* (Eds.): *Rationality, Institutions & Economic Methodology*, London (Routledge) 1993.

MÄKI, U.: "Economics With Institutions: Agenda for Methodological Inquiry", in: U. MÄKI *et al.* (Eds.), (1993): *Rationality, Institutions & Economic Methodology*, London (Routledge) 1993, pp. 3-44.

MÄKI, U. (1993b): "Social Theories of Science and the Fate of Institutionalism in Economics", in: U. MÄKI *et al.* (Eds.), (1993): *Rationality, Institutions & Economic Methodology*, London (Routledge) 1993, pp. 76-109,.

MAINZER, K.: *Thinking in Complexity. The Complex Dynamics of Matter, Mind, and Mankind*, Berlin etc. (Springer) 1997.

MAYER, T.: *Truth versus Precision in Economics*, Aldershot (Edward Elgar) 1993.

MEYER, W.: "Geschichte und Nationalökonomie: Historische Einbettung und allgemeine Theorien", *ORDO*, 40 (1989), pp. 31-54.

NADEL, H.: "La régulation et Marx", in: R. BOYER., Y. SAILLARD (Eds.): *Théorie de la régulation. L'état des savoirs*, Paris (La Découverte) 1995, pp. 40-48.

NIENHAUS, V.: *Kontroversen um Markt und Plan*, Darmstadt (Wissenschaftliche Buchgesellschaft) 1984.

NORTH, D. C.: *Institutions, Institutional Change, and Economic Performance*, Cambridge, Cambridge (Cambridge University Press) 1990.

NOUSSAIR, C. N. *et al.*: "An Experimental Investigation of the Patterns of International Trade", *American Economic Review*, 85 (1995), pp. 462-491.

PENZ, R.: *Legitimität und Viabilität. Zur Theorie der institutionellen Steuerung der Wirtschaft*, Marburg (Metropolis) 1999.

REDMAN, D. A.: *Economics and the Philosophy of Science*, New York/Oxford (Oxford University Press) 1993.

RESCHER, N.: *The Coherence Theory of Truth*, Oxford (Clarendon Press) 1973.

RÖPKE, W.: "Klein- und Mittelbetrieb in der Volkswirtschaft", *ORDO*, 1 (1948), pp. 174.

ROSENBERG, A.: "What is the Cognitive Status of Economic Theory?", in: R.E. BACKHOUSE (Ed.).: *New Directions in Economic Methodology*, London/New York (Routledge) 1994, pp. 216-235.

SCHWEITZER, F.: "Ökonomische Geographie: Räumliche Selbstorganisation in der Standortverteilung", *Selbstorganisation. Jahrbuch für Komplexität in den Natur-, Sozial- und Geisteswissenschaften*, 9 (1998), pp. 97-126.

STREIT, M.: "Ordnungsökonomik - Versuch einer Standortbestimmung", Diskussionsbeitrag 04-95, Jena (Max-Planck-Institut zur Erforschung von Wirtschaftssystemen) 1995.

Part D

Approaches to Transformation

Chapter 10

Regulation Theory and Post-Socialist Transformation

BERNARD CHAVANCE, AGNÈS LABROUSSE

I. Introduction

The Regulation approach, developed for two decades by a group of French scholars, can be defined as an institutional and evolutionary research programme focused on economic change in a historical, theoretical and comparative perspective. The main works have been devoted to capitalist economies, but in the 1980s a number of contributions addressed the problem of socialist systems. During the 1990s, in common with all economic and social theories, the Regulation approach has faced the challenge of understanding the unprecedented experience of post-socialist transformation. Its contributions in this field have shown interesting parallels with other theories or schools of thought that go beyond the traditional orthodoxy/heterodoxy split. Post-socialist transformation is a process of major discontinuous institutional and systemic change, to some extent similar to the historical precedents of war, revolution or major social crises. But it represents a radical new phenomenon, analysis of which should revitalize current theories of economic systems and of the evolution of capitalism.

Regulation-inspired works on socialist systems (reviewed in Chavance 1995a), generally based on extensive historical research, focused on specific configurations of institutional forms, especially the wage labour–nexus and the monetary aspects of economic processes. The traditional socialist development mode had been characterized by an extensive accumulation regime and a "regulation through shortage" based on original investment cycles. The socialist system was understood to differ from capitalism in an "institutional base" consisting of two pillars, the one-party regime and the domination of state ownership of capital, both cemented by the Communist ideology focused on the contrast between capitalism and socialist systems (Chavance 1994b). The process of centralized planning, producing the "mobilized economy" (Sapir 1992) or the shortage syndrome was dependent on that institutional base, and it also persisted in reformed socialism where the megahierarchy was weakened and disaggregation of central targets abolished or limited, as in Hungary after 1968, Poland in the 1980s, or China since the 1980s. In reformed socialism, the high level of organizational and institutional homogeneity in the traditional system gave way to a "mixed socialist economy" where various ownership forms were allowed to coexist with state ownership and where market coordination was activated and "bureaucratic coordination" (Kornai) became indirect, while remaining dominant.

One contribution of regulationist research has been to give an early diagnosis, at the beginning of the 1980s, of the mounting tensions in Central Europe and the Soviet Union as an expression of a qualitative turn, in other words the onset of a "great crisis" in the socialist economies, presenting new challenges compared to the earlier "regulation through shortage" that had allowed significant (although decreasing) growth since the 1950s. But this evolution was limited to the European socialist countries, affecting non-reformed and reformed economies alike, radically contrasting with the high growth trajectory of the Chinese economy which was gradually reforming and opening up. Beyond this general contrast, the variety of national trajectories of institutional change, development modes, social trends and crisis processes was stressed, thus insisting on the diversity of the initial conditions for the post-socialist transformation that began in 1990-91. In the section that follows some regulationist views on the post-socialist transformation are presented. In the final section the specific case of East Germany is discussed.

II. Regulation Theory and Post-Socialist Transformation: Basic Concepts

1. Criticism of standard and neoliberal views on "transition"

The collapse of socialist systems was welcomed by authors referring to the Regulation approach, but without any expression of "capitalist triumphalism" (P. Wiles) during the early years of post-socialist transformation. Implicitly or explicitly, these scholars adopted a kind of Keynesian stance: capitalism is the worst of all systems, except for the rest; they added that just as many national forms of capitalism have historically existed so new ones are probably possible, so let us hope (or work) for less unpleasant, unjust or undesirable ones.

Many specialists in socialist economies were dubious about the radical or "big bang" approaches to post-socialist transformation, as they presented strong parallels with the early "edification of socialism" strategies that had had extremely adverse effects on the societies undergoing such radical change. Ironically, both institutionalists and Austrian economists shared this criticism of such a "constructivist" approach.

The standard view of "transition to the market economy" has been criticized from the outset for its teleological and deterministic approach, its disregard for the variety of initial conditions, its simplified and reductionist view of the "market economy" and its naïve optimism as to the speed and the results of the process. In the light of the early post-socialist depression, and of the unexpected difficulties and the length of the process of change, a number of these criticisms were subsequently voiced by proponents of the standard approach based on contemporary neoclassical foundations or by the international organizations promoting it, but without considering substantial modifications in the methodological and theoretical foundations that had permitted such a mistaken view. Interestingly, criticism of the standard programme demonstrated convergence on the part of schools or individual authors representative of a whole range of trends –Austrian, institutional of varying kinds, post-Keynesian of differing shades and evolutionist (A. Badhuri, M. Ellman, J. Kornai, P. Murrel, D. North, K. Poznanski, D. Nuti, D. Stark, to name a few). A table of synthesis by Robert Delorme (1995 pp. 21-22) gives a view of some general splits (see table 10.1 below).

Table 10.1: The divide between standard and evolutionary perspectives

ITEM	STANDARD PERSPECTIVE	EVOLUTIONARY PERSPECTIVE
I. THEORETICAL STATUS OF CHANGE		
• Character of change	• Novelty excluded;	• Novelty included;
• Source of change	• Novelty is exogeneous.	• Novelty is endogeneous.
II. BASIC SUBSTANTIAL ASSUMPTIONS		
• Unit of analysis	• Individual agent. Transaction;	• Individual-institutional interaction;
• Behaviour	• Substantive rationality. Utility and profit maximization. No creative behavioural capacity. Agents as givens. Explanation of human behaviour centred on the environment and situation of the agent. Neglect of internal structures of agents;	• Procedural rationality. Behavioural spectrum permitted. Creative innovative behaviour. Concern for the internal structures of agents;
• Coordination	• General equilibrium of competitive markets. Information through prices;	• Coordination through institutional and organizational forms. Experiencing and learning;
• Normative principle	• Pareto criterion of allocative efficiency. General equilibrium of competitive markets as paradigmatic example of a well-functioning economy;	• Economic performance depending on capacity to cope with change and on capacity to satisfy human needs defined through democratically agreed procedures and respecting basic universal values;
• Character of evolution	• Evolution as displacements over time of equilibrium states.	• Process.
III. PREVAILING MODE OF REASONING	• End-state driven; • Pre-established model as norm of reference • Transition: viewed as a move from one coherent order to another; • Grand schemes, "designer capitalism"; • Actors motivated from the point of view of the system when architectural design, "getting incentives right"; • Design of a future shaping the present; • The economy as a self-contained system; • Universality; • Typological reasoning. Homogeneity; • Priority to determinateness. One best way, single exit. Priority to analytical tractability; • More "either-or" than "more or less" view of economic problems; • De facto exclusivism.	• Open-ended; • History-bound process. Paths of extrication with differences in kind; • Transformation: plurality of logics of action; • Transformative processes of trial and error, experiencing and learning; • Actors motivated by the urgency of their practical situations; • Future shaped by the pragmatics of the present; • Economy embedded in the socio-historical fabric; • History- and context-dependency; • Population thinking. Diversity; • Indeterminateness permitted. Pragmatism. Priority to the empirically robust and meaningful (hard facts); • More "more or less" than "either-or" view of economic problems; • Inclusivism: subsumes standard approach, accepting its local relevance, for suitably identified situations.
IV. TYPICAL ANALYTIC OUTCOME		
• Expected result	• Equilibrium. Macroeconomic convergence;	• No asymptotic state. Plurality of capitalist trajectories;
• Likelihood of Expected result	• High, provided decision makers, notably governments, follow economists' advice.	• Varying across countries. Evolution does not necessarily entail progress. Decline, political setbacks are possible.

Generally, regulationist analyses (e.g. Boyer 1993; Chavance 1994a; Sapir 1996) stressed the problems of viability, legitimacy and adaptive efficiency in institutional and systemic change, as opposed to a one-sided view of allocative efficiency and the belief that monetary stabilization and swift privatization would naturally lead to economic restructuring, self-correcting convergence to the desired institutional model and a virtuous growth path.

2. A diversified and institutionally embedded coordination set-up

By contrast with dualistic models of coordination based on market/state or market/hierarchy dichotomies, the Regulation approach stresses the variety of coordination modes in both the social and technical division of labour (Boyer 1997; Delorme 1996; Chavance/Magnin 1996). The capitalist and the socialist systems families are not based on single alternative and antagonistic modes of coordination, such as market versus plan, but on different complex coordination set-ups consisting of evolving blends or mixes of a number of similar coordination modes, such as state, markets, networks and hierarchy (micro-hierarchy within firms being distinguished from state megahierarchy found only in socialism). This approach allows a more complex view of the genuine contrasts, but also of common features, between both families of systems; it also leads to an analysis of the variety of national paths in the coordination field, or of changing set-ups through the process of socialist reforms and of post-socialist transformation.

Systemic change thus does not represent a shift from a pure system or order based on a single coordination mechanism, to another one, for instance, a transition from a centrally planned economy to a market economy, but a complex process of change in the coordination mix, with mainly a change of dominance in the mix. The original combination of changing ownership forms and coordination mechanisms[1] defines post-socialist economies, in the

1 The Kornaian notion of an "affinity" between ownership forms and coordination mechanisms, reminiscent of the criticism of socialism by von Mises, may be correct at an abstract level. But in the analysis of the process of middle-term systemic change, it should be qualified, as experience has shown no direct or unambiguous relation between privatization and "marketization". In countries like Poland or Slovakia, the change in the coordination set-up has been much faster than privatization proper. In China, the coordination mix has undergone profound transformations, without any formal privatization (to this day) of the industrial state

first decade of transformation, as a group of "post-socialist mixed economies", a subcategory of capitalist systems, themselves generally understood as "mixed economies".

3. Understanding the variety of national or regional paths of systemic change

A serious question appears as a puzzle for a general theory of transformation in a standard perspective. Why are institutional and growth national trajectories so different in the post-socialist world? The conventional approach, in conformity with its deterministic view of the "transition" stresses idiosyncratic delays and obstacles on this allegedly paved way, and classifies various countries as more-or-less advanced or late on this single path, according to criteria of liberalization, stabilization and privatization. But the contrast between actual post-socialist development paths and such a view of the race to the market economy, shows that the interplay between institutional change and the development mode is much more complex. Why do the Czech Republic and Russia, the two countries that implemented a fast mass privatization programme which was supposed swiftly to provide the right incentive structure, offer such diverging patterns? Why does Poland now seem to be on a middle-term virtuous growth path, while the Czech Republic and Hungary may be locked onto a less positive track? Why has China, the experience of which runs counter to all standard wisdom (for example, the impossibility of gradual reform in the socialist economy, the priority of privatization over restructuring, the necessity for simultaneous liberalization in various fields) been an example for twenty years of "circular and cumulative causation" as far as general growth and institutional change are concerned? The Regulation approach suggests that a more specific analysis of the nationally diverse historical inheritances, mutations in evolving institutional configurations, interplay between formal and informal rules, mode of relation between the state and the economy, interaction between the polity and the economy, can give a clue to the variety of post-socialist transformation paths and emerging forms of capitalism (Boyer 1993; Delorme 1996; Chavance/Magnin 1997; Chavance 1998).

---

sector, and alongside the extension of many non-conventional forms of ownership. In Russia, fast mass privatization has not led to the expected positive consequences in the coordination sphere, on the contrary, the forms and conditions of this privatization have contributed to the perverse evolution in that sphere.

Three stylized trajectories of transformation, each actually encompassing diverse national paths, can be distinguished: Central European, post-Soviet and Asian (see Table below). Three interdependent levels determine the different post socialist pathways: the specific "extrication path" away from socialism, that is, the mode in which the institutional base and the forms of successive political evolution are destructured; the peculiar economic role and intervention of the state; and the general shape of macroeconomic evolution. The former GDR could be added as a fourth type, *sui generis*. The resulting emerging forms of capitalism present important differences, at the intermediate stylized level, or at the more disaggregated national levels (Chavance/Magnin 1997).

Table 10.2: Stylized trajectories of post-socialist transformation (1989-1998)

	Central-European path	Post-Soviet path	Asian path
Extrication path away from socialism (disintegration of institutional base)	Sudden break (destruction of political pillar)	Sudden break (destruction of political pillar)	Gradual evolution (erosion of ownership pillar, ideological euphemization)
Political evolution	Democratic consolidation, political "alternation"	Sham democracy	Authoritarianism (one-party system) with elements of informal pluralization
State legitimacy	Quite strong	Weak	Rather strong
Administrative and taxation capacity of the state	Quite strong	Weak	Rather strong, corruption not-withstanding
Relations between political and economic elites	Noticeable differentiation	Considerable overlap	Considerable overlap, partial differentiation
New formal rules	Quite hard	Soft	Limited formalism, but semi-hard rules
Privatization of the economy	Rather fast, quite legitimate	Rather fast (Russian case), weak legitimacy	Gradual, strong expansion of "non-state", but not exactly private, forms
Social protection	Socialized (quickly externalized from enterprises)	Strongly internalized in large enterprises	Internalized in large enterprises, gradual externalization
Growth	Depression for 3 to 5 years (fall of GDP of 1/4 or 1/3), followed by significant but still fragile growth	Extended and cumulative depression (fall of about 1/2 of GDP)	Sustained and lasting growth, with inflationary tensions (GDP multiplied by 2 or 3 in more than a decade)
Unemployment	Stabilized around 10-15%	Weak (<10%) but poorly registered, growing	Weak but poorly registered, growing, high hidden unemployment in the countryside

Source: Chavance (1998)

III. Regulation Theory and East German Transformation

What about the fourth stylized trajectory of transformation, the East German *sui generis* type, and the Regulation theory? First of all, it is remarkable that, as far as we know, there has been no regulationist study of the East German transformation except for a very brief survey by Robert Boyer (Boyer, 1993; 1997). The West German economic model fascinates the French, and especially Regulationists, so that it is not the East German transformation as such which initially interests a regulationist theoretician but the German unification as a cause of decline and as revelatory of a more structural crisis in the West German model (cf. Streeck 1996).

Consequently, there is no claim here to develop a regulationist view of the East German transformation. The aim of this section is, rather, to question the regulationist approach in the light of the East German experience. To do so, we focus on a few themes which seem very important within this experience: institutions, economic systems and paths of change. Consideration of these topics, the core of regulationist theory defined in the reference works of Boyer (1986) and Boyer and Saillard (1995) provides relevant tools to understand the specific character of the East German case as well as points of comparison with other countries experiencing transformation. But since the initial Regulation research programme was mainly developed in order to explain the crisis of the Fordist economic system, it has to be partly adapted to the specific features of the East German transformation. A closer examination of the East German case requires the application of some of the very recent developments of the Regulation perspective and even reveals some areas of imprecision in its theoretical core (1). Following this critical examination of regulationist tools, we sketch a grid of the East German transformation based on the regulationist approach and supplemented by other institutionalist theories. Stress is laid on a multi-level approach of institutions and organizations which interact and change along different paths. This approach seems fruitful to deal with abrupt change as well as path-dependency and hybridization phenomena (2).

1. Regulation Theory: Tools to conceptualize the East German transformation

How can the French Regulation theory be helpful in analysing the distinctive East German case? This very specificity is principally related to the transfer of the West German institutional system (the so-called *Institutionentransfer*). Whereas the other Eastern bloc countries devoted considerable time to drafting new constitutions, writing new laws and building new institutions, East Germany adopted the West German institutions with only minor modifications. This complex and extensive legislative framework was applied overnight to economic actors and organizations adapted to, and embedded in, a socialist system. The East German transformation is a specific institutional and systemic change (a) which combines different paths of change (b) and necessitates clarification of the articulations between institutions, organizations and economic agents (c). Let us now examine the regulationist "tool-kit" in the light of these three thematic fields.

a) Institutional and systemic change

As an institutionalist theory the Regulation theory provides concepts to analyse transformation process and especially the consequences of the German institutional transfer. Regulation theory emphasizes the complementarity of institutions or at least a minimal degree of compatibility between them (Delorme 1996). According to the Regulation approach, it is not possible to assess the operational efficiency of an institution separately from its connections with the other institutions of the system considered. It is only *ex post facto* that coherence, whenever this is the case, can be observed after a process of trial and error. The regulationist perspective thus rejects the idea according to which the most effectively performing institutional set can be imported in order to warrant economic efficiency (see Boyer in this book). It rejects the conception of a single best way. Every institutional set-up is embedded in a national historical system. One might find an interesting application of this thematic of systemic coherence in the East German case. Even if West German institutions have proved to be relatively consistent and efficient since the Second World War, it is erroneous to believe that the West German institutional system was *per se* adapted to East Germany and particularly to the transformation process (cf. Grabher 1997). Another aspect of this question is the clash that emerged between formal and informal institutions in East Germany. Whereas the formal institutions are a blueprint of the West German ones, the informal institutions have been deeply shaped by the socialist system and its "regulation through shortage". The so-called "chaos qualification", for instance, comes to mind, that is, the development of *ad hoc* solutions and routines in order to cope with chronic shortages (Senghaas-Knobloch 1992; Marz 1992). These two types of institutions correspond to two different economic systems and are often inconsistent with each other. Moreover, the introduction of West German formal institutions has led to a wide destruction of organizations which did not immediately dovetail with the West German institutional system and, with this, a strong depreciation of economic and human capital.

b) Paths of change

Like other institutionalist theories, Regulation theory is a theory of economic change. Yet the Regulation perspective of economic evolution is deeply marked by Marxian thought. Emphasis is placed on a typology of crisis (small and structural or great crisis). Far from the irenic neoclassical standpoint, regulationist theoreticians underline the role of conflicts of interests which may lead to compromises. It should be stressed at this point that the first objective of the Regulation school was to explain the rise and subsequent crises of modes of development (that is, combinations of a regime of accumulation and a type of regulation) in the long run. Thus it has mainly dealt with endogenous changes, whereas the East German transformation is characterized by an abrupt and exogenous break. However, there is a growing attention to "imported change" and notably to the role of organizational transplantation. Boyer, for instance, has studied the transplantation of organization models across countries and proposed the notion of hybridization between two organizational models (Boyer 1996). It seems to us that this notion of hybridization can be transposed to the institutional sphere, since formal and informal institutions are embedded in two different economic systems: in the East German transformation formal institutions are embedded in the Western system, whereas the informal institutions are widely the result of the socialist system. These concepts seem to be appropriate to the East German case: the transplantation of both the West German institutional system and organizational transplants led to hybridization phenomena, as we shall see below.

The Austrian and Freiburg Schools dealt with the role of spontaneity and intentionality according to various approaches. Menger distinguished between pragmatic institutions resulting from deliberate intention and organic institutions corresponding to social structures which are not intentionally created (Menger 1883). Hayek strictly opposes constructivism and spontaneous order (Hayek 1996). Eucken contrasts order with process, viewed as relatively complementary (Eucken 1952). These themes have not been systematically treated by the Regulation school. The Regulation school falls into the category of political economics. Its founding works (Aglietta 1976; Boyer/Mistral 1978; Lipietz 1979) particularly underline the role of government intervention, social struggles and the political institutionalization of compromises between collective actors. However, an evolution towards the study of spontaneous process seems to be taking place according to Sapir (1996). Boyer and Orléan's article about the evolution of conventions (Boyer/Orléan 1991) is one such example. This is probably related to the development of the "French convention economics" (Orléan 1994), whose representatives are closely connected with regulationist theoreticians.

Here, too, the East German case is specific because it presents an extreme combination of different types of socioeconomic change (intentional and spontaneous, endogenous and exogenous). The complex interplay between continuity and break, intentional and unintentional change, reversibility and irreversibility (see Boyer/Chavance/Godard 1991) is a major characteristic of transformation processes and is sharply marked in the East German transformation. These various paths of change correspond to different institutional and organizational levels.

c) Articulations between institutions, organizations and economic actors

Michel Aglietta, one of the founding fathers of Regulation theory, introduces the concept of mediation in the meaning of institution (Aglietta 1997), but he does not distinguish between formal and informal institutions nor between institutions and organizations. Almost invariably, he adds the adjectives "institutional", "organizational" and "conventional" to the word "mediation". But "institution", "organization" and "convention" are not defined. The mediations, or the five institutional forms, of Regulation theory are a *de facto* mix of formal and informal institutions, of organizational relations and of different kinds of conventions. Admittedly, the notion of institution is very polysemic in the economic literature: no wonder Regulation theory does not really prove an exception. Yet in recent contributions Robert Boyer (1997 and chapter 2 below) has demonstrated a growing interest in this issue. Beyond the five institutional forms, he introduces a multi-level model with a hierarchy between constitutional order, institutional forms, organizations, conventions and individual behaviour.

The inheritance of Marx and Keynes led Regulation theory to a macroeconomic approach. Moreover, this approach can be considered as holistic. The articulation between institutions and organizations follows a mainly top-down logic, especially in the early works of regulationist theoreticians. For example, the enterprise is seen as an expression of the national institutional system. Consequently, Regulationists speak of the German or of the Japanese model of enterprise. Regulation theory is a theory of diversity at the national level (the study of "national trajectories"). However, it seems to us that less attention was paid to the diversity of local arrangements. Economic actors are highly stylized (see the study of the wage relation), so that their heterogeneity tends to be blanked out. Here again, Regulation theory evolves, responding to convention economics and to a more general movement in social sciences concerning the relation between micro- and macroanalysis (Revel 1993; Lepetit 1995). Sapir, for instance, underlines the interdependence of horizontal and vertical decision levels (Sapir 1996; 1999). According to him, different levels are discernible both in a vertical and a horizontal logic, relating to a growing heterogeneity of economic agents, their activity and their mode of insertion in the system. The distinction between formal and informal institutions, between institutions and organizations are crucial in analysing the East German transformation, as well as the complex interplay between macro and micro levels.

2. Sketching an institution- and organization-centred approach to the East German transformation

First, we shall sketch an institution and organization-centred approach to the East German transformation partly based on the regulationist framework. Then we shall try to illustrate this approach by dealing with break, path-dependency and hybridization phenomena.

a) A multi-level transformation

The distinction between institutions and organizations is interesting, as Douglass North observed. According to him, institutions are the "rules of the game". The organizations and their entrepreneurs are the players. Organizations are made of groups of individuals bound together by some common purpose to achieve certain objectives. The interaction between institutions and organizations shapes the institutional evolution of an economy (North 1990). Another detailed point made by North is the distinction between formal and informal institutions: whereas formal institutions correspond to legal constraints, informal institutions "are (1) extensions, elaborations and modifications of formal rules, (2), socially sanctioned norms of behaviour, and internally enforced standards of conduct" (North 1990 p. 40). This distinction is necessary to understand the East German transformation: the formal West German rules which were transferred to East Germany are clearly distinct from the informal East German institutions which are largely inherited from the socialist system. Thus, we temporarily obtain three levels: the level of formal institutions, the organizational level and the level of informal institutions.

Formal institutions are the rule of the game defined by law. They are very general and are mainly concerned with areas defined in the four institutional forms of Regulation theory: competition, money and banking, external relations, labour relations. However, the fifth institutional form in Regulation theory, namely the relationship between economy and state, remains very specific (Delorme 1995). First, the state can be considered as an organization itself constituted of different apparatus. But it is not an organization reducible to other organizations such as the enterprise. It is much more a macro-organization overhanging non-statist organizations. Secondly, a fundamental characteristic of the state is that it generates formal rules. Thirdly, it is also an actor who can intervene directly in the economy. To sum up, the state could be defined as a macro-organization which produces the formal regulatory institutions and intervenes in economic activity. Thus, the state belongs to both the institutional and to the organizational spheres.

Then comes the level of organizations. We will only deal here with enterprises. Organizations are finalized subsystems characterized by a certain degree of closure and openness towards their environment. They are made up, on the one hand, of formal and informal organizational rules, that is to say rules which are only valid in a particular organization[2], and on the other hand of actors organized in networks and hierarchical relations. It should be stressed at this point that economic actors are present at each level: formal and informal institutions are actualized and interpreted by economic actors; the state and the economic organizations are made up of economic actors.

---

2 This hierarchization is rooted in a juridical conception. It is a major source of inspiration for ordoliberalism. There are few developments in Regulation theory about the role of the law in economic processes although the main societies studied by Regulation theory could be defined as *Privatrechtgesellschaften* (Böhm 1966).

Figure 10.1: Articulations between institutional and organizational levels

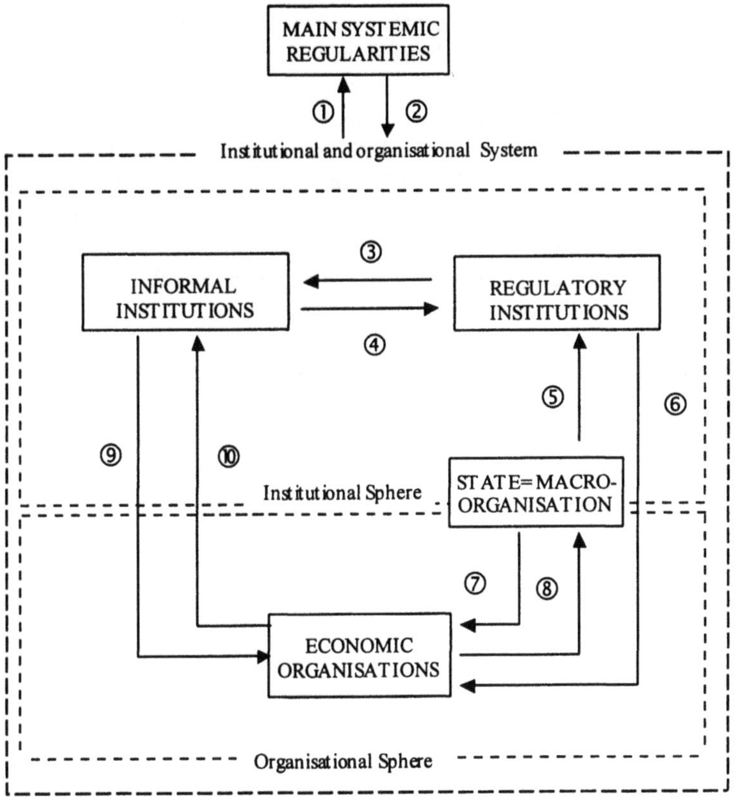

1- The systemic regularities are conditioned by the institutional and organizational system
2 -The systemic regularities reinforce or weaken the institutional and organizational system
3- Influence upon informal rules; legitimacy of informal rules in relation to formal rules, institutionalization of informal rules
4- Reinterpretation of formal rules through informal rules
5- Production of formal rules
6- The formal institutional rules constrain the existence and functioning of organizations
7- Direct intervention of the state upon economic organizations
8- 'Lobbying' in order to influence the formal rules
9- Informal rules playing a central role within the organization
10- Utilization, manipulation and canalization of informal rules to the advantage of the organization

We should add a fourth level: that of the great systemic regularities resulting from the interplay of formal and informal institutions in the functioning of the economic system. The two first systemic regularities are close to the "fundamental relationships" defined in French Regulation theory: money and the wage–labour nexus. The third is property. Their combination defines the economic system.

There is a hierarchy between the different levels, a descending "main line of causality" to use Janos Kornai's concept, although there are mutual influences in several directions between the different levels (Kornai 1992). Figure 10.1 shows the interrelations between them.

A different path and pace of change corresponds to each institutional and organizational level. The comprehensive "overnight" transplantation of formal institutions can be contrasted with the potentially time-expensive change of organizations and informal institutions. It is noticeable that some informal institutions and organizations can evolve rapidly, in some cases, but it is not possible to change them simultaneously from the top and from scratch like formal institutions. Yet, the evolution of organizations and informal institutions cannot be considered separately from the impact of the transplant of formal Western institutions, as it has completely disrupted the game rules East German people were accustomed to. The conditions for enterprise survival and the way the human capital is valued changed completely at an amazing tempo. To sum up, organizations and individuals were compelled to conform to the new rules. But it was not only institutions that were transplanted: German unification also gave rise to a massive influx of Western politicians and civil servants, judges and lawyers, entrepreneurs and managers, consultants and academics. Figure 10.2 shows the different modes of transfer, and Table 10.1 presents a survey of the institutional and organizational transformation of East Germany.

Figure 10.2: The institutional and organizational transfer

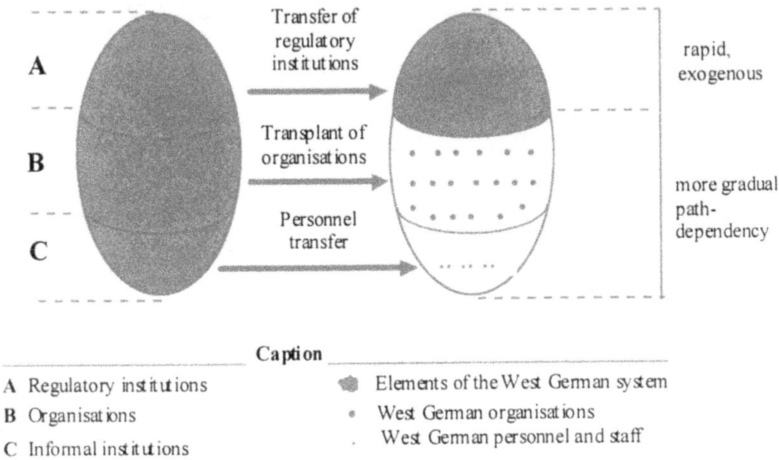

A
B
C
Transfer of regulatory institutions
Transplant of organisations
Personnel transfer
rapid, exogenous
more gradual path-dependency

Caption
A Regulatory institutions
B Organisations
C Informal institutions
Elements of the West German system
West German organisations
West German personnel and staff

b) Rupture, path-dependency and hybridization

Let us now recall very briefly the dislocation of the GDR economic system and its transformation. The fall of the Berlin Wall was in effect the end of the political pillar of the GDR. Yet the period between November 1989 and June 1990 was still open-ended (or at least until April 1990 when the De Maizière government took office) and marked by an intense legislative activity aiming at building a new institutional base. The institutional transfer and the introduction of the *deutsche Mark* (DM) on 1 July 1990 brought this period of institutional innovation to an abrupt end. The introduction of the DM not only meant a revaluation for the East German economy. It should not be forgotten that the forms of money were very different in the socialist system, so that the DM-transfer was central to the systemic change: Three distinct spheres of monetary circulation in the former socialist systems can be distinguished: the scriptural sphere concerning economic organizations, that of fiduciary money (wages) and that

Table 10.3: Institutional and organizational transformation of the East German system

Level of observation	East German System	West German System	Mode and pace of change of the East German System
1. Formal regulatory institutions (legal rules of the game)	• role of formal rules in an authoritarian state	• law (competition, money and banking, external economic relations, industrial relations)	• transplantation of the formal West German institutions incorporating European law
2. State	• party-state, mega-hierarchy	• federalism	• end of the party-state integration to the West-German state
3. Economic organisations (enterprises)			
• relations between state and organisations	• bargaining between the microhierarchy (directors) and the mega-hierarchy	• indirect state intervention and meso corporatism	• extension of the West German state intervention instruments and of West German meso-corporatist actors. Creation of the *Treuhand*.
• relations between organisations	• combines and interentreprise networks (horizontal relationships)	• great variety of entreprises going from small enterprises to big Konzern with tied relationships between banks and industry	• wide disintegration of interentreprise networks
• relations within the entreprise	• micro-hierarchy / tacit compromise between management and workers	• Co-determination (Mitbestimmung)	• organisational change 1) change through evolution of organisational populations - destruction/ survival of GDR entreprises - creation of new entreprises - close-down by the Treuhand - transplants from West Germany 2) change within organisations - exogenous change (break up and buying back by the Treuhand influx of Western managers, transfer of organisation modes and knowhow) - endogenous change (internal adaptation of the organisation)
4. Informal institutions (routines, habits, behavioural patterns)	habits shaped by the socialist system in its formal and informal functioning (second economy)	routines, behavioural patterns particularly informed by the social market economy	slower change • obsolescence, depreciation of a bulk of former informal institutions • learning processes • trial and errors
5. Main systemic regularities			Emergence of several caracteristics of the West German system but with very different forms
• employment regime	• labor shortage	• surplus of labor (unemployment)	• abrupt transition to a surplus of labor (high unemployment level)
• money	• partly passive money with distinct spheres of circulation	• active money	• introduction of the DM (1. July 1990)
• property	• state property	• mainly private property	• slower change of the property system

of external money. Consequently, the *Ostmark* was not a general equivalent. This permitted the existence of different logics among the spheres of circulation which were interconnected to a low degree in comparison with capitalist countries. Thus the introduction of the DM in East Germany is the major factor of the end of systemic coherence (von Hirschhausen 1996) and of the extrication from socialism. After a period of high inflation in East Germany (1990-93), the inflation rate converged rapidly towards the West German level. The employment regime swiftly shifted from a labour shortage, which is a major feature of the socialist system (Kornai 1992) to a labour surplus with a very high level of unemployment. By contrast with the abrupt introduction of the DM, the evolution towards a predominance of private property occurred over several years. The change of property form did not lead to the dislocation of the system but it played a major role in its transformation.

Following the institutional transformation, the organizational change went hand in hand with a deep evolution of organizational populations, networks of organizations (the breaking-up, building and rearrangement of suppliers' and customers' networks, of financing networks, of networks of cooperation and information), as well as an internal restructuring of organizations (sale-oriented organization and production, recentring on the "core business", reduction of redundant personnel and stocks, and so on). It can involve intentional processes (such as the activity of the *Treuhandanstalt*), as well as more spontaneous ones (competition, for instance). It is worth noticing at this stage that the *Treuhand* activity concerned the various types of change mentioned above. The break-up of the combines into separate plants, the down-scaling or closing down of departments such as research and development facilities have had an impact on the population of organizations and on the networks of organizations and on the internal structure of organizations as well. The break-up and down-scaling of the large GDR economic units forms part of an emerging East German economic structure characterized by the dominance of very small enterprises – especially by comparison with the sizes of West German enterprises (cf. Brussig *et al.* 1997). The change through evolution of organizational populations occurred in a context of harsh competition, notably in the industrial sector. Adjustment in industry was, and still is, particularly difficult. In contrast to other branches of the economy, the opening of the domestic market meant that industry was exposed to international competition virtually overnight. At the same time the CMEA collapsed and East Germany lost its traditional markets, which necessitated either a deep and rapid reorientation of the enterprises' customer networks or their exit. Competition, particularly from West Germany, led to widespread deindustrialization, all the more since the industrial structure of East Germany was remarkably similar to that of West Germany as far as the employment repartition in the secondary industries was concerned (DIW, ifW, IWH 1993). Moreover, the shock of revaluation had a very deep impact on the industrial structure inherited from the past. Undoubtedly, the depth of deindustrialization constitutes a characteristic feature of the East German transformation. Change within organizations thus played a crucial but more restricted role. The shock of monetary union and of competition, associated with the integral transposition of West German regulatory institutions, often led to the immediate destruction of East German organizations, impeding an internal adaptation in the mid-term. The internal change of organizations seems therefore mainly to concern sectors depending on local markets (craft industry or construction, for example) and not directly exposed to Western competition.

As for other transforming Eastern European economies, the East German transformation is obviously path-dependent. Path-dependency may be related to the recent evolution but also to the socialist and pre-socialist periods (Chavance 1995). The initial period of the transformation process in a context of hard competition and unfavourable international conjuncture in Western as well as Eastern Europe, combined with the effect of the clash between formal institutions and the inheritance of the former GDR, had deep, lasting and even irreversible consequences on the economic structure of West Germany. The massive deindustrialization that has taken place seems to lock the new *Bundesländer* into a development path characterized by a low degree of industrialization and a very weak export capacity (cf. DIW, ifW, IWH 1999). This is not the mark of a successful transition from an over-industrialized economic structure to a tertiary economy, but the result of an unprecedented phenomenon of close-down in the industrial sector. Besides the development of the services sector is hindered by the low demand for services from industry. Even the high proportion of small enterprises is not to be interpreted as the birth of a new dynamic economy but is partly the product of the break-up of the combines into separate plants or of their shrinking, as already mentioned (Brussig *et al.* 1997). Moreover, business formation is occurring mainly in industries dependent on local markets and on state subsidies and orders. Interestingly, business start-ups are a largely endogenous phenomenon, since 80% of the entrepreneurs are Eastern (Thomas 1996). Globally small enterprises often present management practices and routines (informal institutions) inherited from the socialist system which are reactivated in a new context (Pohlman/Schmidt 1996).

In fact, path-dependency related to the socialist period is largely connected with the role of informal institutions. As North indicates, informal institutions often evolve more slowly than formal institutions. Because of their implicit and tacit nature, it is somehow all the more difficult to make them evolve intentionally since the actors who enforce the informal rules would have to agree on this change in the rules. Informal constraints are thus a pole of stability in a changing environment.[3] This "force of inertia" presents an interesting parallel to the lock-in effect defined in the economics of standards (David 1985; Arthur 1988). In Eastern Germany informal institutions are partly shaped by the socialist system as long as the configurations of the key actors originate from the East.[4] One such example might be the persistent recourse to informal networks of information circulation which was very common in socialist systems. Another might be a kind of "predisposition" towards a flexible working pace: in socialist systems, workers were used to periods of low activity followed by intensive concentrations of work at other periods in order to meet the objectives of the plan (Senghaas-Knobloch 1992). Such a "habit" now contributes to the emergence of forms of flexible working hours in the new *Bundesländer*. The past is not merely a burden but also a resource that economic actors can actualize in a new context. The old personal networks are now, for instance, revived in some enterprises in order to build cooperative relations between enterprises. Unsurprisingly, this is mainly the case in enterprises characterized by relative social continuity and not in those where the management comes from West Germany (Sorge 1996). Thus the role of historical legacies has to be stressed. However, it should not be forgotten that this is a selective and evolving inheritance which is itself partly the result of the transformation process: the "selection" of informal rules occurred under pressure from the transfer of the formal rules and the shock of competition.

3 As Herbert Simon revealed, routines in some fields are necessary to be able to manage change in others (Simon 1979). This is another explanation for the tendency of informal institutions towards stability.

4 Note that informal institutions are enforced by indivual and collective actors, so that the persistence of informal institutions is conditioned by the persistence of configurations of actors sharing the same informal rules.

Regional evolutions show a combination of path-dependency effects corresponding to pre-socialist, socialist and post-socialist periods. Take the agricultural sector. Although West German agricultural policy aimed to propagate the West German model of the small family farm, the dominant model in Eastern Germany remains the big cooperative farm (Lehmbruch/Mayer 1998). Obviously, this can be described as a legacy from the socialist system: an important number of large cooperative farms have been able to gain a foothold in the market, adapting to the new rules but maintaining a large set of traded organizational features which were not necessarily an impediment during transformation but often proved to represent a competitive advantage. However, this general assertion has to be qualified at the regional level. Family holdings are developing in Saxony, especially south Brandenburg. Large farms are overwhelmingly dominant in Saxony Anhalt and Mecklenburg. This local development seems to be related to the old agricultural tradition of family organization in Saxony and to the absence of such a tradition in the north, land of the *Junkers* (Lacquement 1996). The presence of these different traditions and of actors willing to implement them is probably a major explanatory factor of the contrasting regional outcomes of West German agricultural policy in East Germany.

The interplay between formal Western institutions and informal Eastern institutions can produce hybridization processes, that is, functioning processes which are neither reducible to the West German institutional system nor to Eastern legacies. Such emerging novelty may lead to retroaction in the West German system. In order to shed some light on these hybridization[5] and retroaction processes in the East German transformation, let us consider the case of industrial relations. West German industrial relations institutions were introduced in their entirety into East Germany where trade unionism played a totally different role. Not only were formal institutions transferred (the German constitution, the *Tarifvertragsgesetz* and the *Betriebsverfassungsgesetz*), but also West German trade unions staffed by Western officials. However, the West German system of industrial relations is also based on informal institutions. For instance, what is often called the "dual system" of German industrial relations, that is, the combination of collective bargaining by trade unions and workplace representation, corresponds in practice to a system where unions and work councils are

5 We do not deal here with the hybridization process at the organizational level. For a brief sketch of this theme, see Labrousse (1999).

functionally integrated and organizationally interdependent (Streeck 1979). But only the formal part of the system has been transferred. This formal part has combined with East German informal institutions and the emerging outcomes seem to be very different from those in West Germany. Path-dependency at the levels of informal institutions and of organizations modifies the functioning of the industrial relations system. It refers to inherited and resilient aspects of the former system and to the lasting impact of the early months of the transformation process on the new structures.

For instance, a kind of instrumental relationship to trade unions has persisted in East Germany. For most East German workers, the principal value of the former *Freier Deutscher Gerwerkschaftsbund* was its provision of services. Today legal advice for trade union members is one major function of trade unions in East Germany (Hyman 1996). There has been no comparable development of the functional integration of trade unions and work councils in East Germany. A truly "dual system" is emerging. The first work councils faced an environment of radical uncertainty with a precarious future. The fear of closure was the first concern of works councils, managers and workers alike. Subsequently, councils have often collaborated actively with the managers. It is well worth noticing that in the GDR there was commonly a kind of collusion between the management and the work force to develop a "shadow economy" (Senghaas-Knobloch 1992) in order to escape the constraints and inefficiencies resulting from "regulation through shortage".

Eastern Germany is sometimes regarded as a test-bed in which Western companies can experiment with Japanese forms of work organizations to which Western work councils refuse to agree, especially in the motor industry. Meeting the planned output norms in the GDR often necessitated multiple-shift working or even continuous production, whereas such practices were exceptional in the West. Furthermore, Eastern Germany seems to have served as laboratory for the flexibilization of collective bargaining (Hyman 1996). This could be seen as a precedent for subsequent trends in German industrial relations as a whole and could led to retroactions on the West German system.

IV. Conclusion

The Regulation approach should be viewed as a trend in a larger family of socioeconomic theories characterized by historical institutionalism. The central role of economic institutions is interpreted in historical terms, in the sense that the theory deals with historical institutions and organizations, with historically institutionalized actors and processes. Such a view still contrasts with the burgeoning "new institutional economics", which often retains from its neoclassical origins some notion of substantive rationality, the norm of equilibrium and the problematic of allocative efficiency. The institutions considered here do gain a more realistic content compared to the traditional "empty" institutions of standard economics, but they usually remain beyond genuine historical institutions. Nevertheless, it should be added that the frontiers between various strands of institutionalism have become much less rigid and that significant converging trends are at work in this expanding field (Théret 2000).

Capitalism and socialism represent vast systemic families composed of diverse national historical types. While international and global influences have been decisive in the formation, evolution and transformation of these system families, the national histories have been the level of "last resort" where conflicts have developed and compromises agreed or imposed, producing a constantly renewed variety of institutional configurations and organizational arrangements within the general bounds of each systemic family (Chavance *et al.* 1999). This tension is present in the global process of post-socialist transformation, where a strikingly pluralistic process of systemic change has been at work since 1989.

East Germany was, and still is, involved in some elementary process common to other countries experiencing transformation (notably a deep institutional and organizational change and path-dependency effects) but with strongly marked specific features related to its absorption by the West German state. The obvious specificity of the East German transformation, far from excluding the East German trajectory from the field of transformation analysis, could on the contrary enrich an institutionalist approach to transformation such as the regulationist one: it is precisely because it represents an extreme case of transformation that the East German experience is a remarkable laboratory for institutionalist theories about the complex interplay between formal and informal institutions and between institutions and organizations, as well as the creative role of individual and collective economic actors. Regulation theory, supplemented by other institutional approaches, may provide a relevant framework to put this original historical experience in a broader comparative perspective on post-socialist transformation in general.

References

AGLIETTA, M.: *Régulation et crises du capitalisme*, Paris (Odile Jacob) 1997. First edition: Paris (Calmann-Levy) 1976.

AGLIETTA, M., SAPIR, J.: "Inflation et pénurie dans la transition", in: J. SAPIR, V. IVANTER, (Eds.): *Monnaie et finances dans la transition en Russie*, Paris (L'Harmattan/Editions de la MSH), 1995, pp. 3-78.

ARTHUR, B.: "Competing Technologies, Increasing Returns and Lock-In by Historical Events", *Economic Journal*, 99 (1988), pp. 116-131.

BÖHM, F.: "Privatrechtgesellschaft und Marktwirtschaft", *Ordo* 17 (1966), pp. 95-108.

BOYER, R.: *Théorie de la régulation: une analyse critique*, Paris (La Découverte) 1986.

BOYER, R.: "La grande transformation de l'Europe de l'Est: une lecture régulationniste", Working Paper, Couverture orange 9319, Paris (CEPREMAP) 1993. Shortened version in: *Problèmes économiques*, 2374/4 (1994). English transl. in: *Emergo*, 2/4 (Autumn 1995), pp. 25-41.

BOYER, R.: "Vers une théorie originale des institutions économiques ?", in: R. BOYER, SAILLARD, Y. (Eds.): *Théorie de la régulation: l'état des savoirs*, Paris (La Découverte) 1995, pp. 530-538.

BOYER, R.: "The Seven Paradoxes of Capitalism", Working Paper, Paris (CEPREMAP) 1997.

BOYER, R., CHAVANCE, B., GODARD, O. (Eds.): Les figures de l'irréversibilité en économie, Paris (Editions de l'EHESS) 1991.

BOYER, R., FREYSSENET, M.: "Des modèles industriels aux stratégies d'internationalization", provisory version of chapter 10 entitled: "Mise en concurrence et migration des modèles industriels Des transplants japonais à la géographie de l'hybridation", in: *Le monde qui a changé la machine*, forthcoming, Paris, 1996.

BOYER, R., MISTRAL, J.: *Accumulation, inflation, crises*, Paris (Presses Universitaires de France) 1978.

BOYER, R., ORLÉAN, A.: "Why Are Institutional Transitions so Difficult ?", Working Paper, Paris (CEPREMAP), 1991.

BOYER, R., SAILLARD, Y. (Eds.): *Théorie de la régulation: l'état des savoirs*, Paris (La Découverte) 1995.

BRUSSIG, M. *et al.* (Eds.): *Kleinbetriebe in den neuen Bundesländern*, Opladen (Leske and Budrich) 1997.

CHAVANCE, B. (1994a): *La fin des systèmes socialistes: crise, réforme, transformation*, Paris (L'Harmattan) 1994.

CHAVANCE, B. (1994b): *The Transformation of Communist Systems. Economics Reforms since the 1950s*, Boulder (Westview Press) 1994. Original: *Les réformes économiques à l'Est*, Paris (Nathan) 1992.

CHAVANCE, B. (1995a): "Institutions, régulation et crise dans les économies socialistes", in: R. BOYER, Y. SAILLARD (Eds.): *Théorie de la régulation: l'état des savoirs*, Paris (La Découverte) 1995, pp. 417-426.

CHAVANCE, B. (1995b): "Hierarchical Forms and Coordination Problems in Socialist Systems", *Industrial and Corporate Change*, 4/1(1995), pp. 271-291.

CHAVANCE, B.: "Grand-route et chemins de traverse de la transformation post-socialiste", *Economies et sociétés*, XXXII 1/36 (1998), pp.141-149.

CHAVANCE, B.: "The Evolutionary Path away from Socialism: The Chinese Experience", in: E. MASKIN, A. SIMONOVITS (Eds.): *Planning, shortage, and Transformation. Essays in Honor of Janos Kornai*, Cambridge MA (MIT Press), forthcoming 2000. Original in: *Actuel Marx*, 22 (1997), pp. 69-90.

CHAVANCE, B., MAGNIN, E.: "L'émergence d'économies mixtes `dépendantes du chemin' en Europe centrale post-socialiste", in: R. DELORME (Ed.): *A l'Est du nouveau. Changement institutionnel et transformations économiques*, Paris (L'Harmattan) 1996 pp. 125-153. English translation in: *Emergo*, 2/4 (1995), pp. 55-75, and in: A. AMIN, J. HAUSNER (Eds.): *Beyond Market and Hierarchy. Interactive Governance and Social Complexity*, Cheltenham (Edward Elgar) 1997.

CHAVANCE, B., MAGNIN, E.: "Trajectoires post-socialistes et capitalismes occidentaux", in: J.-P. FAUGÈRE *et al.*: (Eds.): *Convergence et diversité à l'heure de la mondialisation*, Paris (Economica) 1997. English version in: *Prague Economic Papers*, 3 (1998), pp. 227-237.

CHAVANCE, B. *et al.* (Eds.): *Capitalisme et socialisme en perspective. Evolution et transformation des systèmes économiques*, Paris (La Découverte) 1999.

CZADA, R., LEHMBRUCH, G. (Eds.): *Sektorale Entwicklungspfade in Ostdeutschland*, Frankfurt/New York (Campus Verlag) 1998.

DAVID, P.: "Clio and the Economics of Qwerty ", *American Economic Review*, Proceedings 75, pp. 332-337.

DELORME, R. (1995a): "L'Etat relationnel intégré complexe", in: R. BOYER, Y. SAILLARD (Eds.): *Théorie de la régulation: l'état des savoirs*, Paris (La Découverte) 1995, pp. 180-188.

DELORME, R. (Ed.) (1995b): "Forms of Organization and Transformation in Central and East European Countries", *Emergo, Journal of Transforming Economies and Societies*, 2/4 (1995).

DELORME, R.: "Un cadre théorique pour les relations entre l'Etat et l'économie dans les économies en transformation", in R. DELORME (Ed.): *A l'Est du nouveau. Changement institutionnel et transformations économiques*, Paris (L'Harmattan) 1996. English translation in: R. DELORME, 1995b.

DELORME, R.: "*Régulation* as an Analytical Perspective. The French Approach", in: A. MIDTTUN and E. SVINLAND (Eds.): *Approaches to and Dilemmas in Economic Regulation*, London (Macmillan) 1999.

DIW, ifW, IWH: *Gesamtwirtschaftliche und unternehmerische Anpassungsprozesse in Ostdeutschland*, Report 13 (1993), pp. 131-158.

DIW, ifW, IWH: *Gesamtwirtschaftliche und unternehmehrische Anpassungsprozesse in Ostdeutschland*, Report 18 (1998), pp. 571-596.

EUCKEN, W.: *Grundsätze der Wirtschaftspolitik*, Tübingen (Mohr) 1952.

GRABHER, G., STARK, D. (Eds.): *Restructuring Networks in Post-Socialism*, Oxford (University Press) 1997.

GRABHER, G.: "Adaptation at the Cost of Adaptability? Restructuring the Eastern Regional Economy", in: G. GRABHER,, D. STARK (Eds.): *Restructuring Networks in Post-Socialism*, Oxford (University Press) 1997, pp. 107-134.

HAYEK, F. von: *Die Anmassung von Wissen*, Tübingen (Mohr) 1996.

HIRSCHHAUSEN, C. von: *Du combinat socialiste à l'entreprise capitaliste*, Paris (L'Harmattan) 1996.

HYMAN, R.: "Institutional Transfer: Industrial Relations in Eastern Germany", *Work, Employment and Society*, 10/4 (1996), pp. 601-639.

KORNAI, J.: *The Socialist Economic System. The Political Economy of Communism*, Oxford (Clarendon Press) 1992.

LABROUSSE, A.: "Der komplexe Wandel von Institutionen und Organisationen in der ostdeutschen Transformation", BISS public, 27 (1999), pp. 105-131.

LACQUEMENT, G.: *La décollectivisation dans les nouveaux Länder allemands. Acteurs et territoires face au changement de modèle agricole*, Paris (L'Harmattan) 1996.

LEHMBRUCH, G., MAYER, J.: "Kollektivwirtschaften im Anpassungsprozeβ: Der Agrarsektor", in: R. CZADA, G. LEHMBRUCH (Eds.): *Sektorale Entwicklungspfade in Ostdeutschland*, Frankfurt/New York (Campus Verlag) 1998, pp. 306-331.

LEPETIT, B. (Ed.): *Les formes de l'expérience. Une autre histoire sociale*, Paris, (Albin Michel) 1995.

LIPIETZ, A.: *Crise et inflation, pourquoi?*, Paris (Maspero) 1979.

LUTZ, B. et al. (Eds.): *Arbeit, Arbeitsmarkt und Betriebe*, Opladen (Leske and Budrich) 1996.

MARZ, L.: "Beziehungsarbeit und Mentalität", in: E. SENGHAAS-KNOBLOCH, H. LANGE (Eds.): *DDR-Gesellschaft von Innen: Arbeit und Technik im Transformationsprozeß*, Forum Humane Technikgestaltung 5, Bonn, 1992, pp. 75-90.

MENGER, C.: *Untersuchungen über die Methode der Sozialwissenschaften, und der politischen Ökonomie insbesondere*, Leipzig, 1883, in: MENGER, C.: *Gesammelte Werke*, Vol. 2, Tübingen 1969.

NORTH, D.: *Institutions, Institutional Change and Economic Performance*, Cambridge (University Press) 1990.

NORTH, D.: "Understanding Economic Change", in: J. NELSON, C. TILLY, L. WALKER, (Eds.): *Transforming Post-communist Political Economies*, Washington D.C. (National Academy Press) 1997.

ORLÉAN, A. (Ed.): *Analyse économique des conventions*, Paris (PUF) 1994.

POHLMAN, M., SCHMIDT, R. (Eds.): *Management in der ostdeutschen Industrie*, Opladen (Leske and Budrich) 1996.

PORTES, R.: "Transformation Traps", *Economic Journal*, 104 (1994), pp. 1178-89.

REVEL, J. (Ed.): *Jeux d'échelles. La micro-analyse à l'expérience*, Paris (Gallimard, Le Seuil) 1996.

SAPIR, J.: *Logik der Sowjetischen Ökonomie oder die permanente Kriegswirtschaft*, Münster/Hamburg (Lit Verlag) 1992. Original: *L'économie mobilisée*, Paris (La Découverte) 1989.

SAPIR, J. (1996a): "Inflation and Transition: From Soviet Experience to Russian Reality", in: S. SEN (Ed.): *Financial Fragility, Debt and Economic Reform*, London (Macmillan, St Martin) 1996.

SAPIR, J. (1996b): *Le chaos russe*, Paris (La Découverte) 1996.

SAPIR, J. (1996c): "Théorie de la *régulation*, conventions, institutions, et approches hétérodoxes de l'interdépendance des niveaux de décisions", Working paper, Paris (CEMI) 1996.

SAPIR, J.: "Penser l'expérience soviétique", in: J. SAPIR, (Ed.), *Retour sur l'URSS. Economie, société, histoire*, Paris (L'Harmattan.) 1997.

SAPIR, J.: "Le capitalisme au regard de l'autre", in: CHAVANCE, B. *et al.* (Eds.): *Capitalisme et socialisme en perspective. Evolution et transformation des systèmes économiques*, Paris (La Découverte) 1999, pp. 185-216.

SENGHAAS-KNOBLOCH, E., LANGE, H. (Eds.): *DDR-Gesellschaft von Innen: Arbeit und Technik im Transformationsprozeß*, Bonn, Forum Humane Technikgestaltung 5, 1992.

SENGHAAS-KNOBLOCH, E.: "Notgemeinschaft und Improvisationsgeschick. Zwei Tugenden im Transformationsprozeß", in: E. SENGHAAS-KNOBLOCH, H. LANGE (Eds.): *DDR-Gesellschaft von Innen: Arbeit und Technik im Transformationsprozeß*, Bonn, Forum Humane Technikgestaltung 5, 1992, pp. 91-101.

SIMON, H.: "Rationality as a Process and as a Product of Thought", *American Economic Review*, 68/2 (1978), pp. 1-16.

SORGE, A.: "Kleinbetriebe: Enstehung, Bedingung und Entwicklungspotentiale", in: B. LUTZ *et al.* (Eds.): *Arbeit, Arbeitsmarkt und Betriebe*, Opladen (Leske and Budrich) 1996, pp. 189-226.

STREECK, W.: "Gewerkschaftsorganisation und industrielle Beziehungen", *Politische Vierteljahresschrift*, 20(1979), pp. 241-247.

STREECK, W.: "Le capitalisme allemand existe-t-il, peut-il survivre ?", in: C. CROUCH, W. STREECK: *Les capitalismes en Europe*, Paris (La Découverte) 1996, pp. 47-76.

THÉRET, B.: "Institutions et institutionnalismes. Vers une convergence des conceptions de l'institution ?", Paper presented to the Conference "Institutions et organisations", GERME-ERSI, Amiens (France), 25-26 May 2000.

THOMAS, M. (Ed.): *Selbständige, Gründer, Unternehmer. Passagen und Paßformen im Umbruch*, Berlin (Berliner Debatte Wissenschaftsverlag)1997.

Chapter 11

The Contribution of Neoliberal Ordnungstheorie to Transformation Policy

HELMUT LEIPOLD

Since the ongoing transition from socialist to democratic and market societies differs considerably from country to country it is not easy to identify the actual or even possible contribution of neoliberal *Ordnungstheorie* and its concept of a market order to transformation policy. Common to all countries is only the experience that transformation has proved much harder and more costly than most economists and politicians expected in 1990. The initial political euphoria was matched by overoptimism about the economic prospects, in Eastern Germany and Eastern Europe alike. Eastern Germany is a special case because it experienced, with currency unification and political reunification, a radical and specific "shock therapy". The transplantation of the West German order was motivated by political considerations.

Neoliberal principles and ideas remained without direct political influence. At most, they were an implicit part of the transplanted concept of the social market economy. In the other reform countries, neoclassical economics and not the neoliberal *Ordnungstheorie* provided the general framework for the transformation debates and policies.

In view of the fact that the contribution of the neoliberal *Ordnungstheorie* to the practical transformation policy was quite modest, this chapter will confine itself to a discussion of some theoretical advantages and disadvantages of neoliberal *Ordnungstheorie* in relation to the possible foundation of transformation policy. The question is not only of academic interest. In many countries the transformation process is far from complete. Hence Kant's dictum that "nothing is more practical than a good theory" still applies to transformation policy, irrespective of the stages of transitional development. The main criterion for a "good theory" should be its ability to give a realistic explanation of the impact of economic institutions and policy measures. The question is to what degree neoliberal *Ordnungstheorie* can meet this requirement.

For clarity and brevity the arguments will be presented in the form of propositions.

Proposition 1: Neoliberal *Ordnungstheorie* provides a well-founded analysis of the impact of institutions, that is, of the functioning of a market economy.

Economic neoliberalism was developed by German economists in the 1930s and 1940s. It is not possible to outline the ideas of all neo- or ordoliberal economists, such as W. Eucken, F. Böhm, W. Röpke, A. Rüstow or A. Müller-Armack. In what follows I will be concerned with the work of Walter Eucken, who is generally acknowledged as the leading representative of this school (see also Leipold 1990; 1991).

Eucken's theoretical starting-point was the famous *Methodenstreit* between Menger and Schmoller. The central issue was the debate about the possibility or impossibility of general economic theories versus theories which can be applied only to particular historical periods and places. Eucken gave his answer to this issue in his book, *Die Grundlagen der Nationalökonomie* (1950), by presenting a theory of economic order. Eucken, who started his academic work under the influence of the Historical school, could neither be satisfied with the historical reasoning because it lacked a sound theoretical background nor with the abstract classical and neoclassical theory. Instead, he proposed an integration of historical and theoretical enquiry, where the method to be used would be based on the idea that economic behaviour always occurs within orders or rules. Consequently, an explanation of complex economic processes would have to start with an analysis of the economic order. In this context economic order consists of the totality of forms and rules in which processes occur. This totality is composed of a few elementary forms: the forms of planning and coordinating economic processes (that is, centralized or decentralized planning systems as the basic economic systems) which can be characterized according to different distribution forms, market forms and monetary systems. The economic order embraces – in other words – the total set of institutions or rules which condition economic behaviour.

Since economic activity is dependent on the whole set of institutional conditions, *Ordnungstheorie* allows a comprehensive institutional analysis. It provides a theoretical basis for the study and comparison of the impacts of different historical and present arrangements.

A second and more practical inspiration of the neoliberal school derived from the economic and social crisis of the Weimar Republic, especially the high unemployment and inflation rates during the crisis of the world economy and the predominance of monopolized markets by cartels, which were estimated at 2,500 in Germany alone in 1925. A rethinking of the classical liberal theory consequently seemed necessary. Eucken's study,

Grundsätze der Wirtschaftspolitik (1952) deals with these practical issues. It answers the question, which economic order should be chosen?

Proposition 2: Neoliberal *Ordnungstheorie* provides a reliable basis for establishing and securing an efficient and socially acceptable market economy.

Based on the comparison of different economic systems and a profound analysis of the economic policies of past decades, Eucken opted for a competitive market order. Analysing the past, Eucken realized that neither the policy of *laissez faire* nor the policy of specific interventions could be an effective solution. In his view, the most important failure of the *laissez faire* idea was the fact that economic order was not considered to be a special responsibility of the state. Order was shaped, to a large extent, by private interests and power. The main failure of the period of specific interventions was that economic policy resulted in a chaos of measures, which contradicted each other, leading to the fettering of the state by economic interests. Therefore, he recommended a competitive market economy, in which the state has to shape the institutional framework. For this purpose, he elaborated a set of principles of economic policy. Three groups of principles can be distinguished:

- The constituent principles of a competitive market system, which represent the aims of the order policy in the narrow sense.
- The regulating principles, which should serve to correct market failures and represent the aims of the so-called *Ablauf-* or *Prozeßpolitik*, that is, current economic policy.
- The principles of state action, dealing with the permissible competence of the state in a market order.

The principles can be explored here only briefly (see Table 11.1). A market system is dependent on prices as its means of coordination. But a workable price system does not emerge spontaneously. It needs to be supplemented by carefully considered institutions. The fundamental question is, how can the price mechanism be made to function properly? According to Eucken this should be the basic principle of the economic policy. The state has to remove all barriers to free trade, to allow competitive markets and scarcity-oriented prices. In addition, competition has to be secured by an anti-monopoly (cartel) policy. Every policy which does not succeed in creating a competitive price system is bound to fail. All other principles have to support this basic requirement.

Table 11.1: Eucken's Principles for a market economy

Constitutive Principles
1. The fundamental principle of striving for competitive prices
2. Principle of the primacy of stable prices and value of money
3. Principle of open markets
4. Principle of preference to private property rights
5. Principle of freedom of contract
6. Principle of avoidance of limits on liability (concordance of rights of disposal and liability)
7. Principle of continuity (steadiness) of economic policy
Regulating Principles
1. Principle of dissolving or supervising monopolies
2. Principle of correction and redistribution of market income
3. Principle of correction of external effects
4. Principle of correction of abnormal supply reactions
Principles of the state and its policy
1. Principle of the limitation of the power and influence of interest groups
2. Principle of the priority of order policy over a policy of interventions at particular points.

Sources: W. Eucken (1952) pp. 254-337; see also Grossekettler (1989).

This explains the primacy of price stability. Stability of prices or money is to be given priority, because inflation has negative effects on competition. It distorts economic calculation and leads to an unsocial distribution of income. Therefore, the monetary institutions should be constructed in such a way as to keep the value of money as stable as possible. A practical device is the independence of Germany's federal bank from political direction, which allows a tight monetary policy for price stability.

The principle of keeping markets open requires an active policy of freedom of trade, which applies both at home and to international trade. Opening markets and borders is the most efficient measure to initiate and safeguard competition as the only reliable yardstick for evaluating and judging the efficiency of home production.

The principle of keeping markets open can be realized only in combination with private property rights and private entrepreneurship. Private ownership of the means of production represents only a legal device for free choice of trade and free development of entrepreneurial potential. For Eucken (1952 p. 275) the interrelation between private property and competition, that is, dynamic market processes, is most important: "Just as private ownership of the means of production is a precondition for competition, so competition is a precondition if the private ownership of the means of production is not to lead to economic and social abuses." If control by competition is missing, private property has to be restricted. Neoliberal economists always stress the complex interplay between private property rights, entrepreneurial activities and workable competition.

The development and functioning of capital markets are especially dependent on exclusive and transferable property rights. According to Röpke (1950 p. 51) the development of a free capital market is an essential precondition for the development of a properly functioning market system. Capital markets direct scarce resources to the most successful entrepreneurs and stimulate careful management. Without these markets, the problems of restructuring the economies, that is, dissolving and transforming the existing firms and establishing new ones, cannot be realized.

Freedom of contract is a natural prerequisite for market relations. This freedom has to be restricted to prevent abuse by cartel agreements and other constraints on competition. Unrestricted liability is seen as the complement to making profits. Laws restricting liability are viewed with scepticism, because they encourage risky undertakings and concentration.

Finally, Eucken demands continuity of economic policy. This should improve the climate for long-term investments. Steadiness of economic policy is the opposite of what Eucken calls a policy of interventions at particular points, in other words, a short-sighted and volatile policy, neglecting the universal interdependence of economic processes.

The regulating principles result from the idea that certain market shortcomings remain, even if the basic principles are implemented in the best possible way. They are designed to correct certain market failures, such as natural monopolies, externalities, the abnormal behaviour of supply in labour markets or in agriculture, and the adjustment of income distribution by a progressive tax system which should not endanger the propensity to invest. Moreover, Eucken also called for an effective social policy. But throughout, he emphasized the primary importance of the basic principles of the economic-constitutional framework.

The ultimate aim of *Ordnungspolitik* is to stimulate competition and to restrict both private and state power. The significance of Eucken's constituent and regulating principles for curbing the general tendency to constrain competition is obvious. But the limitation of state power is equally important. It is the restriction of economic policy to the establishment of general procedures and institutions that ensures the independence of the state from the pressures of interest groups. It is a fact that growth in government activities corresponds to a loss in state authority. Hence, the state should concentrate on order policy and attempt to limit the influence of interest groups.

Crucial to the idea of interdependence of specific orders is the relation between the economic and political system. In the long run, a market economy can function only when basic personal rights are guaranteed. Among basic economic rights are the following: freedom of choice, contract, property, occupation and competition. Political rights are manifest in free democratic elections of the government, in freedom of religion, speech, coalition and the press, and last but not least in academic freedom. A democratic state, in which the rule of law prevails, is thus an essential component of the market economy. Neoliberal *Ordnungspolitik* lays special emphasis on this interdependence and especially on rules restricting political and economic power. Eucken's political principles are also meant to further this objective.

To sum up, the principles of economic policy should, first, ensure the development and functioning of a competitive and socially acceptable market system, and secondly, restrict the competences of the state, thereby strengthening it in the execution of its proper role in a market economy. Thus, the neoliberal concept amounts to the deliberate shaping of a competitive economic order, which is seen as being the best means both of achieving prosperity and of restricting the power of government.

Proposition 3: The neoliberal school contributed the essential theoretical foundation to the concept of a social market economy as the official *leitbild* of the German economy.

The concept of a social market economy, which was mainly implemented by L. Erhard in the late 1940s and early 1950s, shares with the neoliberal programme the conviction that a competitive market system represents the best order to achieve and secure individual freedom and economic welfare, and that such an order needs to be safeguarded by a consistent constitutional framework. It goes beyond the neoliberal concept in trying to reconcile a free market economy with the demands of social justice. The German economist A. Müller-Armack (1974 p. 163), who coined the term "social market economy", describes the basic conceptual idea as the attempt to reach a true synthesis of market economy and social security, where the market forces support the extension of social security, and social security, in turn, guarantees the functioning and the existence of the market.

The adjective "social" has different meanings. First, it stresses the fact that a market economy in itself is already "social" in comparison to a centrally planned economy. It creates liberty and choice not only for entrepreneurs but also for consumers and employees. By its effectiveness and high increases in productivity, it promotes economic welfare and thus the cake to be distributed. This last point indicates the second meaning, namely the necessity of an active social policy as a demand to the state to redistribute the primary market incomes, to organize a social security net and to create a social order concerning the relations between employers and employees. This explains the third meaning. The social market economy offers a possibility of reconciling different ideological and theoretical concepts. The relevant positions at least in the Western world are the liberal and neoliberal concepts, the social ethics of Christian churches (Catholic and Protestant social doctrines) and social democratic ideas. Müller-Armack (1982) always stressed the idea of "social irenics". He regarded the reconciliation of diverging values and interests as a prerequisite for the social acceptability and stability of an economic order.

The practical policy programme also reflects these ethical and philosophical dimensions. The social market concept fully incorporates the neoliberal principles of economic policy and adds the social policy, stabilization policy and structural policy.

The main aim of social policy is to protect people from insecurities in the case of retirement from work, either through illness, unemployment, age or accidents. Therefore the state has to organize social security systems and to provide certain social services. Social market economists prefer a contribution-based financing of the social security systems. They also recommend a modest redistribution of market incomes. The market distributes incomes according to performance and cannot take social aspects into account. Thus primary distribution should be corrected through a redistribution policy, by means of a progressive tax system and subsidies, compensation payments and so on.

Social policy should be assisted by a stabilization and a structural policy. Market processes often result in cyclical movements and, along with them, in macroeconomic imbalances. This smoothing of cyclical movements demands a business cycle policy. Time and again, there are bottlenecks or excess capacities in special sectors. In case of long-term problems in adjusting production, a structural policy should provide relief measures to branches or specific regions. Social market theorists do not support industrial policies, that is, selective state interventions to promote economic development.

Sound interventions have to be regulated in a manner compatible with the market forces. The social market concept stresses the "principle of conformity with the market", which applies to all economic and social policies mentioned so far. According to Röpke (1982), who specified the conformity principle, interventions should work with the price mechanism and not against it. Non-conformable measures are those which attempt to influence prices and quantities directly. A rent-freeze, for example, which affects the housing market is non-conformable, while rent assistance to the poor is in conformity with the housing market. The market-conformity principle stresses a technical device of state intervention in the market. But it gives no conclusive answer to the question as to what extent the state should intervene and what decisions should be reserved for the market and private initiatives. The founding fathers of the social market concept were aware of this problem. As Müller-Armack (1989 p.84) noticed at an early date, it is certainly easy and tempting for politicians to step over the threshold at which disturbance of the market begins. But it would be impossible to determine in advance at which point state interventions do this. Without doubt, a comprehensive welfare state was not, or is not, intended by the social market theorists. The basic idea was the establishing of a market economy tempered by social safeguards which are consistent with liberal and free-market principles. Reality, however, did not follow this intention. During the 1970s and 1980s the West German economic and social order developed into a comprehensive welfare state.

Proposition 4: The economic and social order introduced in the course of political unification in Eastern Germany together with the subsequent economic policy only partly corresponded to neoliberal principles.

It does not require detailed empirical proof to show the discrepancies between the real development of economic and social policies during recent decades and neoliberal principles. The neoliberal economist Hamm (1989 p. 189) deplored this development before 1990 in trenchant terms:

"The German economy has reached a stage at which the thoughts and actions of people have already been largely corrupted by the welfare state. Reflecting on one's own performance is regarded as outmoded; claims and rights to financial aid from the community dominate behaviour patterns. Because of the reduction in the number of achievers and highly motivated people, welfare state policy is leading into a blind alley – the welfare state is destroying its own base."

This statement already indicates the conviction that the undifferentiated imposition of the comprehensive welfare state order and its many sclerotic regulations with unification on 3 October 1990 only partly corresponded to neoliberal principles. Neoliberal economists lament that the opportunity to reconstruct Germany's economic and social order in the process of unification has not been seized. This especially applies to the social security system, the organization of labour markets and other overregulated areas (cf. Apolte/Cassel/Cichy 1994). Admittedly, unification policy was dominated by political and not by economic considerations. It was not the moment to discuss the concept of the social market economy which was at that time highly acceptable to most East Germans. Nevertheless, the first experiences of the German economic and currency union of 1 July 1990 gave ample evidence that the complete transplantation of the whole network of regulations, taxes and social costs would be too much for East German firms to bear. The currency union revealed in no uncertain terms the true degree of backwardness of the GDR economy. Compared to the West German economy the work productivity of the GDR at the end of the eighties reached at most 33%. With the introduction of the German Mark and the politically determined economically unrealistic conversion rates the East German economy was suddenly subjected to fierce international competition. Because of the conversion of the continuous payments (wages, prices) at the rate 1 D-mark to 1 Ost Mark, the East German firms were forced to pay their relatively high costs in D-marks and to adjust their equally high prices to the relatively lower prices of international and, above all, West German competition. The pressure of costs continued to increase enormously because of the high wage settlements after 1990 and the high social costs which resulted from the political unification in October 1990. The firms were unable to stand up to these challenges on account of their low work productivity, deficient product quality and lack of marketing capabilities. In addition, there was a loss of established markets in Eastern Europe countries.

The needs of the moment, or rather the requirements of the changed political circumstances, demanded privatization of state-owned assets as rapidly as possible. The task of privatization was given to the *Treuhandanstalt* which thus had enormous influence on the economic reconstruction of East Germany. This is not the place to evaluate the privatization policy which was a unique task. This policy was certainly very expensive, and some mistakes were made. However, it correctly paved the way for the rapid denationalization of the East German economy and for setting free private initiative (cf. Leipold 1996).

The German government tried to cushion the shock of political and especially of the economic and currency unification by massive transfers from the West, a large part of which served to finance the social security system. In addition, the government granted substantial investment subsidies and practised different forms of active labour market policies. Although this policy was extremely expensive, it could not stop the rise of unemployment. Even with faster growth and high investments, unemployment will remain high and massive transfers will be necessary for a longer period (for further details, see Gutmann 1995).

Proposition 5: In other Central and Eastern European countries the transformation process started under different initial conditions. In most Central European countries only the initial reform packages were in accordance with neoliberal principles.

The economic reform policy of Poland, introduced on 1 January 1990, may serve as an example. The reform package consisted of

- a set of macroeconomic stabilization measures, setting a restrictive course for the monetary policy and reducing budgetary subsidies;
- the liberalization of domestic consumer and wholesale prices;
- the liberalization of domestic and international trade from controls;
- the establishing of a unified exchange rate and current account convertibility;
- the abolition of existing laws regulating the freedom of entrepreneurship and the freedom of contract;
- the announcement of the beginning of privatization of state ownership.

The reform package can be read as a pure reflection and implementation of the constituent principles of Eucken's economic policy mentioned above. Actually the reform policy was not directly inspired by German neoliberal theory. L. Balcerowicz, who stayed several weeks in Marburg in 1998 and V. Klaus were familiar with neoliberal *Ordnungstheorie* and the economic policy of L. Erhard, which had resulted in the rapid West German economic recovery during the post-war years. Yet the early attempts at reform in Poland and the Czech Republic reflected and followed the standard reform prescription as it was advised by neoclassical economists and by international institutions (IMF and so on).

The high degree of similarity between neoliberal and neoclassical theory is not surprising. As has been mentioned, Eucken's main idea was to identify particular historical conditions as variable combinations of pure forms of economic order and to study the impact of these specific combinations with the help of general economic theory, which in his time was in essence neoclassical theory. Compared with neoclassical economics, neoliberal *Ordnungstheorie* has perhaps two advantages: first, it tries to specify the applicability of standard economic theories. Secondly, it tries to pay attention to the interdependence between the whole set of economic order and adequate economic policy measures. Neoliberal *Ordnungstheorie* and neoclassical economics are thus complementary.

Besides the prescription of the same set of policy measures for all reform countries, representatives of neoliberal and neoclassical economic theory shared the belief that the creation of an institutional framework for a market economy should be a state task which could be realized quickly and easily. The reality of transformation policy did not confirm this optimism. In Poland as well as in other reform countries transition soon began to stall (Winiecki 1995). The political process evolved according to its own dynamics. In Poland, for example, the resistance of politicians, voters and pressure groups to the promotion of a competitive market order manifestly increased from 1993 onwards. More and more markets have been shielded from external competition. Privatization laws and policy were postponed. At the same time, one could observe a rebirth of industrial policy in the shape of sectoral policies and trade protection. The main aim was and is the defence of the economic and social *status quo* and the slowing-down of competition and the adjustment to market processes. The increased pressures of various groups, especially from interests stemming from the state-owned sector which was still huge although decreasing, for financial support and for the defence of the domestic markets reflect the normal course of economic policy in democracies. Transformation policy in the young democracies seems to confirm the experiences in Western welfare states. In both cases, growing divergences between neoliberal (and neoclassical) economic prescriptions and realities are evident. The reasons for these divergences are to be found in the incentives and pressures of political competition, which Eucken and other representatives of neoliberal *Ordnungstheorie* did or could not foresee.

Proposition 6: The main shortcoming of neoliberal *Ordnungstheorie* is that it has not sufficiently analysed the political process and the role of informal rules. Hence the programme of *Ordnungspolitik* remains a somewhat idealized concept.

The neglect of analysis of the political process is understandable from a historical and methodological point of view. Eucken wrote his main work during the regime of Hitler. He had to be cautious in writing about political issues. Nevertheless, he was a supporter of a free and democratic society. Secondly, for methodological reasons Eucken viewed the political order – *inter alia* – as data for economic analysis. "Data" means that the evolution and change of the political, social or legal order and their impact on economic behaviour cannot be explained by the tools of economics. This view reflects Eucken's scepticism of historical theory and the myth of the inevitability of social development. It also encourages modesty concerning the applicability and ability of economic theory to explain historical and cultural influences on economic behaviour. Nevertheless, Eucken always emphasized that historical and cultural conditions should be taken into consideration in economic policy. But his morphology was, and is, only capable of identifying pure forms of economic orders. It cannot explain the embeddedness of the economic order and institutions into the historical and cultural structures and hence it can not understand the prevailing structures of consciousness (values, belief systems). Both neoliberal *Ordnungstheorie* and neoclassical economics have no instruments to analyse the interaction between informal and formal rules. Informal rules, which include tradition, customs, religious beliefs and other spontaneously grown rules are not seen as variables which influence economic and political processes.

Transformation policy provides ample evidence that informal rules matter. It is the mixture of formal and informal rules together with the political and moral enforcement conditions that shape economic performance. Reform countries, especially in Eastern Europe, that adopt the legal framework of Western societies show different performance characteristics because of different informal norms and the lack of a civil society and the rule of law. As North (1994 p. 366) remarks, the transfer of formal rules of Western societies is not a sufficient condition for good economic performance (see also Leipold 1997).

The consideration of the importance of informal rules suggests that instead of building a market order exclusively from above, governments in reform countries should try to provide – admittedly by *fiat* – legal possibilities that allow people to choose among alternative institutional arrangements, that is, to allow competition of jurisdictional rules. It would give governments and private individuals a chance to learn about new market institutions, try them out and select those that are in harmony with the prevailing informal rules. A convincing theoretical legitimation for this proposal is offered by Hayek, not Eucken (cf. Pejovich 1994).

The neglect of the importance of informal rules and thus of history and path-dependence, together with the neglect of the properties of political processes, explains why neoliberal (and neoclassical) reform prescriptions are only concerned with the transformation of formal (in other words, legal) rules, which has to be brought about by a powerful state. The idea of a powerful state that is supposed to restrict itself to the setting-up and control of a competitive market order is only an another version of the benevolent despot model of neoclassical welfare economics.

Especially in the new democracies, which experience a dramatic change of formal rules and informal values, the postulated primacy of *Ordnungspolitik* is not sufficient to restrict the expansion of state interventionism and the influence of organized interest groups.

The explanation offered by the new political economy and constitutional political economics provides a more realistic understanding of the incentives and dynamics of the behaviour of politicians, bureaucrats and interest groups (Buchanan 1984; Leipold 1989). Analysis of the interrelations between economic and political interests and processes also reveals the causes of constitutional shortcomings. Hence the constitution of the state and other collective organizations deserves as much consideration as the constitution of markets to which Eucken directed attention. The concern of constitutional political economics with the question of how to impose constitutional fiscal constraints on government and other public institutions seems to be an important issue of transformation policy and thus a necessary complement to the neoliberal concern with the imposition of constraints on market processes. To propose such constitutional fiscal constraints is one thing, to implement them is quite another and a far more utopian story.

References

APOLTE, T., CASSEL D., CICHY E. U.: "Die Vereinigung: Verpaßte ordnungspolitische Chancen", in: G. GUTMANN, U. WAGNER (Eds..): *Ökonomische Erfolge und Mißerfolge der deutschen Vereinigung - Eine Zwischenbilanz*, Stuttgart/Jena/New York (Gustav Fischer Verlag) 1994, pp. 105-128.

BUCHANAN, J. M.: "Constitutional Restrictions on the Power of Government", in: J. M. BUCHANAN, R. D. TOLLISON (Eds.), *The Theory of Public Choice*, Vol. 2, Ann Arbor (University of Michigan Press) 1984, pp. 439-452.

EUCKEN, W.: *Grundlagen der Nationalökonomie*, 6th ed., Berlin/Göttingen/ Heidelberg (Springer Verlag) 1950. First edition: 1939.

EUCKEN, W.: *Grundsätze der Wirtschaftspolitik*, Tübingen (J.C.B. Mohr) 1952.

GROSSEKETTLER, H. G.: "On Designing an Economic Order. The Contributions of the Freiburg School", in: D. A. WALTER (Ed.): *Perspectives on the History of Economic Thought*, ldershot (Elgar) 1989, pp. 28-84 (= Twentieth-Century Economic Thought, Vol. 2).

GUTMANN, G. (Ed.): *Die Wettbewerbsfähigkeit der ostdeutschen Wirtschaft. Ausgangslage, Handlungserfordernisse, Perspektiven*, Berlin (Duncker & Humblot) 1995.

HAMM, W.: "The Welfare State at Its Limits", in: A. PEACOCK, H. WILLGERODT (Eds.): *Germany's Social Market Economy: Origins and Evolution*, London (Macmillan) 1989, pp. 171-194.

LEIPOLD, H.: "Ordnungspolitische Konsequenzen der ökonomischen Theorie der Verfassung", in: D. CASSEL, B.-T. RAMB, H. J. THIEME (Eds.): *Ordnungspolitik*, Munich (Franz Vahlen) 1989, pp. 257-284.

LEIPOLD, H.: "Neoliberal Ordnungstheorie and Constitutional Economics. A Comparison between Eucken and Buchanan", *Constitutional Political Economy*, 1/1 (1990), pp. 47-65.

LEIPOLD, H.: "The Neoliberal Concept of Economic Order", in: C. T. SAUNDERS (Ed.): *Economic and Politics of Transition*, Houndmills/London (Macmillan) 1991, pp. 73-85.

LEIPOLD, H.: "Privatisation Policy and Restructuring - The German Example", in: R. OVIN (Ed.): *Reform and Economic Policy Coordination*, Maribor (Studio Linea) 1996, pp. 87-106.

LEIPOLD, H.: "Der Zusammenhang zwischen gewachsener und gesetzter Ordnung: Einige Lehren aus den postsozialistischen Reformerfahrungen", in: D. CASSEL (Ed.): *Institutionelle Probleme der Systemtransformation*, Berlin (Duncker & Humblot) 1997, pp. 43-68.

MÜLLER-ARMACK, A.: *Genealogie der Sozialen Marktwirtschaft: Frühschriften und weiterführende Konzepte*, Bern/Stuttgart (Haupt) 1974.

MÜLLER-ARMACK, A.: "Social Irenics", in: LUDWIG-ERHARD-STIFTUNG (Ed.): *Standard Texts on the Social Market Economy*, Stuttgart/New York (Gustav Fischer Verlag) 1982, pp. 347-359.

MÜLLER-ARMACK, A.: "The Meaning of the Social Market Economy", in: A. PEACOCK, H. WILLGERODT (Eds.): *Germany's Social Market Economy: Origins and Evolution*, London (Macmillan) 1989, pp. 82-86.

NORTH, D. C.: "Economic Performance through Time", *American Economic Review*, 84/ 3 (1994), pp. 359-368.

PEJOVICH, S.: "The Market of Institutions versus Capitalism by Fiat", *Kyklos*, 47 (1994), pp. 519-529.

RÖPKE, W.: *Ist die deutsche Wirtschaftspolitik richtig? Analyse und Kritik*, Stuttgart/Cologne (Kohlhammer) 1950.

RÖPKE, W.: "The Guiding Principles of the Liberal Programme", in: LUDWIG-ERHARD-STIFTUNG (Ed.), *Standard Texts on the Social Market Economy*, Stuttgart/New York (Gustav Fischer Verlag) 1982, pp. 187-191.

WINIECKI, J.: *Institutional Barriers to Self-Sustaining Growth: Poland's Incomplete Transition*, Warsaw (Adam Smith Research Centre) 1995.

Chapter 12

Ordnungstheorie and Theory of Regulation: How Productive Are They? A Virtual Panel Discussion

HANS-JÜRGEN WAGENER[1]

Moderator: Ladies and gentlemen, for two days we have been listening to fascinating expositions of two continental economic theories and lively discussions about their respective theoretical value. Clearly, the two are as different as they could be:

• They originated in different periods of the twentieth century. *Ordnungstheorie* was developed in Nazi Germany by Eucken (1940) with the historical experience of the Great Depression and its cures, corporatism, state planning and Keynesian interventionism. Regulation theory evolved in post-war France during the 1970s with the historical experience of the *trente glorieuses*, the post-war boom, and its abrupt ending in 1973.

• They have different theoretical backgrounds. Eucken was inspired by liberalism which, at the time, brought him in close contact with neoclassical and (neo)Austrian theory. As a matter of fact, his microeconomics was basically neoclassical and his monetary theory informed by the Austrians, since the neoclassical approach had little to offer in this respect. Of course, Eucken was well acquainted with, although not enamoured of, the German Historical school. The Regulationists are secularized Marxists with a solid mainstream training. When it comes to economic policy questions, an inclination towards post-Keynesianism can be detected. However, it rarely does come to questions of economic policy, since their perspective, like that of Marx, is the long run.

• They have different philosophical backgrounds. In Eucken we may trace two sources, Husserl's phenomenology and the ethical idealism of his father, Rudolf Eucken, that led him to believe in a predetermined natural order which mankind was free either to implement or not, hence his idea of

[1] I thank Frank Bönker and Stefan Voigt for valuable comments and criticism of an earlier draft. This chapter was written while I was fellow of the Institute for Advanced Study Berlin. The stimulating atmosphere of this institution may be responsible for the unconventional form of the paper. Any shortcomings are mine.

ordo. With the Regulationists the position is somewhat more difficult. They have freed themselves from Marxian historical determinism and from the static fetters of structuralism. But, of course, those are their roots.

• A major difference may be seen in the vitality of the respective research programmes measured by the number of contributing scholars. *Ordnungstheorie*, as a rule, is described with reference to Eucken, and Eucken only. As a research programme inspiring a larger group of theoretically oriented scholars, it never really took off. The volume commemorating 20 years of theory of Regulation (Boyer/Saillard 1995) has contributions by 46 authors, perhaps not all of them members of the school. Although Boyer is always a central reference-point, he is one among many others who contributed to the core propositions.

• In stark contrast to theoretical vitality are political vitality and relevance. It is hardly surprising that *Ordnungstheorie* has been described as a theory of economic policy rather than a theory of the economic system (Grossekettler 1989). *Ordnungspolitik* is said to have been the dominating idea behind the post-war West German social market economy. It has a clear-cut liberal message: institutions and policy measures should not interfere with the price mechanism and should actively further competition. Of course, this message can be derived from other theoretical complexes as well (see e.g. Ribhegge 1991). Nothing similar can be said about the Regulationists. Their programme knows nothing of normative ideal states but wants to explain historical development. The long-term dynamic turns short-term economic policy into a voluntaristic illusion (Delorme/André 1983).

At the same time, both have several things in common:

• They are aware of the problematic relation between singular historical events and universal general theories. And they both try to solve it by having morphological analysis precede theoretical analysis. But here their ways already part again: Eucken's morphology defines a limited set of options of order – a consequence of his philosophy; the regulationists' morphology defines a limited set of focal points, institutional forms which can be filled in an unlimited number of ways, even if the number of viable configurations may be restricted – more a question of evolution than one of choice.

• Both approaches are engaged in comparative systems analysis. The Ordo people concentrated more in the past on the functioning of socialist economic systems, while the Regulationists preferred to compare different brands of capitalism in order to isolate individual modes of regulation. The systemic approach yields a common focus of interest, the idea of complementarity of elements of order that Eucken calls interdependence of orders and that the Regulationists call coherence of a regime.

• Both have an ambiguous relation to the neoclassical mainstream. They are unanimously critical of the ahistorical approach that abstracts from institutions. But when it comes to explaining the micro-working of markets, for instance, neoclassical logic is unavoidable: monopoly theory is not falsified by particular historical situations. However, it cannot explain why and how specific institutions perhaps enabling monopoly evolve and, if it tries (Demsetz 1982, for example), it is not particularly successful. It is also at pains to explain macro-phenomena like crises, persistent unemployment and monetary shocks. This is the field of research of Regulationists who think that such macro-phenomena cannot be properly analysed by methods based upon methodological individualism. Eucken, on the other hand, accepts the neoclassical doctrine. Once the proper order and competition have been installed, markets will work smoothly and neoclassical theory is applicable.

• A third commonality may be found in their increasing interest in Hayekian ideas. Seen from a distance, the Freiburg school looks like the school of Eucken and Hayek, since they both taught there, although at different points in time. Hayek's name appears in the editorial circle of the journal, Ordo. In his inaugural lecture of 1962, Hayek explicitly took up the legacy of Eucken (Vanberg 1999 p. 8). However, Ordoliberalism and neo-Austrian liberalism have different theoretical backgrounds. A major difficulty in the further development of Ordnungstheorie obviously lies in Eucken's philosophical assumptions. They exclude evolution and make his approach a static one of "no future". In a way, it is a justification of Fukuyama's notorious end of history after the collapse of Soviet Communism. Hayek is a possible way out, allowing for fairly similar policy conclusions, in particular abstinence from policy interventionism, but he also calls for abstinence from order policy. Hayek did not add to *Ordnungstheorie* but has supplied a theoretical substitute. It may be too early to speak of a "new Freiburg school", but the efforts of V. Vanberg and M. Streit (see the textbook, Kasper/Streit 1998, where Hayek is by far the most frequently cited author) go in that direction: the integration of *Ordnungstheorie*, Hayekian evolutionism, institutional and constitutional economics which they call *Ordnungsökonomik* [the economics of order]. The Regulationists, too, are on speaking terms with Hayek. His idea of spontaneous order and his theory of complex phenomena attract their attention, despite his typically Austrian methodological individualism. And, evidently, Hayekian therapeutic nihilism in economic policy based on limited human knowledge corresponds partly with the voluntaristic illusion of economic policy based on structural determinism already mentioned.

- A final commonality is their national limitation. Until now, neither approach has succeeded in transcending national borders. The history of the general equilibrium approach, however, which was developed in the French milieu of the 1870s and became a truly international research programme only in the 1950s, shows that time is needed to convince the profession of the productivity of a paradigm.

So we have come this far. But, as is well known, the proof of the pudding is in the eating. Transformation from socialist planning to a democratic market system seems to be an ideal test case for economic theories. That is why we have called together a panel to discuss the merits of different approaches *vis-à-vis* the challenges of transformation in the post-Communist states of Central and Eastern Europe. The question put before the panellists is simple: what did their economic theory contribute to the problem of transformation? Let me introduce the panel to you. To my left you have Mme Callot from the University of Paris, a protagonist of the Regulation school. On my right there is, first, Oxford professor, Jack Miller, who was a policy adviser in several transition countries. Since both *Ordnungstheorie* and theory of Regulation are heterodox, representation of the mainstream view on the panel seemed indispensable. Further to the right, there is Prof. Dr. Klaus Barthaus from – as you surely expected – Freiburg University.[2] I think we will start with short statements and then jump right into the discussion. Mme Callot would you like to start?

Mme Callot: I would have preferred to be the last to speak. For, as you will see later, the position of the Regulation school is in a way antithetical to conventional wisdom. So, in the first round I shall just delineate the playing field of transformation. Two things are important in my eyes. First, the theory of Regulation sees modes of regulation as historically contingent. This implies that central planning is not *per se* inefficient. It may be helpful to get modern economic growth started. It is, however, evidently incapable of coordinating and mobilizing highly diversified production and consumption, where market competition is more effective. The inability to introduce markets into the socialist economic system – which the Chinese reforms prove are not impossible – is a major cause of crisis in the system. From this

2 Names of the panellists are purely fictional. Although their affiliation has been chosen to be representative, it is open to question whether they are truly representative members of the respective schools.

it follows that transformation should not follow the eraser strategy of abolishing all old institutions and beginning from scratch but should carefully differentiate between institutions that were functional only in the old mode of regulation and those that can be converted or adapted to the new situation. Delorme (1995 p. 19) has aptly described this approach: "Instead of building from *tabula rasa* on the ruins of socialism, it would seem more reasonable to accept the idea of building with the ruins of socialism." It also follows that the market is not a definite and once and for all optimal configuration but is, rather, embedded in a web of constitutional rules, legal and social institutions which are constantly adapting the market to its environment and are themselves permanently adapting to their own environment. The second important thing is the focal-points of transformation policy, which follow from the morphology of Regulation Theory. These include the wage–labour nexus, the monetary regime, the forms of competition, the relationship between state and economy and the mode of international economic integration. The concrete institutions defining these institutional forms are the object of transformation policy. Since they are specific in time and space, we refrain from making definite prescriptions. However, it is pretty clear that to have a functioning market mechanism you need a well-developed monetary system with a stable currency and a reliable and low-cost payments system, as well as a legal system providing a low-cost framework for making and controlling contracts.

Moderator: Thank you very much. Certainly there are already a number of questions. But before proceeding to the discussion we will hear Dr. Barthaus.

Barthaus: I will refrain from commenting on the introductory remarks of our chairman, although some of his qualifications seem controversial to me. Allow me to make just one comment. As an economist, you see Eucken alone as the founding figure of the Freiburg school. However, it is a school of law and economics. Hence, F. Böhm is of equal importance, and the title of his 1937 book *Die Ordnung der Wirtschaft als geschichtliche Aufgabe und rechtsschöpferische Leistung* [Economic Order as Historical Task and Achievement of Legislative Creativity] contains the programme of the school. I must admit, the promising start did not experience an equally promising follow-up and law and economics developed independent of *Ordnungstheorie*. Only recently, as you mentioned, has constitutional economics returned to Freiburg. In answering the question, "What did *Ordnungstheorie* contribute to the transformation problem?" I can be fairly brief. Grossekettler (1989 p. 38) begins his survey: "Imagine that someone were to give you the task of developing the economic institutions for a modern industrial state from scratch". He could not anticipate, of course, that such a situation would arise the very year these words were published. The design of an economic order lies at the core of *Ordnungstheorie*. Eucken, in his book on economic policy (Eucken 1952) tackled the issue exhaustively, referring at the time, of course, to the transition from the Nazi war economy to a democratic competitive system. He has listed the core elements of a market system that he called constitutive principles, indicating that they had all to be in place for a market to function. Nevertheless, his ordering of the elements reflects their relative importance. They have been mentioned already several times, so I need not repeat them.

I think the constitutive principles are self-explanatory. As has been said, they are necessary but not sufficient elements. To protect the market from degeneration, as under *laissez faire*, Eucken adds four regulative principles, as he calls them, in which we may see the protective belt of the market order:

- private property has a natural tendency to monopolize, hence competition has to be protected by active economic policy;
- the results of the market process need not be in accordance with prevailing ideas of justice, hence incomes policy can make corrections within very strict limits;
- external effects are not properly taken account of by individual decision-makers, hence an appropriate policy has to correct for the shortcomings of the market;
- falling supply curves, especially in the labour market, may impede market coordination, hence labour market regulation may be advisable under certain conditions.

Again, the first principle is the most important and constitutes, so to say, the essence of *Ordnungspolitik* as a permanent task of the state.

Asked what should be done to transform socialist planning into democratic competition, the answer is clear from the *ordo* point of view: free prices, stabilize money, privatize state-owned enterprises, open all markets, install an appropriate legal system, and see to it that government policy is predictable. All this has to happen immediately and at the same time. Not all of the market-protecting measures need to be implemented in the first round: if opening the external market is insufficient to guarantee competition, as it is in big countries, competition policy will be very important in the highly-monopolized post-Communist economies. In a transitional phase, the new markets will produce unemployment, hence some social security measures are indispensable right from the start. For the rest, regulative measures are of secondary importance.

Miller: I suppose it is now my turn and that I have been invited to speak for the incriminated neoclassical mainstream. I would love to, if only to play devil's advocate, if only I knew the exact nature of the beast called "the neoclassical mainstream". I have a strong suspicion that it is a fiction. Of course, General Equilibrium Theory will not tell us anything about how to manage transformation. It was not meant to do so, even if Frank Hahn (1992) tried, rather unsuccessfully to my mind, to convince the Czechoslovaks to the contrary. If there is such thing as a mainstream, neoclassical or not, it will be found in introductory economics textbooks that do, indeed, exhibit an astonishing uniformity of topics, definitions and theories: students of economics all over the world are nourished on more-or-less the same diet. But it would be going too far to claim that this diet is a coherent body of theory deduced from the same axiomatic principles.

What does this body of knowledge tell us about transformation? Just take the advice given by Western experts in the early 1990s (e.g. Sachs 1990): liberalization, stabilization, privatization. As the previous speaker stated, introduction of the market price mechanism is the first priority. This includes the introduction of a convertible currency, in order to have free trade. The second part of the programme is the elimination of all restrictions on private economic activity or freedom of trade. Since privatization cannot be done at once, disciplining state-owned enterprise is a third element. And fourthly, price stability has to be achieved by tight monetary and fiscal policies. All that has to be done rapidly and at once. For Eucken – whose programme I met here for the first time – was right: these measures form a seamless web. Piecemeal engineering is not the way to manage great transformations. This is what later came to be known in a more elaborate form as the "Washington consensus", which I would claim is the economics consensus.

Interjection from the floor: What about the notorious mainstream imperialism, Professor Miller. You suppress alternative projects. There have been other recommendations, in particular those moulded by Keynesianism stressing the importance of demand management and distributional measures. And they were wise recommendations considering the enormous costs of the stabilization programmes. What you called an economics consensus *was* a Washington consensus, since it was backed by Washington money, which forced the transition states to accept the advice and persuaded the advisers to give it.

Miller: Indeed, there was Keynesian agenda for transformation in the early years. But little has been heard from that quarter subsequently. For it is quite plain that those countries which managed their transformation most successfully were those that implemented the necessary reform steps most thoroughly – necessary, that is, in the context of liberalization, stabilization, privatization. Certainly, the operation was costly, for instance, in terms of unemployment. But this we knew in advance. You cannot have structural change without turnover of labour, part of it experiencing unemployment. The Czechs, not known to be particularly Keynesian, seemed to escape this, but in fact they only delayed restructuring.

Mme Callot: Let me add that this is a good example of the historicity of regulation modes. Keynesian policies were appropriate during the *trente glorieuses*, but in the 1970s and 1980s they became less and less effective. Therefore it would have been rather anachronistic to base transition on Keynesian policies. It does not imply that we would subscribe to the Washington consensus without qualification. The World Bank's chief economist Jo Stiglitz (1999) has recently provided a forceful critique, which is perfectly in line with our point of view. Stiglitz stresses social and organizational capital, the mode of regulating the modern corporation, where ownership and control are separated, the difference between individual restructuring and systemic reorganization, the institutional complementarities, the critique of utopian social engineering which manifests itself in shock therapy. All this boils down to a protracted and carefully managed change, not from above, but from within, taking into account the specific historical and social situation.

Miller: Two brief answers. As to Keynesianism, you could interpret the facts in a different way and it is not so easy to discriminate between the two propositions, although I prefer the second since it is theory-based and more general. Modern macroeconomics has shown that Keynesian counter-cyclical policies are ineffective. When there was more-or-less stable growth they did little harm. When the weather became rough they proved not to work. So there was recourse to flexible markets, which seemed to be able to cope with crises better than government policies. Secondly, I have read Stiglitz's paper, too and did so with great pleasure. For his conclusion is precisely the liberal one I just mentioned: "economic development and transition is more a matter of institutional transformation than of day-to-day economic management" (Stiglitz 1999 Part 2 p. 13), meaning by institutional transformation an evolutionary process. However, this did not fit the situation in 1989. The old system of planning had collapsed because state authority had disappeared. So, any reference to China is off the mark. A new system had to be put on track, even if one knew that it would not run smoothly right from the start. Indeed, shock privatization, for instance, was not necessary. But it was not part of the economists' consensus. Remember Jeff Sachs's recommendations I quoted earlier. Nor was it implemented in all transition countries, least so in successful ones like Poland and Hungary. But if you leave transformation to the old cadres - cultural and organizational capital is embodied - and the passage of time, you will not get it done. Stiglitz's inclusiveness, popular participation and involvement are principles we all share. In the face of an external shock and widespread ignorance about the options, extraordinary politics can be defended.

Barthaus: Let me come back to the Keynesian experiment, which was one of Eucken's points of departure. Government full employment policies may be effective if the government employs idle workers in, for instance, what is now called public employment companies. It puts the market price mechanism out of operation, initially only in a small part of the economic system. But the process has a tendency towards self-enforcement. All economic orders between competition and central administration are unstable and these two can be seen as attractors. There is also an evolutionary asymmetry: deviations from competition tend to converge on the other extreme. To reverse it, a deliberate pro-market choice for is required. The Keynesian experiment failed because it used monetary policy as an instrument and not direct government employment measures. It did so because direct government measures were well understood to be incompatible with the market order.

Mme Callot: Dr. Barthaus, "the inevitable logical principle of the excluded middle", as Boyer (1995 p. 29) has called it, does not stand up the empirical test: between Soviet-type planning and American-type *laissez faire*, neither of which are pure extreme systems, there is a non-empty space of mixed orders, of which the German social market economy is just one such. By way of the logical principle of the excluded centre, Eucken's compromise between theoretical universalism and historical contingency, the great antinomy, adds to the universalist neoclassical theory a likewise universalist theory of order with only one degree of freedom - the choice between serfdom and freedom. The interjection from the floor should, however, not be dismissed so easily. Regulationists are very much in accord with post-Keynesians (e.g. Kregel/Matzner/Grabher 1992) when it comes to describing the constitution of markets which the mainstream consistently neglects to do. Markets are not the result of *laissez faire*, but of deliberate political and social action. Here is a task for the state. And markets will not function under *laissez faire* rules. For unconstrained market participants will conceal information, restrict access and avoid competition. To create the necessary stimuli is another task for the state, as our friends from Freiburg also stress. Finally, the market as mode of coordination needs social acceptance: institutionalizing social security and solidarity will be indispensable. Managing a smooth transition from plan to market, which means, to quote Boyer again (1995 p. 28), "maintaining an equilibrium between productivity and increasing standards of living, whilst at the same time maintaining full employment", appropriate policy measures and time are needed. The mainstream "big bang" philosophy assumes that deregulation creates markets and markets produce equilibrium. Gradual approaches, we are convinced, allow for smooth transitions, greater fairness and broader acceptance.

Miller: Empirical evidence seems to contradict your conviction. Those countries that took to rapid changes have fared much better than others that tried to distribute the pain over a longer period of time. For let it be clear once more – the idea that the structural changes needed in Central and Eastern Europe could be managed without social costs, such as unemployment, real wage losses and a higher degree of inequality, is an illusion. The valley of tears has to be crossed. And the valley of tears is not Schumpeter's optimistic creative destruction. What's the difference? Well, Schumpeter's creative destruction implies that a higher productive process is bidding with bank credit for productive factors, which are released by marginal firms that can no longer cover the temporarily higher factor prices. Unemployment of factors of production is a transitory situation. In the socialist countries, most conspicuously in Eastern Germany, the marginal firms were producing goods that could not stand the market test, virtual production as it was called (Gaddy/Ickes 1998). Exposed to the market, this virtual production collapsed, but there were no higher productive activities immediately round the corner. In referring to Schumpeter, I admit that conventional theory is not very explicit about the transition path from the old to the new equilibrium. But since we all agree that stable money is at the basis of markets, there is no alternative to monetary and fiscal discipline and that right from the beginning.

Moderator: From these short statements it looks to me as if the term "consensus" would also apply to the views of our three contestants. Am I right in concluding that, apart from speed, there are no alternatives to a basic transformation programme informed by alternative theories?

Mme Callot: You are definitely incorrect! Let me enumerate some points of disagreement:

- First of all, you should notice the difference in interpreting your initial question, "What did the respective theory contribute to the problem of transformation?" For my two colleagues, it was self-evident that the question aims at a reform policy programme – policy advice as a natural task of theory. There are other tasks. The theory of Regulation concentrates on the causes of the old system's collapse and the conditions for the emergence of a new system.
- To continue, neoclassical theory cannot motivate a preference for the capitalist mode of regulation. It is indifferent to the choice of economic order.
- It has no theoretical basis for the necessary and sufficient institutional forms. Its transformation programme is sheer pragmatism. That private property rights, a functioning monetary system and deregulation of price formation and economic activity are needed to introduce a market system is pretty obvious.
- Merely by enacting deregulating laws or distributing private property rights at random or making the currency more-or-less stable you do not get functioning capitalist institutions. There is no invisible hand which guarantees their beneficial working. How and when matters.
- The role of institutions, in particular of the state, has been utterly neglected by the Washington consensus. How could it have been otherwise with a theory that disregards institutions?

Barthaus: I can agree with several of the points of your critique. However, I would like to point out a major difference between *Ordnungstheorie* and the theory of Regulation. You consider private property rights and deregulation evident elements of a functioning market system. In the morphological scheme of the five institutional forms of a regulation mode there is no place for them, whereas they figure prominently in Eucken's constitutive principles. Private property rights as the basic institutional form enabling individual decisions and the independence of the individual from the state are a central issue in transformation from plan to market. In Regulation theory, they are a blank space on the map – perhaps a Marxian legacy.

Mme Callot: Of course, they are not. I would call this a misunderstanding. Where is the proper place for property rights? Evidently in the mode of interaction between the state and the economy, which is a missing link in Eucken's *Ordnungstheorie*. Delorme (1995) has described it in a hierarchy of four levels: nature of interaction, rules of interaction, forms of interaction and actors, respectively. The nature of interaction between state and economy defines the game to be played, and this is done by the definition of individual rights relating to the market and to property and by constraining the activity of the state. Whether the nature of interaction is laid down in a kind of social contract, as constitutional economists - not unknown in Freiburg - would have it, remains to be seen from historical evidence. But the intended switch of the nature of interaction between state and economy lies at the root of transformation from plan to market.

Barthaus: Let me come back to the aforementioned theoretical indifference between plan and market that Eucken shared, although he had an absolute preference for a competitive order. Hayek, however, has made a forceful defence of economic arguments as to why socialist planning cannot function efficiently, and history has not disproved him. Eucken's preference for a competitive order is based on a non-economic, but I think none the less relevant argument: it is freedom. Freedom does not come without a price and some people in transforming countries feel nostalgic for the cosy paternalism of the old system – paternalism, as Kornai (1992) has told us, being a constitutive principle of it. However, Kant already pointed out that paternalism is inconsistent with freedom (I happen to have the quotation with me and would like to share this beautiful Kantian extract with you):

"A government constructed on the principle of benevolence towards the people as of a father towards his children, that is, a paternalistic government (*imperium paternale*), where the subjects are forced to behave passively as minors who cannot discern what is really useful for or detrimental to them and expect their happiness from the judgement of the head of state and from his benevolence in order that he desires so – this is the greatest conceivable despotism (constitution that cancels all freedom of the subjects who then have no rights whatsoever)". (Kant 1968 pp. 145-6)

But tell me, what is the Regulationists' answer to the question: why capitalism?

Mme Callot: Very simple: history. Regulation modes emerge from the coevolution of many institutions in the five institutional forms.

Barthaus: Transformation from plan to market in Central and Eastern Europe is not a question of evolution, but clearly one of institutional choice. If the right constitutional choices are made, the economies will achieve the track of self-sustained growth.

Mme Callot: This was not the idea of Freiburg's Hayek. His was the concept of spontaneous order.

Barthaus: Here is a widespread misunderstanding. Hayek speaks of spontaneous order, not of spontaneous institutions. By order, he understands structural relations between system elements which allow for more-or-less stable expectations about the behaviour of the system. The market equilibrium, if it existed, would be an ideal case of such an order that evolves spontaneously. The institutions of capitalism that support the generation of such order were invented unintentionally, indeed. But once they are in place, countries that had hitherto adhered to inefficient rules and regulations could adopt them. There was a long path from medieval money to the *Bundesbank* statute, a path transition economies need not travel once more when they introduce a monetary regime fit for a market order. As you remember, Hayek in *Law, Legislation and Liberty* went still further and constructed (*horribile dictu*) "a model constitution", which could guide countries "without a tradition even remotely similar to the ideal of the Rule of Law which the nations of Europe have long held" or could be useful in the context of "the contemporary endeavours to create new supra-national institutions" (Hayek 1979 p. 108). Clearly, as far as rules are concerned, learning is possible.

Mme Callot: All right, here we are fairly close together with the exception of the possibility of easy adoption or imitation of alien institutions. Institutional forms are complex entities whose elements must fit into each other. I think I have heard something of the kind from your side, too – interdependence of orders. To introduce a complex income tax scheme, for instance, you need able accountants, controllable income statements, a reliable tax administration, a certain degree of loyalty towards the state, in short, a whole number of institutional elements which may be viable in present-day Hungary, but not in Tadjikistan. Even if the capitalist mode of production has been more-or-less globalized by now, this does not necessarily entail a convergence of institutional forms.

Miller: But is that not simply a question of the level of abstraction? We may agree about the fact that the Hungarian tax system will exhibit Hungarian peculiarities. From a welfare theoretical point of view it is important that taxes should distort neither price information nor entrepreneurial motivation. This is the advice to be given, which means that turnover taxes should be replaced by value-added taxes, for instance.

Mme Callot: The difference between the Anglo-American mode of corporate governance via the capital market and the continental network model is not a question of the level of abstraction. It has something to do with history. It is not enough to distribute private property rights no matter how and expect that a contractarian process *à la* Coase will find the most productive allocation. The sociohistorical context in which the institution of private property is embedded has to be taken into account.

Miller: Here, I would claim that there is a misunderstanding here – a misunderstanding of Coase: it was exactly he who introduced the decisive variable, namely transaction costs. He did not say that freedom of contract and markets tend to find the best solution. By now it is good economics to know that transaction costs make a difference and institutions can be understood as rational devices to reduce these costs.

Mme Callot: *Voilà*, you are in a rationality trap. Transaction costs are a catch-all phrase for factors other than direct costs of production, like technical progress. Until now, at least, they have not been properly operationalized. Hence, concrete historical institutions cannot be evaluated as to whether or not they really minimize transaction costs; they are assumed to be optimal, given the preferences and the constraints, transaction costs included. Regulationists would rather hold that institutions emerge from conflicts, crises, and wars and that their stability is much less due to optimality than to the frequency of support, cost-effectiveness being, of course, not the least effective argument in favour of an institution. Since the body of existing institutions is a coherent whole, at least for a certain slice of time, it does not make much sense to look for individual optimality rather than for total coherence, which is less the result of *ex ante* optimization than of learning. The viability of a mode of regulation is then the decisive quality.

A voice from the floor: How do you know?

Mme Callot: From the same source that Eucken gets his morphology and his constitutive principles - history. If we hold, to give an example, that the *Bundesbank* could not have functioned so well in the French context, where distributional conflicts were mediated by inflation, that is, by an accommodating bank, this could not be deduced from any axiomatic principle, but only inferred from historical observation. The French system of conflict management is evidently viable, as is the German, but they are different.

Miller: But this is not generally true. For the French very quickly grew accustomed to the *Bundesbank* policy once it was imposed on them by the European Central Bank. If viability is, as you say, historically contingent, I wonder how you can possibly advise transition governments about what an appropriate set of measures could be and in what sequence they should be implemented.

Mme Callot: Indeed, we have to be very careful in giving advice. What has worked successfully in South America may not work at all in Russia. On the other hand, we are much more specific than the Washington consensus, to which I have the impression, the Freiburg people also subscribe. It is not enough to install an independent central bank and a apply stabilization policy to get a functioning monetary system. It starts from such basic things as a payments system, the current absence of which in Russia has disastrous consequences.

Miller: You can count on the IMF and World Bank specialists to tell anyone who wants to know everything about the technicalities of payment systems and the like. The problem is that no payments system will function properly unless it is fuelled by good money. So the central issue is to have good money and then see to it that the banking system fulfills its role, namely to provide for a payments system and to mediate between savings and investment.

Barthaus: This brings us back to the issue of the role of the state that Mme Callot mentioned earlier. Liberalization, stabilization, privatization –we are certainly not far apart on the importance of these steps. But how should they be accomplished? It is a basic tenet of *Ordnungspolitik* that you need a strong state to get the transformation task done. The lesson of *laissez faire* should not be forgotten. Individual interests, if left unchecked, will create power positions that are harmful to the welfare maximizing functioning of the market system. The rule of law, competition, good money – no transformation country will enjoy these institutions without a strong state to provide for them and to guard them, even if "good money" means that the government has to keep its hands off the central bank. A strong state does not imply a powerful interventionist state, but one that sticks to the accepted constitutional rules.

Miller: That reminds me a little of the Communists who told us that their system would work perfectly if only the people would behave appropriately: how can you embezzle people's property if you belong to the people? Well, experience has shown that people did. The state, too, is run by self-interested people. So, it seems wise to give it as little control over the national income as possible.

Mme Callot: You miss the point, which is that there is a close relationship between the action of government and the emergence of markets or, more generally, between collective action and the emergence of individual entrepreneurship. If there is no rule of law, individual entrepreneurship may emerge, but it will be of the mafia-type. If the state does its job badly, markets will be thin, transaction costs will be high, there will be paralysing uncertainty. If there is no careful provision for competition, fairness, the protection of nature and the like, the market system will run into a crisis – monopolistic power concentration, social unrest, environmental degradation.

Miller: It sounds like a transformation paradox: transformation is caused by the crisis or the collapse of the old system and its state, but it can be performed successfully only by a strong state. If the state is still strong, why shouldn't it stick to the Ancien Régime after all? If the state is new, how can it be strong? You expect Herculean efforts of the state: to be new born and to strangle pythons immediately.

Mme Callot: Nobody has claimed that transformation is an easy task. Evidence from Eastern Europe shows that the establishment of markets is closely related to the effective working of the state. If you want to explain why Poland has successfully transformed while Russia, to date, has not, state authority seems to be the most important single factor. If you want to explain why Poland was able to institutionalize a strong state after the collapse of Communism and Russia was unable to do so, you will have to resort to history.

Miller: Well, is it really true that the mainstream has totally neglected the role of the state? Take stabilization or privatization: the necessary measures evidently require a strong state. This is exactly one of the reasons for the big bang recommendation: in the first phase of transformation, the so-called honeymoon period, a relatively weak, but determined government can get incisive policies accepted which, under normal conditions, would require a lot of political authority. You will have noticed, by the way, that after the

IMF and the World Bank people at Washington from had given advice about the emergency tasks of stabilization and liberalization, they turned their attention to institutions and to the state in particular. The academic journals, too, have taken up the topic (e.g. Shleifer 1997) and it has become clear, to use the metaphor of Frye and Shleifer (1997), who compared the working of exactly the Polish and the Russian states, that the invisible hand is the hand of the strong state, while the weak state has the grabbing hand. The approach is very pragmatic. But I have the impression that *Ordnungstheorie* and theory of Regulation are pragmatic in this respect, too. For, as far as I can see, there is no elaborate economic theory of the state to hand.

Barthaus: This is a critique of *Ordnungstheorie* which is often heard and it is indeed true. If there is such a thing as the *theoriebildende Kraft des Faktischen* (theory-generating power of facts: Sutela 1999), and this is the basic philosophy of both *Ordnungstheorie* and theory of Regulation as I understand them, then we may perhaps be able to distil by means of isolating abstraction the economic theory of the state from the ample stock of facts that the transition process in Central and Eastern Europe has produced. Of course, the facts do not make a theory and a sea of facts is no guarantee of a coherent theory, but the many recent approaches to the subject, like new institutional economics, new political economy, and constitutional economics, are promising in this respect.

Mme Callot: I can follow the argument for a comprehensive decision for early reform. But there seems to be linked to it some kind of time inconsistency, from which follows what I would like to dub "reform Leninism". The transition from plan to market goes hand in hand with the transition from autocratic to democratic rule. If the individual processes of change have different time patterns, as Delorme (1995) convincingly states, with changes in the political sphere happening rather quickly, changes in public opinion demanding an intermediate length, and changes in the patterns of perception being a rather long-term affair of generational change, then the avant-garde reform, the resolute reform team at the highest central level, will have to make the constitutive early decisions without proper democratic resonance – "extraordinary policy" as L. Balcerowicz has called it. Gradual transformation is much more in line with the cognitive development of the citizens and with the inclusion, popular participation and involvement, which World Bank chief economist Stiglitz (1999) has called for, in 1999 not in 1989, as was mentioned earlier.

Barthaus: Whether he would have appreciated the idea of reform Leninism is open to question, but there is no doubt that, in 1948, Ludwig Erhard and the Allied reform team of the Allies were avant-gardist and took decisions for liberalization and stabilization which would hardly have passed the democratic test in a hypothetical German parliament of the period (the German parliament came into existence only a year later). I think it is typical of great transformations that they are initiated in such a way. It distinguishes them from ordinary reform measures, which in a democracy must pass parliament – as the Balcerowicz plan in Poland has, to be sure – and which encounter there well-prepared and well-organized vested interests – such as the Balcerowicz plan has not.

Moderator: Dear colleagues, our discussion must come to an end. The time allotted to us has elapsed. I will give each of you the opportunity for a final word. Jack, will you start, please?

Miller: I have listened to a range of fascinating views, ideas, problems coming up in the context of transformation. The events in Central and Eastern Europe will provide rich material for further analysis and for the formation of theory. But let me return to basics. The reform politician of 1989-90 was looking for a systematic answer to the questions:

1. What is a consistent transformation programme?
2. What is the proper sequence of its implementation?
3. Under which auxiliary conditions can it be successful?

The same questions were repeated in each individual field of the transformation programme, privatization, for instance:

1. What is a consistent programme for privatization?
2. By which methods can it be implemented?
3. Under which side-conditions do we get an efficient property rights structure?

Our friends from Freiburg seem to have approached these questions in general. Competition policy, for instance, is important. But what does it imply in concrete terms? When it comes to a lawsuit on the misuse of market power, the judge, I bet, will fall back on modern industrial economics in order to evaluate the facts. From what I have heard, our friends from Paris leave the reform politicians pretty much in the dark. I have not still seen any acceptable alternative to what I earlier called the "economics consensus". It was important, first, to free prices, open up the economy, stabilize money, establish fiscal discipline, which meant simultaneously tightening the budget constraints of state-owned enterprises and getting privatization started. The sooner these things were done the better. A real wage cut of 25 per cent or so was inevitable in this context. Let me remind you of the fact that, at the very last moment, the central planners of the GDR informed their government that a 25 per cent cut in real wages was needed for the GDR to survive economically. In other words, we have to distinguish between the outcome of the old system and the consequences of the reform policies. After these emergency measures in the first phase, the problems of institution-building and institutional reform, of restructuring and political reform, were paramount in a second phase. These things take time and will still occupy us for some while to come.

Barthaus: The orthodox reform policy programme meets little criticism from the Freiburg school. The distinguishing feature, however, is its theoretical foundation. Little has been said on that account. There is a solid body of theoretical literature on stabilization, to be sure and this perhaps makes for the central position of stabilization in mainstream transformation policy. As for privatization, for instance, little can be said with theoretical authority about motive, method and effect. *Ordnungstheorie* identifies the aims of transformation. Practical reform policy based on the theory will stress a coherent body of institutions and organizations that guarantee autonomous individual decision-making and competition at the same time. The orthodox programme forms part of this, but it lacks the property of order, a consistent whole. *Ordnungspolitik*, in particular the German model of the social market economy, has attracted the attention of the transition countries. Indeed, the reformers have come and studied the individual institutions, from the body of economic laws, the constitution of the *Bundesbank*, the cartel administration to the unemployment and benefits administration and the health system. The debate over reform within Germany shows that these institutions and organizations are far from optimal. Therefore it cannot be a question of importing them lock, stock and barrel. But the fact that they are working as a whole testifies to their viability. You see that this type of reasoning is very close to the Regulation approach, the difference lying in the theory-based ordo-imperative "beware of a degeneration of competition". A focus on historical models rather than theory has surely been strengthened by the desire to join the European Union. The conditions of access serve as an anchor for transformation policy, which is lacking in those countries with no chance of joining in the medium term.

Mme Callot: Metaphorical language in economics more often than not camouflages either ignorance or disagreeable facts. Adam Smith's invisible hand is a conspicuous example of the former, the EU anchor belongs to the latter. I would not go as far as my former Paris colleague, Marie Lavigne (1999 p. 346), who wrote of Central and Eastern Europe: "The important point is that this 'second world' is also a 'second-class world' dominated by the international institutions and the international groupings of which the countries in transformation are so eager to become part", but there is little choice as to the economic order if they want to obtain IMF support or to join the EU. This has something to do with the concrete institutional form of the international system of integration, of which, indeed, the transformation countries want to become part after decades of isolation in the Communist bloc, at great cost in welfare. Regulation theory is a theory of macrodevelopment integrating economic history and economic theory in order to explain endogenously the sequence of periods of crises and periods of relatively stable equilibrium. It is not a theory of economic policy, which *Ordnungstheorie*, for instance, aims to be. Therefore, the test of the pudding is not whether we are able to give appropriate transformation advice, but rather whether we are able to explain what happened. Here its Marxian parentage cannot be overlooked. Marx was very reluctant to give policy advice, witness his hesitations over writing any comment on the Gotha programme and or posting the letters to Vera Sassulitsch. It is a historical irony that each policy measure in the Soviet-type economies had to be backed by a quotation from the master. As a research programme aiming at the analysis of modern capitalism and its development, the theory of Regulation has to be evaluated against its ability to explain not only the workings of the market mechanism, but of the whole set of institutions linked to it. Without proper analysis of this whole, institutional reform will easily degenerate into institutional *bricolage*. Equilibrium theory analyses the conditions under which the market *can* work; Regulation theory concentrates on the conditions under which the market *does* work. The first is an abstract theoretical approach, the second a historical and theoretical approach. What we have learned from history is that institutions and the whole order of capitalism evolve over a long period of time. So, whatever immediate measures are taken to get the process started, the great transformation will be accomplished only after a long transition period full of contradictions, pain, and crises. Since the initial conditions are important for the path of transition, not only the legacies of the past, but certainly also the immediate measures of reform policy deserve the attention of the analyst.

Moderator: I do not think the panellists' final words require any further comment on my part. The show will go on, in other words, the three research programmes will continue their efforts. It just remains for me to thank the panellists and the audience for a most enjoyable hour.

References

BÖHM, F.: *Die Ordnung der Wirtschaft als geschichtliche Aufgabe und rechtsschöpferische Leistung*, Stuttgart (Kohlhammer) 1937.

BOYER, R.: "The Great Transformation of Eastern Europe: A 'Regulationist' Perspective", *Emergo*, 2/4 (1995), pp. 25-41.

BOYER, R., Y. SAILLARD (Eds.): *Théorie de la régulation. L'état des savoirs*, Paris (La Découverte) 1995.

DELORME, R.: "An Alternative Theoretical Framework for State-Economy Interactions in Transforming Economies", *Emergo*, 2/4 (1995), pp. 5-24.

DELORME, R., ANDRÉ C.: *L'État et l'économie*, Paris (Seuil) 1983.

DEMSETZ, H.: *Efficiency, Competition and Policy: The Organization of Economic Activity*, Oxford (Basil Blackwell) 1982.

EUCKEN, W. *Die Grundlagen der Nationalökonomie*, Jena (Gustav Fischer) 1940.

EUCKEN, W. *Grundsätze der Wirtschaftspolitik*, Tübingen (Mohr) 1990. First edition: 1952.

FRYE, T., SHLEIFER A.: "The Invisible Hand and the Grabbing Hand", *American Economic Review, Papers and Proceedings*, 87 (1997), pp. 354-8.

GADDY, C. G., ICKES W.: "Russia's Virtual Economy", *Foreign Affairs*, 77/5, (1998), pp. 53-67.

GROSSEKETTLER, H. G. "On Designing an Economic Order. The Contributions of the Freiburg School", in: D. A. WALKER (Ed.) *Perspectives on the History of Economic Thought*, Vol. 2, Aldershot (Elgar) 1989, pp. 38-84.

HAHN, F.: "The Relevance of General Equilibrium Theory for the Transformation of Centrally Planned Economies", *Prague Economic Papers*, 1/2 (1992), pp. 99-108.

HAYEK, F. A.: *Law, Legislation and Liberty. Vol. 3 The Political Order of a Free People*, London (Routledge & Kegan Paul) 1979.

KANT, I.: "Über den Gemeinspruch: Das mag in der Theorie richtig sein, taugt aber nicht für die Praxis", in: *Werke*, Vol. 9, Darmstadt (Wissenschaftliche Buchgesellschaft) 1968, pp.123-72.

KASPER, W., STREIT M. E.: *Institutional Economics. Social Order and Public Policy*, Cheltenham (Edward Elgar) 1998.

KORNAI, J.: *The Socialist System. The Political Economy of Communism*, Oxford (Clarendon Press) 1992.

KREGEL, J., MATZNER E., GRABHER G. (Eds.) *The Market Shock: An Agenda for Socio-Economic Reconstruction of Central and Eastern Europe*, Vienna (Austrian Academy of Sciences) 1992.

LAVIGNE, M.: "The International Framework: How Autonomous is the Transition States' Economic Policy?", in: H.-H. HÖHMANN (Ed.): *Spontaner oder gestalteter Prozeß? Die Rolle des Staates in der Wirtschaftstransformation osteuropäischer Länder*, Baden-Baden (Nomos), pp. 334-47.

RIBHEGGE, H.: "Der Beitrag der Neuen Institutionenökonomik zur Ordnungspolitik", *Jahrbuch für Neue Politische Ökonomie*, 10 (1991), pp. 38-60.

SACHS, J.: "Eastern Europe's Economies. What Is to be Done?", *The Economist* 13 January 1990, pp. 19-24.

SHLEIFER, A.: "Government in Transition", *European Economic Review*, 41 (1997), pp. 385-410.

STIGLITZ, J. E.: *Whither Reform? Ten Years of the Transition*, Annual Bank Conference on Development Economics, Washington (World Bank) 1999.

SUTELA, P. "Zur theoriebildenden Kraft des Faktischen: Russische und estnische Erfahrungen", in: H.-H. HÖHMANN (Ed.) *Spontaner oder gestalteter Prozeß? Die Rolle des Staates in der Wirtschaftstransformation osteuropäischer Länder*, Baden-Baden (Nomos), 1999, pp. 80-95.

VANBERG, V. (Ed.): *Freiheit, Wettbewerb und Wirtschaftsordnung*, Freiburg (Haufe), 1999.

List of Authors

FRANK BÖNKER is Senior Lecturer in Economics, European University Viadrina in Frankfurt/ Oder, and Research Fellow, Frankfurt Institute for Transformation Studies (FIT), Germany.

ROBERT BOYER is Professor of Economics, Ecole des Hautes Etudes en Sciences Sociales (EHESS), Director of the Research Unit "Régulation, ressources humaines et économie publique" of the "Centre d'Etudes Prospectives d'Economie Mathématique Appliquées à la Planification" (CEPREMAP), Centre National de la Recherche Scientifique, Paris, France.

SYLVAIN BROYER is Doctoral Student at the Institute of Economic Theory and Analysis (GATE), University Lyon II, France.

BERNARD CHAVANCE is Professor of Economics, University of Paris 7 and Director of the "Groupe d'Etudes et de Recherche sur la Régulation et les Mutations des Economies" (GERME), Paris, France.

ROBERT DELORME is Professor of Economics at the University of Versailles-Saint Quentin and Researcher at the "Centre d'Etudes Prospectives d'Economie Mathématique Appliquées à la Planification" (CEPREMAP), Centre National de la Recherche Scientifique, Paris, France.

CARSTEN HERRMANN-PILLATH is Professor of Economics, University Witten/Herdecke, and Director of the Institute for Comparative Research into Culture and Economic Systems, Germany.

AGNÈS LABROUSSE is Doctoral Student at the "Centre d'Etudes des Modes d'Industrialisation" (CEMI), Ecole des Hautes Etudes en Sciences Sociales (EHESS), Paris, France and at the Centre Marc Bloch, Berlin, Germany.

HELMUT LEIPOLD is Professor of Economics, Philipps University, Marburg, Germany.

JEAN-FRANÇOIS VIDAL is Professor of Economics, University of Sceaux, France.

LIST OF AUTHORS

HANS-JÜRGEN WAGENER is Professor of Economics, European University Viadrina, and Director of the Frankfurt Institute for Transformation Studies, Franfurt/Oder, Germany.

JEAN-DANIEL WEISZ is Doctoral Student at the Centre d'Etudes des Dynamiques Internationales (CEDI), University Paris-XIII, France and at the Centre Marc Bloch, Berlin, Germany.

MICHAEL WOHLGEMUTH is Research Assistant, Max-Planck Institute for Research into Economics Systems, Institutional Economic Unit, Jena, Germany and Visiting Scholar, Department of Economics, New York University, United States.

List of Illustrations

- Diagrams

- Figures

- Graphs

- Tables

Index of Names

INDEX OF NAMES

INDEX OF NAMES

Index of Subjects

INDEX OF SUBJECTS

Studies in Economic Ethics and Philosophy

P. Koslowski (Ed.)
Ethics in Economics, Business, and Economic Policy
X, 178 pages. 1992
ISBN 3-540-55359-2 (out of print)

P. Koslowski and Y. Shionoya (Eds.)
The Good and the Economical: Ethical Choices in Economics and Management
X, 202 pages. 1993
ISBN 3-540-57339-9 (out of print)

H. De Geer (Ed.)
Business Ethics in Progress?
IX, 124 pages. 1994
ISBN 3-540-57758-0

P. Koslowski (Ed.)
The Theory of Ethical Economy in the Historical School
XI, 343 pages. 1995
ISBN 3-540-59070-6

A. Argandoña (Ed.)
The Ethical Dimension of Financial Institutions and Markets
XI, 263 pages. 1995
ISBN 3-540-59209-1 (out of print)

G.K. Becker (Ed.)
Ethics in Business and Society Chinese and Western Perspectives
VIII, 233 pages. 1996
ISBN 3-540-60773-0

P. Koslowski (Ed.)
Ethics of Capitalism and Critique of Sociobiology. Two Essays with a Comment by James M. Buchanan
IX, 142 pages. 1996
ISBN 3-540-61035-9

F. Neil Brady (Ed.)
Ethical Universals in International Business
X, 246 pages. 1996
ISBN 3-540-61588-1

P. Koslowski and A. Føllesdal (Eds.)
Restructuring the Welfare State Theory and Reform of Social Policy
VIII, 402 pages. 1997
ISBN 3-540-62035-4 (out of print)

G. Erreygers and T. Vandevelde
Is Inheritance Legitimate? Ethical and Economic Aspects of Wealth Transfers
X, 236 pages. 1997
ISBN 3-540-62725-1

P. Koslowski (Ed.)
Business Ethics in East Central Europe
XII, 151 pages. 1997
ISBN 3-540-63367-7

P. Koslowski (Ed.)
Methodology of the Social Sciences, Ethics, and Economics in the Newer Historical School From Max Weber and Rickert to Sombart and Rothacker
XII, 565 pages. 1997
ISBN 3-540-63458-4

A. Føllesdal and P. Koslowski (Eds.)
Democracy and the European Union
X, 309 pages. 1998
ISBN 3-540-63457-6

P. Koslowski (Ed.)
The Social Market Economy Theory and Ethics of the Economic Order
XII, 360 pages. 1998
ISBN 3-540-64043-6

Amitai Etzioni
Essays in Socio-Economics
XII, 182 pages. 1999
ISBN 3-540-64466-0

P. Koslowski (Ed.)
Sociobiology and Bioeconomics The Theory of Evolution in Biological and Economic Theory
X, 341 pages. 1999
ISBN 3-540-65380-5

J. Kuçuradi (Ed.)
The Ethics of the Professions: Medicine, Business, Media, Law
X, 172 pages. 1999
ISBN 3-540-65726-6

Studies in Economic Ethics and Philosophy

S. K. Chakraborty and S. R. Chatterjee (Eds.)
Applied Ethics in Management
Towards New Perspectives
X, 298 pages. 1999
ISBN 3-540-65726-6

P. Koslowski (Ed.)
The Theory of Capitalism in the German Economic Tradition.
Historism, Ordo-Liberalism, Critical Theory, Solidarism
XII, 577 pages. 2000
ISBN 3-540-66674-5

P. Koslowski (Ed.)
Contemporary Economic Ethics and Business Ethics
IX, 265 pages. 2000
ISBN 3-540-66665-6

L. Sacconi
The Social Contract of the Firm.
Economics, Ethics and Organisation
XV, 229 pages. 2000
ISBN 3-540-67219-2

M. Casson and A. Godley (Eds.)
Cultural Factors in Economic Growth
VIII, 244 pages. 2001
ISBN 3-540-66293-6

Y. Shionoya and K. Yagi (Eds.)
Competition, Trust, and Cooperation
IX, 252 pages. 2001
ISBN 3-540-67870-0

B. Hodgson
Economics as Moral Science
XIV, 380 pages. 2001
ISBN 3-540-41062-7

www.ingramcontent.com/pod-product-compliance
Ingram Content Group UK Ltd.
Pitfield, Milton Keynes, MK11 3LW, UK
UKHW021858190726
13853UKWH00003B/1324

* 9 7 8 3 6 6 2 0 4 4 7 3 5 *